D1725122

Martin Bojowald

The Universe: A View from Classical and Quantum Gravity

Related Titles

Erdmenger, J. (ed.)

String Cosmology
Modern String Theory Concepts from the Big Bang to Cosmic Structure

2009
ISBN: 978-3-527-40862-7

Ng, T.-K.

Introduction to Classical and Quantum Field Theory

2009
ISBN: 978-3-527-40726-2

Forshaw, J., Smith, G.

Dynamics and Relativity

2009
ISBN: 978-0-470-01459-2

Fayngold, M.

Special Relativity and How it Works

2008
ISBN: 978-3-527-40607-4

Morsch, O.

Quantum Bits and Quantum Secrets
How Quantum Physics is Revolutionizing Codes and Computers

2008
ISBN: 978-3-527-40710-1

Scully, R. J., Scully, M. O.

The Demon and the Quantum
From the Pythagorean Mystics to Maxwell's Demon and Quantum Mystery

2007
ISBN: 978-3-527-40688-3

Liddle, A.

An Introduction to Modern Cosmology

2003
ISBN: 978-0-470-84835-7

Fayngold, M.

Special Relativity and Motions Faster than Light

2002
ISBN: 978-3-527-40344-8

Martin Bojowald

The Universe: A View from Classical and Quantum Gravity

WILEY-VCH

WILEY-VCH Verlag GmbH & Co. KGaA

The Author

Martin Bojowald
Pennsylvania State University
Physics Department
University Park
USA

Author photography on backcover:
© Silke Weinsheim
(Silke.weinsheim@web.de)

Library of Congress Card No.:
applied for

British Library Cataloguing-in-Publication Data:
A catalogue record for this book is available from the British Library.

Bibliographic information published by the Deutsche Nationalbibliothek
The Deutsche Nationalbibliothek lists this publication in the Deutsche Nationalbibliografie; detailed bibliographic data are available on the Internet at http://dnb.d-nb.de.

Composition le-tex publishing services GmbH, Leipzig
Printing and Binding Markono Print Media Pte Ltd, Singapore
Cover Design Adam-Design, Weinheim

Printed in Singapore
Printed on acid-free paper

ISBN 978-3-527-41018-7

Contents

for A

You delight in laying down laws,
Yet you delight more in breaking them.
Like children playing by the ocean who build sand-towers with constancy and then destroy them with laughter.
But while you build your sand-towers the ocean brings more sand to the shore, and when you destroy them the ocean laughs with you.

Kahlil Gibran: *The Prophet*

1 The Universe I

Observations of the stars and planets have always captured our imagination, in myths as well as methods. While the myths have lost much in importance and relevance, traces of inspiration from the stars can still be seen in modern thinking and technology.

1.1 Newtonian Gravity

A concept in physics that, despite its age, enjoys enormous importance, yet has often been remodeled, is the one of the force. In colloquial speech, it is associated with violence or, in personal terms, egotism: "It is a necessary and general law of nature to rule whatever one can."[1] We speak of forces of nature that cause inescapable calamity, or the personal force of despots small and large. Physicists, as human beings, know about these facets of force, but they have also developed it into a powerful and impartial method for predicting motion of all kinds.

Force In physics, forces play a more important role than realized colloquially, a role perhaps most powerful and at the same time, most innocent. Forces in physics are embodied by laws of nature. If there is a force acting on some material object, this object has to move in a certain way dictated by the force. The action is most powerful because it cannot be evaded, and it is most innocent because it happens without a purpose, without a hidden (or any) agenda. This concept of a force, although it has been somewhat eroded by quantum mechanics[2] is one of the central pillars of our description of nature. It serves to explain motion of all kinds, to predict new phenomena, and to understand the workings of useful technology.

Forces govern what we do on Earth, but the sky is no exception. A force causes acceleration, or a change of velocity, in a precise and qualitative way: For any given object, the acceleration due to a force is proportional to the magnitude of the force;

1) Thucydides: *History of the Peloponnesian War.*
2) Nothing in physics, it appears, is permanent or unquestionable, but the demise of one piece of thinking often begets a new, stronger term. The ocean laughs with us, and in its playful way guides us to higher learning.

The Universe: A View from Classical and Quantum Gravity, First Edition. Martin Bojowald.

force equals mass times acceleration. This is a mathematical equality which tells us that its two sides, force and the product of mass times acceleration, are two sides of the same coin. Knowing the force in a given situation is as much as knowing the value which the product of mass and acceleration takes, for they can never be different.

> **Newton's second law** In formula, $F = ma$, where F stands for the value of the force acting on an object, m for the object's mass, and a for the acceleration caused. This equality is often called Newton's second law. (The first law, some kind of precursor of the second law, states that the action performed on an object equals the reaction caused. Action, in this sense, has only received vague definitions in physics, and so the role of the first law waned in the course of history.)

One might think that the concept of a force seems to be redundant because we could always replace it with the product of mass times acceleration. However, the distinction of the two sides of an equation is important. We interpret the force, one side of the equation, as causing an action on the object, while the acceleration, on the other side of the equation, is the effect. Although both sides always take the same numerical value, their physical interpretations are not identical to each other. While we see and measure the effect of a force by the acceleration it causes on objects (or by deformations of their shapes), the value of the force itself is derived from the expectation we have for the strength of a certain action. This expectation may come from the physical strength of our arms which we know from experience, or it may be based on a detailed theory for a general physics notion such as the gravitational force. In this way, the distinction of the two sides of a mathematical equation becomes an important conceptual one: ingrained features of an actor in the play of nature, and effects that the action causes. Both sides of Newton's second law can be as different as psychology and history.

As in a literary play, one often understands the characters by watching their actions and reactions. After some time, one begins to form a theory about their nature; their actions reflect back on them by the way we judge them and expect how they may behave in future scenes. In physics, forces are first understood by the effects they cause, such as making an apple fall to the ground; physicists then begin to judge the force, and try to understand what makes it pull or push. One forms a theory about the force, that is, more mathematical equations that allow one to compute the value of the force from other ingredients, independent of the accelerations caused on objects. This theory, in turn is tested by using it to compute forces in new situations and comparing with observed accelerations. New predictions can be made to see what the same force may do under hitherto unseen circumstances. The theory is being tested, and can be trusted if many tests are passed.

Acceleration Forces can be determined by the accelerations caused (or deformations, as a secondary effect of acceleration of parts of elastic materials). We feel this interplay physically when we attempt to exert a force: our muscles contract, and our body begins to move. For acceleration or general causes of forces, we have a more direct sense than for the forces themselves, reflected in the way physics determines forces by measuring accelerations of objects. We notice acceleration whenever our velocity changes. Here, we have a second relationship involving acceleration: it is the ratio of the change of velocity in a given interval of time by the magnitude of that interval, assumed small. The condition of smallness is necessary because the value of acceleration may be different at different times; if we were to compute acceleration as the ratio of velocity changes by long time intervals, we would at best obtain some measure for the average acceleration during that time. By taking small values of time intervals (and assuming that the acceleration over such short time intervals does not change rapidly), we obtain the acceleration at one instant of time.

If a is the acceleration and v the velocity, we write $a = \Delta v/\Delta t$, denoting by Δv the change of velocity during the time interval Δt. (The Greek letter Δ, Delta, is used to indicate that a difference has been taken, here the difference of velocity values at the beginning and end of the time interval considered.) This notation does not show that a precise value for acceleration is obtained for small time intervals. To show the limiting smallness, a limit is often quite literally included in the formula as $a = \lim_{\Delta t \to 0}(\Delta v/\Delta t)$, where small values of Δt are defined as those being close to zero: we try smaller and smaller $\Delta t \to 0$, until the ratio $\Delta v/\Delta t$ reaches a stable value. At zero, the ratio is not defined, or ill-defined as mathematicians like to say, because Δt as well as Δv become zero. (If the time interval shrinks to zero size, the velocity does not change over it.)

The ratio of two vanishing numbers is undefined because its value depends on how those numbers, in a limiting process, approach zero. Dividing by zero is a dangerous operation and often produces meaningless infinities, also called divergences. If we have to divide a cake among four people, the pieces are smaller than those dividing it among two people. If there is only one person, the cake remains undivided. Fractional people are difficult to consider in this picture, but if we just continue the pattern, the ratio will continue to rise above the original size of the numerator, that is, the cake, and can be made large by making the denominator small.

If Δv approaches zero twice as fast as Δt, we write $\Delta v = 2\Delta t$ and cancel out Δt in the ratio, obtaining the value 2. If Δv approaches zero half as fast as Δt, we obtain the value 1/2. If Δv approaches zero as fast as the square of Δt, we can cancel out one Δt in the ratio but are still left with one such factor, making the whole ratio vanish for small Δt: $\lim_{\Delta t \to 0} \Delta t = 0$. If Δv approaches zero as fast as the square root of Δt, only factors of $\sqrt{\Delta t}$ can be canceled and we are left with $\lim_{\Delta t \to 0} 1/\sqrt{\Delta t}$. The denominator becomes smaller and smaller, and dividing by a smaller number makes the ratio larger. In the limit of vanishing Δt, the ratio is larger than any number, it is infinite, or diverges.

Continuum A limiting value depends on how the limit is taken, for instance, on the behavior of numerator and denominator in a fraction. The velocity depends on time through the way the object is moving, and so small changes of Δv are tied to small changes of Δt; they cannot be chosen arbitrarily. There is a functional relationship (or a function, for short) between velocity and time, a relationship $v(t)$ which tells us what velocity the object has at a given time. Whenever there is such a relationship between two quantities, we can consider small changes starting from one point and compute limiting ratios such as a above. The limiting ratio is called the derivative of v by t. Computing derivatives is a common operation in mathematics, and so a notation shorter than writing the limit explicitly has become customary. Instead of writing ratios of differences, we write ratios of so-called differentials: $a = \mathrm{d}v/\mathrm{d}t$. The limiting procedure is understood by replacing the edgy difference symbol Δ by the smoother letter d. This notational change reflects the procedure of going from differences of values taken at two separate times (ignoring the slight changes the function $v(t)$ may take between those times) to differentials that consider the change between times closer to each other than can be imagined.

Differentials and the limiting procedures used in them implement the idea of the continuum, an abstract concept of structureless space. Although the notion is not intuitive, it simplifies several mathematical constructions and equations. Derivatives and the continuum view therefore pervade many branches of mathematics and physics. Nevertheless, such limits may not be correct to describe nature at a fundamental level. We cannot be sure that time durations shorter than any number we can imagine do exist, or that objects can move in increments smaller still than any small number. We will come back to the physics behind these questions in our chapter on quantum gravity; for now, we must leave the answers open. In physical statements, we will forgo using the continuum limit and derivatives, and keep working with differences and the symbol Δ, even if it provides only approximations to instant mathematical changes of a curve.

Derivative The French amateur mathematician Pierre de Fermat, a lawyer by his main profession, is the first person known to have grasped the meaning of differentials, although he did not go as far as introducing them in precise terms. He paved the way for the development of calculus with its many applications in geometry and theoretical physics. In consistent form, differentiation and integration were introduced by Isaac Newton and Gottfried Leibniz, whereas Fermat's assumptions and methods remained, from a strict point of view, self-contradictory. What was missing for Fermat was the notion of an infinitesimally small quantity, a quantity, as it were, squaring the circle by being smaller than any number, yet not zero.

Fermat's crucial idea was to compute the position of an extremum of a curve, that is, a maximum, minimum or saddle point, by taking seriously the intuition that the curve at such a point is momentarily constant. When we walk over a hill, our altitude is unchanging as we pass the top. The curve may not be constant for any nonzero change of its argument, but it is "more constant" than at any other place.

Fermat formulates this observation by considering the difference of the values the function takes at two nearby places, separated by a small displacement h: $\Delta_h f(x) = f(x+h) - f(x)$. If he lets h become ever smaller, setting it to zero, nothing is won because he simply subtracts one number from itself. To prevent triviality, Fermat inserts another step into his calculation: Realizing that the difference $\Delta_h f(x)$, for small h, is proportional to the value of h, he tries to factor out one uninteresting contribution of h by writing $\Delta_h f(x) = d_h f(x) h$. The factor $d_h f(x)$ still depends on h, but the dependence is removed by setting $h = 0$. (We obtain the derivative $\mathrm{d}f(x)/\mathrm{d}x = d_0 f(x)$.) Zeros of $d_0 f(x)$ correspond to places x where the original function $f(x)$ is "more constant" than elsewhere because the difference $\Delta_h f(x) = d_h f(x) h$ vanishes for $h = 0$ not just by the explicit factor of h but also by the vanishing $d_0 f(x)$.

Fermat himself provides the following example: Consider a rope of length $2L$, by which we are to enclose a rectangle of maximum area by kinking it at three places by right angles, and then connecting the endpoints by another right angle. There is one variable in this problem: We can choose a point on the rope, placed at position x between one endpoint and the midpoint at distance L. Kinking the rope at x, L and $L + x$ by right angles then forms a rectangle of side lengths x and $L - x$. Its area is $A(x) = x(L - x) = Lx - x^2$, a quadratic function.

Following Fermat's procedure, we compute $\Delta_h f(x) = A(x+h) - A(x) = L(x+h) - (x+h)^2 - Lx + x^2 = Lh - 2xh - h^2$. As understood by Fermat, we can factor out one contribution of h: $\Delta_h f(x) = (L - 2x - h)h = d_h f(x) h$ such that $d_h f(x) = L - 2x - h$ and $d_0 f(x) = L - 2x$. The latter function of x vanishes for $x = L/2$, the value for which the rectangle of side lengths $x = L/2$ and $L - x = L/2$ is a square.

Showing that the value found by Fermat's procedure is a maximum or a minimum requires some extra work, which can be done by considering extreme values of x of small or large size. If such x reduce the value of the function, the value found for x is a maximum, otherwise a minimum. More problematic in Fermat's procedure is the fact that it uses two contradictory assumptions about h. We are first asked to factor out one contribution of h, in the difference $\Delta_h f(x) = d_h f(x) h$, then set h to zero. If h vanishes, however, the difference $\Delta_h f(x)$ vanishes automatically and cannot contain any nontrivial information about the original function. Fermat's intuition is based on the observation that $\Delta_h f(x) = d_h f(x) h$ vanishes "more rapidly" as h shrinks to zero if not only the explicit factor of h but also its coefficient $d_h f(x)$ vanish at some specific x. However, how can one make the notion of "vanishing more rapidly" precise? The invention of calculus after Fermat, aided much by his work, provides a satisfactory procedure by defining $d_0 f(x)$ as the limit $\lim_{h \to 0}(\Delta_h f(x)/h)$ in which we recognize the limiting quotient defining the derivative of f at x.

Fermat developed his method further to compute the slope of tangents to curves, amounting to nonvanishing derivatives of the function defining the curve. Fermat did not publish his methods, which was not usual at his time, not only for an amateur mathematician. Instead, we know of his ideas by letters he sent to some of the eminent thinkers among his contemporaries. (Nowadays, this methods of publication is no longer encouraged. Important work could land in spam filters.) The method of finding maxima of a function was sketched by Fermat in a letter to René Descartes in 1638, a fitting recipient because the way Fermat proceeded to solving several geometrical problems relied on Descartes' algebraization of geometry, an essential mathematical contribution which we will soon encounter.

Motion We now have two equations involving an object's acceleration: Newton's second law equating it to the force (divided by the mass) and the relation to the change of velocity. Just as there is a difference of perception of a force and its effect, there is an important difference between the two equations. We feel acceleration through the change of velocity it implies, and we relate it to an acting force only by experience or after some thinking. Similarly, the equation that relates acceleration to the change of velocity is considered much more evident than Newton's second law. It is not even a law, relating two seemingly different quantities in a strict way, but rather a definition. When we talk about acceleration, we mean nothing else than the change of velocity in a given time. Unlike with Newton's second law, the equality does not give us insights into cause-and-effect relationships. It does not make sense to speak of the change of velocity as the cause of an acceleration, or vice versa; under all circumstances, both quantities are one and the same, not just regarding their values but also their conceptual basis. Equating acceleration to the ratio of velocity change by time interval defines what we mean by acceleration in quantitative terms. It is a definition to clarify the meaning of a colloquial word, namely, acceleration, rather than a law that would give us insights into the interplay of different objects in nature.

Velocity and acceleration describe the motion of an object without regard to the forces that cause or influence the motion. Such quantities and the equations they obey amongst themselves are called kinematical, from the Greek word for motion. Laws that involve the force, or in general, the cause for motion, are called dynamical, from the Greek word for force. Kinematical equations, or kinematics for short, are easier to uncover and to formulate than dynamical ones because they do not require as much knowledge of interactions in nature. Nevertheless, they are important because they form the basis of dynamical laws so that the latter can refer to the effects caused by forces on motion. Moreover, kinematics is often far from trivial, and does not exhaust itself in definitions. In the context of special relativity, we will have an impressive example for how nontrivial, even counterintuitive, the kinematics of a physical theory can be.

A second kinematical relationship arises when we think about the meaning of velocity. We say that we have a nonzero velocity when we are moving, that is, when our position changes in time. Just as acceleration is quantified as the ratio of velocity change by time interval, we define the value of velocity as the ratio of position change by time interval: $v = \Delta x/\Delta t$ where x is now the position, for instance, taken as the distance traveled along a road, or the yards run along a stadium round.

If we know how an object is moving at all times, we are given the functional relationship $x(t)$. From this position function, we compute the velocity function $v(t)$ by taking ratios (and limits) of changes of position, and then the acceleration. If we also know the object's mass, we can use the computed values of acceleration to compare with an idea we might have about the acting force and see if the values match. In this way, theories for different forces can be tested.

Integration The opposite of differentiation is integration. In mathematics, we differentiate the values of a function by computing how much it changes under small variations of its argument. Integration is the reverse procedure: we try to reconstruct the function from its minuscule changes between nearby points. Regarding motion, this procedure allows us to go from acceleration to velocity and then to position.

Once an understanding of a certain force has been established, the reverse of differentiation becomes more interesting. We start with an equation for the force, telling us what values the force has at different positions or at different times. Newton's second law then identifies the acceleration of an object moving subject to the force. Knowing the acceleration, we would like to determine the corresponding velocity and position as functions of time, to give us a direct view of the way we expect the object to move. The process of taking ratios of differences can be reversed by choosing a starting value, that is, of the velocity, say, if we are to determine $v(t)$ from $a(t)$, and then adding up all differences as given by $a(t)$.

> Starting from $v(t_0)$ at some time t_0, we write $v(t_0 + \Delta t) = v(t_0) + \Delta v(t_0) = v(t_0) + a(t_0)\Delta t$, $a(t_0)$ computed from the force at time t_0, and repeat until we reach the desired stretch of time.

Acceleration is defined not by the differences themselves but by their values obtained in the limit of small time intervals. If the time intervals are zero, it takes an infinite number of them to reach any time other than the one chosen as the starting point. The mathematical procedure to compute the velocity from acceleration thus requires us to sum up an infinite number of terms, all so small that the final value of the sum is still finite. The procedure, reversing the computation of a derivative, is called integration, and the value of the sum is called the integral of $a(t)$.

Once we have a formula for the force independent of the one relating it to acceleration, we can compute the velocity by integration. Velocity, in turn, is defined as the ratio of position and time changes, taken in the limit of small time intervals. Once the velocity is known, we can, again choosing a starting value of x at some time t_0, compute the position function $x(t)$ by the same mathematical procedure, integrating the velocity. In this mathematical way, force laws are used to compute all details of the motion of objects, making predictions about the effects of forces.

Locality Physicists prefer local products. If we know an object's position $x_0 = x(t_0)$ and velocity $v_0 = v(t_0)$ at one time t_0, we can compute how it moves from then on subject to a given force. It is not necessary to know the complete prehistory before t_0. Newton's laws do not ask for records of preexisting conditions: the force $F(t)$ is local; it depends only on a single instance of time, for instance, as the product of position and velocity at just one time t and perhaps constant parameters such as the mass. Friction by motion through a medium such as air is one example, often described by a force $F(t) = Kv(t)$ with a constant K. To see how to move on from t_0 according to a local force $F(t)$, we only refer to the current state (x_0, v_0). Equations relating different variables at a single moment in time (or a single place in space) are called local.

Motion is not always local. Cars at an intersection may go straight or turn, even though they are in the same spot with about the same speed. In addition to the cars' positions and speeds, we must know more, the prehistory and memory of their drivers. Such nonlocal motion would be much more complicated to compute than local motion, sensitive to just a few parameters.

Sometimes, nonlocal laws can be made local by introducing auxiliary variables: parameters that are not essential to describe the motion but play a helpful mathematical role. In the example of traffic, an auxiliary variable is realized by turn signals, or at intersections with turn lanes. Cars turning in different ways occupy separate lanes; the lane position is one new variable to describe motion along the road in local terms. (Then, we must know which lane a driver will choose.)

Algebra and geometry Both differentiation and integration have geometrical interpretations, which is another reason for their widespread use. The derivative of a function is related to its change, or when plotted as a graph to its slope. The integral of a function, less obviously so, determines the area under the graph: Figure 1.1.

If integration is the opposite of differentiation, we can proof certain properties of integration by showing that they are undone by differentiation. As an example, we claim that the integral $I_{x_0}^{x}(f)$ of a function $f(x)$, starting at one point x_0 and proceeding to another x, is equal to the area under the curve between points x_0 and x. If we can show that the derivative of the area is equal to the original function, our claim is proven.

Calculus The derivative of $I_{x_0}^{x}(f)$ is the limit of $[I_{x_0}^{x+h}(f) - I_{x_0}^{x}(f)]/h$ for h small. If $I_{x_0}^{x}(f)$ is equal to the area under the curve from x_0 to x, $I_{x_0}^{x+h}(f) - I_{x_0}^{x}(f)$ is the area from x to $x + h$. If h is small, this area is close to $hf(x)$, the area of a rectangle with width h and height $f(x)$. Dividing by h and letting h go to zero produces the original function f evaluated at the endpoint x, as was to be shown. Integration can undo differentiation only when we take limits of quantities like h going to zero, one of the reasons for the mathematical power of infinitesimals. When we want to leave the question of limits open, in equations of physics, we will write an explicit sum $\mathbf{Sum}_x f(x)\Delta L(x)$ for the integral, summing over all points x separated by $\Delta L(x)$ in some interval. The continuum version uses smoothed-out notation, $\int f(x)dx$.

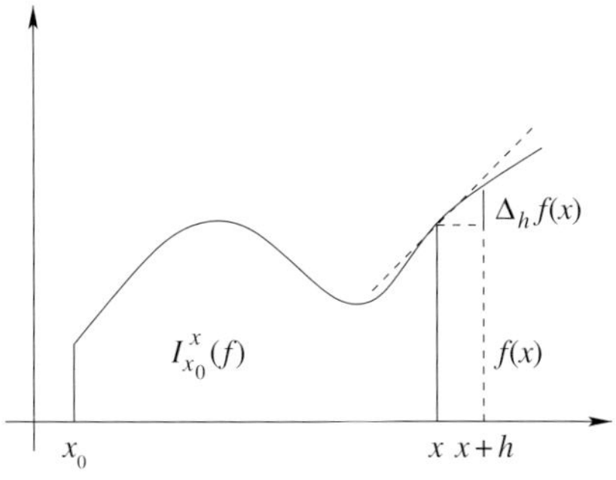

Figure 1.1 Differentiation and integration. The derivative computes the slope $\Delta_h f(x)/h$ of a function $f(x)$, the integral $I_{x0}^x(f)$ the area under the curve between x_0 and x. The derivative of the integral is the original function: $\Delta_h I_{x0}^x(f) = I_{x0}^{x+h}(f) - I_{x0}^x(f)$, for small h, is an area of the size $f(x)h$.

Differentiation and integration, or calculus as the subject of their study is called, have two sides: an algebraic one and a geometric one. The algebraic side appears in the definitions given, according to which the values are ratios of numbers in a derivative or sums of terms in an integral. It also shows its formal face when differentiation and integration are applied to special classes of functions. Computing a derivative according to the definition, from quotients of changes of a function, can be tedious. It is often easier to compute derivatives for certain functions that appear often, and then combine those results according to general rules until one finds the derivative of a particular function of interest. Once the derivatives of a large class of functions are known, also integration is facilitated because one no longer needs to sum up terms as the definition suggests; rather, one can use differentiation tables to look for a function whose derivative agrees with the function to be integrated.

Polynomials The derivative of the linear function $f(x) = Ax + b$ is $df/dx = \lim_{h\to 0}[f(x+h) - f(x)]/h = A$, a constant. The derivative of any monomial $f(x) = Ax^n$ with an integer n can be computed from $f(x+h) - f(x) = A(x+h)^n - Ax^n = Anx^{n-1}h + \ldots$, where $\ldots$ indicates terms that contain at least two factors of h. Dividing by h and letting h go to zero shows that the derivative is Anx^{n-1}. Such different functions can be combined by using the sum and the product rule: The derivative of the sum of two functions is the sum of the derivatives, and the derivative of the product fg is $f(dg/dx) + (df/dx)g$.

The first statements is rather easy to see. For the latter, we note that $f(x+h)g(x+h) - f(x)g(x) = f(x+h)[g(x+h) - g(x)] + [f(x+h) - f(x)]g(x)$. Dividing by h and letting h go to zero produces the two derivatives, and also sends the first factor of $f(x+h)$ to $f(x)$. Dividing by a function is the same as multiplying with its inverse, $1/f(x)$. Such derivatives can be computed using $1/f(x+h) - 1/f(x) = -[f(x+h) - f(x)]/[f(x)f(x+h)]$. Once again, dividing by h and letting h go to zero shows that the derivative of $1/f(x)$ is $-(\mathrm{d}f/\mathrm{d}x)/f(x)^2$.

Given an operation, interesting objects are always those that remain unchanged, or are reproduced to their original form after a finite number of repetitions. The operations are then easy to perform, even if the invariant or reproduced objects may be difficult to find or construct. Knowing that an operation does not change an object, or can be applied several more times to go back to the original form, has something reassuring.

Exponential The exponential function $\exp(x)$, shown in Figure 1.2, is defined as the function equal to its own derivative, with the condition that $\exp(0) = 1$: $\mathrm{d}\exp(x)/\mathrm{d}x = \exp(x)$. The function is always positive and ever-increasing because its derivative is positive, too. There is a number $e \approx 2.72$ such that $\exp(x) = e^x$ is its power by x.
From the invariant exponential function, we obtain reproducing functions by inserting suitable numbers in front of x. We have $\mathrm{d}\exp(kx)/\mathrm{d}x = k\,\mathrm{d}\exp(kx)/\mathrm{d}x$. If $k^n = 1$, the original function is reproduced after n steps of differentiation. For real numbers k, $k^n = 1$ can be achieved only for $k = 1$ (if n is odd) or $k = \pm 1$ (if n is even). However, if we allow complex numbers, more interesting possibilities arise. For instance, the imaginary unit is defined as the number i such that $i^2 = -1$. The functions $\exp(\pm ix)$ are reproduced after four steps of differentiation. Common symmetric combinations are the functions $C(x) = \frac{1}{2}[\exp(ix) + \exp(-ix)]$ $S(x) = -\frac{1}{2}i[\exp(ix) - \exp(-ix)]$, reproduced after four steps of integration and intercommuted by a single differentiation: $\mathrm{d}S(x)/\mathrm{d}x = C(x)$ and $\mathrm{d}C(x)/\mathrm{d}x = -S(x)$.

Trigonometry Algebra and geometry overlap in many areas, also regarding trigonometric functions. The sine and cosine functions allow us to compute angles from ratios of side lengths in a right triangle. The functions can be introduced as coordinates of a point moving along the unit circle, a geometrical realization. The coordinates, by reference to a right-angled triangle in Figure 1.3a with the origin of coordinates and the point on the circle as two corners and the third on the horizontal axis, are given by $(\cos(\sphericalangle), \sin(\sphericalangle))$ if $\sphericalangle$ is the angle by which the point has moved from the horizontal axis.

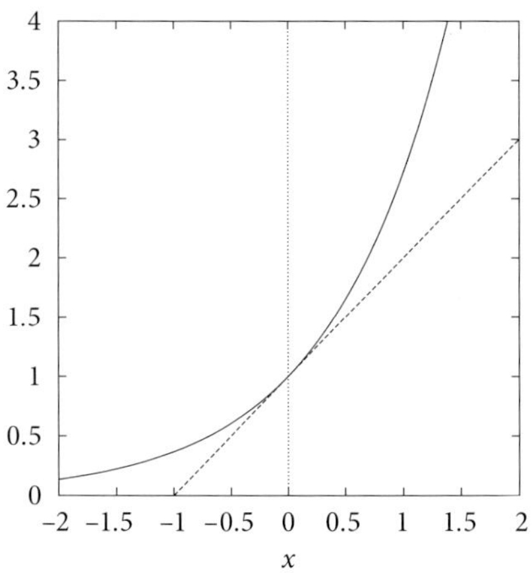

Figure 1.2 The exponential function $\exp(x)$, monotonic (ever-increasing) and always positive. Its slope grows by the same rate as the function increases. At $x = 0$, the slope is $45°$ (the dashed tangent is diagonal) and $d\exp(x)/dx = 1$. For all x, $d\exp(x)/dx = \exp(x)$.

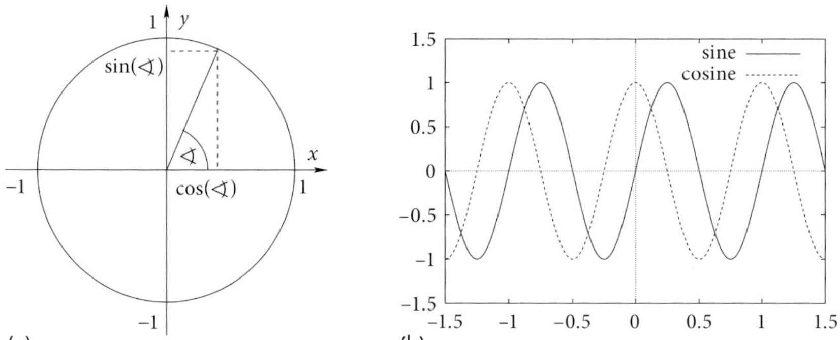

(a) (b)

Figure 1.3 Sine $\sin(\bigcirc x)$ and cosine $\cos(\bigcirc x)$, periodic functions. (a) Their definitions as coordinates of points on the unit circle of radius one. (b) Plots as functions of fractions $x\bigcirc$ of the full circumference.

The closure provided by the circle shows that sine and cosine are periodic functions (Figure 1.3): They return to their original values when we move around the whole circle. We write $\sin(\sphericalangle + \bigcirc) = \sin(\sphericalangle)$ and $\cos(\sphericalangle + \bigcirc) = \cos(\sphericalangle)$: if we add the full circumference of the circle to the arguments, the functional values do not change. It is often convenient to measure the circumference by the actual distance traveled along a unit circle, with radius one, in which case $\bigcirc = 2\pi$. Angles, fractions of the circumference, are then expressed as multiples of π, called radians, for instance, $\pi/2 = \frac{1}{4}\bigcirc$ for a right angle. However, one could also use standard degrees, such that $\bigcirc = 360°$. We will leave the option open by writing $\bigcirc$ unspecified.

The geometrical identification of sine and cosine with coordinates of points on a circle helps us to compute derivatives of these functions. To derive changes under small variations of the angle, some trigonometric identities are useful.

Trigonometric identities Figure 1.4 shows two rays at the angles $\sphericalangle_1$ and $\sphericalangle_2$, intersecting the unit circle. Adding the two angles means that we turn the second ray by an amount $\sphericalangle_2$, starting not at the horizontal axis but at the ray of $\sphericalangle_1$. The ray with angle $\sphericalangle_1 + \sphericalangle_2$ then intersects the unit circle at point q, and the coordinates of q are $x = \cos(\sphericalangle_1 + \sphericalangle_2)$ and $y = \sin(\sphericalangle_1 + \sphericalangle_2)$ as the lengths of the two dashed lines in Figure 1.4. In order to relate these functions to those of just one of the angles, we first draw several auxiliary lines, dotted in Figure 1.4. We draw a line starting at q and intersecting the ray of angle $\sphericalangle_1$ with a right angle, at a new point r. Turning the diagram by an angle $\sphericalangle_1$, so that the $\sphericalangle_1$-ray becomes horizontal (the dashed x'-axis), we see that this dotted line, according to the definition of the sine function, has length given by $\sin(\sphericalangle_2)$. This line, therefore, should be useful in relating the trigonometric functions of added angles to functions of individual angles. We draw two more auxiliary lines from r, one vertical until it intersects the horizontal axis and one horizontal until it intersects the dashed vertical line from q. We can now recognize the angle $\sphericalangle_1$ at two different places, both with the same size. At p, we have the original ray of angle $\sphericalangle_1$ with the horizontal axis. At q, as indicated in the figure, we have the dotted line from q to r, which by construction is perpendicular to the first ray, and the dashed vertical line perpendicular to the horizontal axis. Each of these two lines is perpendicular to one of the two lines making the angle $\sphericalangle_1$, and so the angle between them must be of the same size: it agrees with the previous one if we just rotate the two new lines by 90°.

At this stage, we have related the angles $\sphericalangle_1$ and $\sphericalangle_2$ to each other: The dotted line from q to r has length $\sin(\sphericalangle_2)$, and it has an angle $\sphericalangle_1$ with the vertical dashed line. Moreover, the length of the line from p to r, which has an angle $\sphericalangle_1$ with the horizontal axis, is $\cos(\sphericalangle_2)$. In the triangle with corners q, r and s, with s the intersection of a vertical line from q and a horizontal line from r, we can then write the length of the line from r to s as $\sin(\sphericalangle_1)\sin(\sphericalangle_2)$ and the length of the line from q to r as $\cos(\sphericalangle_1)\sin(\sphericalangle_2)$. The distance from s to the vertical axis, as recognized earlier, is $\cos(\sphericalangle_1 + \sphericalangle_2)$, and we can write it as the distance from r to the vertical axis minus the distance from r to s; see Figure 1.5. The distance from r to the vertical axis is the distance from r to p, which is $\cos(\sphericalangle_2)$, times $\cos(\sphericalangle_1)$. The distance from s to the vertical axis is thus

$$\cos(\sphericalangle_1 + \sphericalangle_2) = \cos(\sphericalangle_1)\cos(\sphericalangle_2) - \sin(\sphericalangle_1)\sin(\sphericalangle_2) . \tag{1.1}$$

As also shown in Figure 1.5, we recognize $\sin(\sphericalangle_1 + \sphericalangle_2)$ as the distance from q to the horizontal axis, or as the distance from q to s, $\cos(\sphericalangle_1)\sin(\sphericalangle_2)$, plus the distance from s to the horizontal axis, the latter the same as the distance $\sin(\sphericalangle_1)\cos(\sphericalangle_2)$ from r to the horizontal axis. Equating these two expressions, we have

$$\sin(\sphericalangle_1 + \sphericalangle_2) = \cos(\sphericalangle_1)\sin(\sphericalangle_2) + \sin(\sphericalangle_1)\cos(\sphericalangle_2)\,. \tag{1.2}$$

The trigonometric identities allow us to demonstrate that the sine and cosine functions are reproduced by four steps of differentiation, intercommuting them after a single step. The interplay of algebra and geometry facilitates the calculations, an example for the usefulness of being able to take both the algebraic and geometric viewpoints.

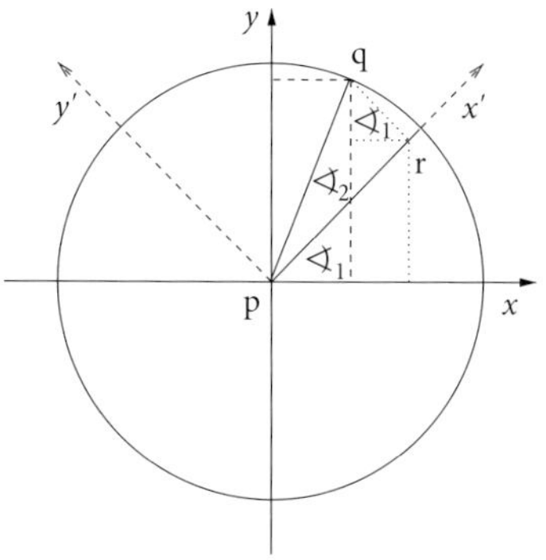

Figure 1.4 A diagram demonstrating the addition theorem of trigonometric functions. The lengths of the dashed lines are the sine and the cosine of $\sphericalangle_1 + \sphericalangle_2$; the dotted lines are auxiliary. The angle $\sphericalangle_1$ appears twice in the diagram and helps to relate $\sin(\sphericalangle_1 + \sphericalangle_2)$ and $\cos(\sphericalangle_1 + \sphericalangle_2)$ to the trigonometric functions of $\sphericalangle_1$ and $\sphericalangle_2$.

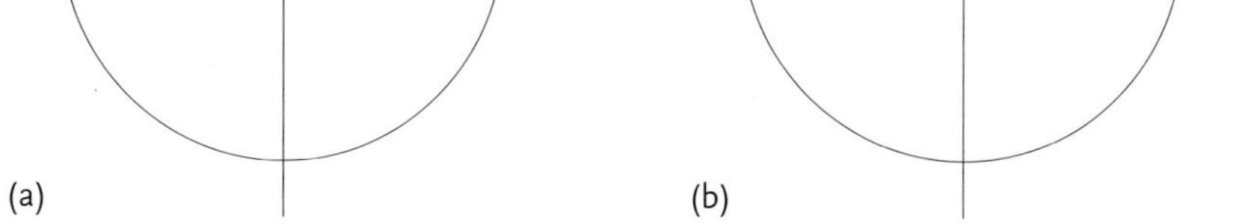

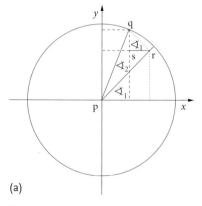

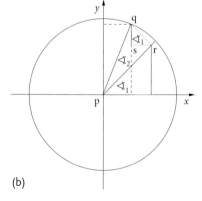

Figure 1.5 The same diagram as in Figure 1.4 with additional solid lines showing how the lengths of the dashed lines, which are $\sin(\sphericalangle_1+\sphericalangle_2)$ and $\cos(\sphericalangle_1+\sphericalangle_2)$, respectively, can be calculated by adding smaller pieces related to the trigonometric functions of $\sphericalangle_1$ and $\sphericalangle_2$. (a) illustrates how to obtain $\cos(\sphericalangle_1+\sphericalangle_2)$ by subtracting the distance from r to s from the distance from r to the vertical axis. In (b), we obtain $\sin(\sphericalangle_1 + \sphericalangle_2)$ by adding the distance from q to s to the distance from s (or r) to the horizontal axis.

Euler's identity We write $\sin(x+h) - \sin(x) = \sin(x)\cos(h) + \cos(x)\sin(h) - \sin(x)$. For small h, $\cos(h)$ is close to one, $\sin(h)$ close to the length of a circle segment with angle h, the fraction $2\pi h/\bigcirc$ of the full circumference. The first and last terms cancel, while the middle term is proportional to $2\pi h/\bigcirc$. Dividing by h gives $\mathrm{d}\sin(x)/\mathrm{d}x = 2\pi\cos(x)/\bigcirc$. Similarly, $\cos(x+h) - \cos(x) = \cos(x)\cos(h) - \sin(x)\sin(h) - \cos(x)$ produces $\mathrm{d}\cos(x)/\mathrm{d}x = -2\pi\sin(x)/\bigcirc$ from the middle term. The trigonometric functions show the same behavior of change as the functions we defined using the complex exponential $\exp(2\pi i x/\bigcirc)$. We write $\exp(2\pi i x/\bigcirc) = \cos(x) + i\sin(x)$, Euler's identity. As a special case, $e^{i\pi} = -1$ with $x = \frac{1}{2}\bigcirc$.

Algebra and geometry The interplay of algebra and geometry has been important ever since the beginnings of modern physics; it even played a large role before physics as its own field of study was established. The philosopher René Descartes was the first to algebraize geometry in a form still used today. By describing the shape of a geometric object in terms of coordinates, that is, unique numbers assigned to all points in a plane or in space, most geometrical questions, such as those about the length of a curve, relationships between angles and side lengths of triangles, or the areas of figures, can be formulated in algebraic terms and then be answered with established methods to solve equations. Doing geometry is formalized to the extent that a direct geometrical picture, the ability to visualize geometrical objects and relations between them, becomes secondary. Only in this way has it been possible to drive mathematics to higher and higher levels of abstraction, leading for instance to the concepts of higher-dimensional and curved space. Despite their abstraction and the loss of direct imagination, these concepts can be real, as evidenced by modern theories of physics, in particular relativity.

Using the bridge between algebra and geometry both ways, geometrical visualization can also be employed to understand and derive algebraic relationships. Algebra and geometry are two sides of a coin, both essential. Intuition remains important to investigate geometry, pose interesting questions and interpret algebraic results; it was not Descartes' intention to cut all this short by dry, formalistic manipulations. As geometry advanced over the centuries, more complicated figures had come to focus which were difficult to visualize even for trained and experienced geometers, at a time well before computer-aided visualization tools became available. There are examples in books and letters in which properties of curves were analyzed correctly even though accompanying sketches of them showed the wrong behavior.

The Cartesian leaf, named in a not-so-obvious way after René Descartes, is the curve made of all points (x, y) that obey the "implicit" equation $x^3 + y^3 = pxy$ with a given value for p. The curve, shown in Figure 1.6, is of historical interest because it showed the importance of Fermat's methods of computing tangent lines to curves. Unlike Fermat, Descartes was unable to solve this problem.

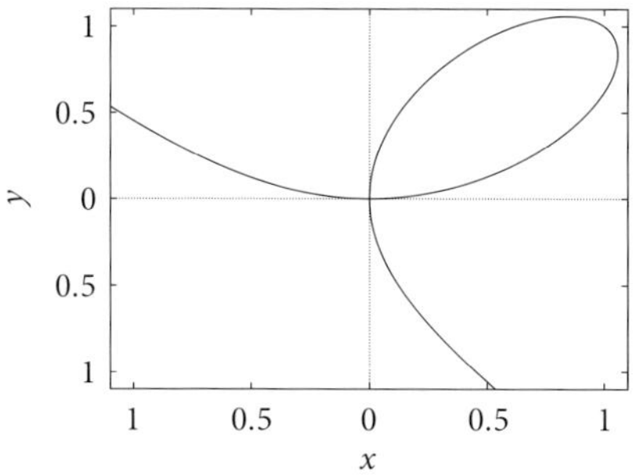

Figure 1.6 The Cartesian leaf: the set of solutions of the equation $x^3 + y^3 = 2xy$.

Numbers Neither Descartes nor Fermat had a correct understanding of the shape of the Cartesian leaf, as shown by the incorrect diagrams found in their correspondence. Nevertheless, the results obtained by Fermat for tangent lines to the curve were correct, a fact which shows the importance of algebra in situations in which our intuition no longer suffices to visualize mathematical constructions. A similar sentiment was voiced about 200 years later by Carl Friedrich Gauss in a letter to Friedrich Wilhelm Bessel in 1830, in which he declared arithmetic to have primacy over geometry, saying that only the former's laws are "necessary and true." This correspondence happened under the impression of new versions of geometry, discovered by Gauss and others, that did not obey all the traditional rules laid out by the ancient Greeks, known as Euclid's axioms. By using algebra, Gauss was able to perform geometry even in settings outside of the traditional realm of this field. Geometry and its rules were generalized, while algebra remained the same. By now, Gauss' sentiment toward algebra as primary compared to geometry is no longer considered valid. Our imagination may fail, but not geometry. And just as geometry, also algebra can be formulated in different systems, depending on which of its rules one would like to be obeyed.

Multiplication The algebraic operations we know can be realized in many different ways. Multiplication is consistent in a set of just two elements, which we may call "O" and "M," if we declare their products to be $O \cdot O = O$, $O \cdot M = M$, $M \cdot O = M$ and $M \cdot M = O$. Taking these products does not result in any object other than O and M; the multiplication operation is closed. Moreover, one can verify the laws $a \cdot b = b \cdot a$ (commutativity) and $a \cdot (b \cdot c) = (a \cdot b) \cdot c$ (associativity), well-known from our usual algebra. In fact, the product rules provided are satisfied if we identify O with the number 1 and M with -1.[3] The rules we defined are not new, but we do obtain a new algebra by restricting all numbers to just ± 1. (Other restrictions do not give rise to closed multiplication.)

Commutativity and associativity are commonly used rules when we calculate, sometimes implicitly so: When we write $a + b + c$ the result is clear even if we do not use parentheses to specify which numbers are to be added first. In any case, $(a + b) + c = a + (b + c)$. Although commutativity and associativity are common, not all operations obey these rules. Language is neither commutative nor associative. An interesting example for noncommutativity is the statement[4] "A life is worth nothing, but nothing is worth a life." with two parts of opposite meanings after switching just some words. In mathematical parlance, saying the operation or comparison "is worth" is not always commutative in language, even though its mathematical analog, the equality, is. Also, the law of associativity is not always respected by language. For unique meanings, we set parentheses not in the mathematical way but by using hyphens, for instance, to distinguish "mock turtle soup" from the nonexisting "mock-turtle soup."[5]

A set of numbers (or other objects, for what makes a number a number, in the eyes of a mathematician, is just the kind of rules it obeys) is called a group if (i) one can multiply any two of its elements, (ii) the multiplication of three elements does not depend on whether one multiplies the product of the latter two with the first or the product of the first two with the latter, (iii) there is a unit element which does not change any other number when multiplying by it, and (iv) any element can be inverted so as to provide the unit element when multiplied with this inverse number. The inverse provides the possibility of division, or of subtraction, for we could have stated the same laws with "addition" instead of "multiplication," "sum" instead of "product," "zero" instead of "unit element," and "negative" instead of "inverse." A group, however, does not offer both multiplication and addition at the same time, so its elements cannot be fully respectable numbers in the common sense.

Playing matches at tournaments is a nonassociative operation. We define an operation $W(A, B)$ as the winner of two teams A and B playing against each other, assuming that there is a mechanism to avoid ties, such as penalty kicks. It is not clear what $W(A, A)$ should be, but there is no harm in just declaring it as $W(A, A) = A$. The zero of this operation is a team that always loses (a true zero): $W(A, 0) = A$ for all A. There is no negative team because $W(A, B)$ is either A or B, not 0 unless $A = B = 0$. (A negative team could at best be considered one with a demoralizing spell, turning every opponent into a permanent loser. However, if two losers have to play each other, one should win.) One of the usual laws is violated, but

3) Formulated as addition,
"In the twigs of the cherry tree
I saw two naked pigeons;
the one was the other,
and the two were no one."
(F. Garcia Lorca: Song)

4) André Malraux: *Les Conquérants.*

5) "Then the queen left off, quite out of breath, and said to Alice, 'Have you seen the Mock Turtle yet?' 'No,' said Alice, 'I don't know what a Mock-turtle is.' 'It is a thing Mock Turtle Soup is made from,' the Queen said." Lewis Carroll: *Alice in Wonderland.*

we can still test commutativity and associativity. We don't distinguish the order of both teams, and so W is commutative, $W(A, B) = W(B, A)$. (It may not be commutative when there is a rematch to have both teams enjoy the home advantage.) Associativity is not realized: $W(A, W(B, C))$ does not always equal $W(W(A, B), C)$. If A loses to B, and B loses to C, A may still win against $C = W(W(A, B), C)$, but $W(A, W(B, C)) = A$.

Matrices A set of objects different from the set of usual numbers, but still similar in the rules obeyed, is obtained if we consider the so-called 2×2-matrix algebra with elements

$$M = \begin{pmatrix} a & b \\ c & d \end{pmatrix}$$

for all ordinary numbers a, b, c and d. We define addition and multiplication by

$$\begin{pmatrix} a_1 & b_1 \\ c_1 & d_1 \end{pmatrix} + \begin{pmatrix} a_2 & b_2 \\ c_2 & d_2 \end{pmatrix} = \begin{pmatrix} a_1 + a_2 & b_1 + b_2 \\ c_1 + c_2 & d_1 + d_2 \end{pmatrix}$$

and

$$\begin{pmatrix} a_1 & b_1 \\ c_1 & d_1 \end{pmatrix} \cdot \begin{pmatrix} a_2 & b_2 \\ c_2 & d_2 \end{pmatrix} = \begin{pmatrix} a_1 a_2 + b_1 c_2 & a_1 b_2 + b_1 d_2 \\ c_1 a_2 + d_1 c_2 & c_1 b_2 + d_1 d_2 \end{pmatrix} .$$

Again, the rules of commutativity and associativity are satisfied for addition, as is distributivity defined as $M_1 \cdot (M_2 + M_3) = M_1 \cdot M_2 + M_1 \cdot M_3$. Multiplication, while still associative, is not commutative as one can check. Another rule which we have to forgo for these "numbers," in general, is division. For two nonzero matrices M_1 and M_2, there is not always a third matrix M_3 such that $M_1 M_3 = M_2$. (However, there is always a negative matrix which we can subtract from a given one to obtain zero.)

Mathematicians call an "algebra" a set of objects any two elements of which can be added or multiplied, or a "division algebra" if there is also division (or inversion) among the objects. Commutativity, or even associativity, is often forgone in order to allow a larger variety.

Complex numbers A nonreal set of numbers in which all the familiar algebraic rules are satisfied, is the complex numbers, defined as expressions $a + bi$ where a and b are real numbers and i is another number obeying all the usual rules except that $i \cdot i = i^2 = -1$. No real number satisfies this relation, and so the numbers defined here are new. Using this rule, together with familiar ones such as distributivity, we can compute products such as

$$(a + bi)(c + di) = ac + bci + adi + bdi^2 = ac - bd + (bc + ad)i ,$$

again of the form defined. With this multiplication rule, we can confirm that we divide by any nonzero number by multiplying with its inverse

$$\frac{1}{a+bi} = \frac{a}{a^2+b^2} - \frac{b}{a^2+b^2}i\,.$$

This is a good complex number as long as a or b is nonvanishing, that is, as long as $a+bi \neq 0$. We can divide by any complex number other than zero, and all the usual rules apply.

Numbers that can be added, subtracted, multiplied and divided by can be defined in different ways, but there are not many options, only four. Two of them are the familiar real and less familiar complex numbers; two more are the quaternions and octonions. Complex numbers are pairs of real numbers, a two-dimensional extension. Quaternions are pairs of complex numbers, or sets of four real ones, and octonions again double the dimension. However, as we go up in dimensions, we lose laws. Quaternion multiplication is not commutative, and octonions don't even obey the law of associativity. If we try to extend them further, no number system is possible. Our laws of numbers stand firm; they cannot be realized in arbitrary ways. Real numbers indeed seem real.

Quaternions We keep the i of complex numbers, and include two more, j and k, with the same property that they square to -1. Moreover, we declare $ijk = -1$. With these properties, the multiplication of "four-dimensional" numbers $a+bi+cj+dk$, the quaternions, is determined. Multiplying the equation $ijk = -1$ from the right with k and using $k^2 = -1$, we derive $ij = k$; multiplying with i from the left, we have $jk = i$. If we multiply the latter equation with j from the left, we obtain $-k = ji$, differing from $ij = k$ by its sign. Quaternion multiplication is not commutative: results depend on the order of factors.

Space-time Descartes' algebraization was important, not only for the long-established subject of geometry, but also for the emerging field of physics. It preceded calculus, as seen in the example given by Fermat, a contemporary of Descartes' with whom he kept correspondence. And calculus, in the early days, was very much developed in parallel with laws of motion needed to address questions in physics. Stepping far ahead, much of modern physics would not be possible without an algebraic view on geometrical concepts, such as spaces of dimension larger than three, which cannot be visualized by our minds that evolved and grew up in a three-dimensional environment. Algebra, and with it an operational instead of visual view, has for a long time taken primacy in our understanding of physics, until his insights into relativity forced Albert Einstein to geometrize physics. Nowadays, geometrical formulations, even if the geometry in them may be highly abstracted and hardly recognizable, are considered more powerful and even more fundamental than algebraic ones. Most powerful, however, remains the interplay of algebra and geometry as it is needed to imagine new relations and then do computations to realize them.

One of the first instances in which the new geometrical view of physics became crucial was the understanding of space-time in relativity. In a rather different way, Immanuel Kant already associated the pair of algebra and geometry with the distinction of space and time. Geometry, according to him, is associated with space; we visualize objects in space and determine their relationships when we do geometry. Algebra, on the other hand, he associated with time; we do algebraic operations, such as manipulations of an equation in order to solve it, in sequential order, one at a time. This view of time suggests an operational interpretation, in which what happens, and ultimately the whole universe, amounts to running a computer program, a sequence of operations. It seems a small step to conclude that also our role as human beings is nothing but the outcome of a calculation, just to be fed into another part of the program, an idea which seems rather popular in modern culture. As for understanding the universe (leaving aside humans), such a view is not justified in modern physics, where not just the algebraic ideal of following rules but also intuition as embodied by geometry is central. A unified view of algebra and geometry has emerged in physics, at just about the same time when the two poles of space and time, brought in contact with the mathematical fields by Kant's arguments, have conglomerated into a single whole: space-time.

Personal force Ideas are often driven by and associated with individuals. It was Descartes who had the thought of endowing the empty plane, the stage of geometry, with a structure of its own, laying on it a grid of numbers to name its points; for this idea, we still honor him by calling those numbers Cartesian coordinates. Much later, Einstein began contemplating how these coordinates, that is, numbers to identify points, chosen with almost as much arbitrariness as the names we assign to people (or pets), can, in some cases, be equipped with physical meaning, related, for instance, to distances between points measured by different observers. The coordinates are not unique, and two different observers need not always agree on their values. However, transformation formulas can be provided that allow one to compute the numbers assigned by one observer from those obtained by another one. After that, Einstein developed a framework by which physical predictions can be derived in a way insensitive to the arbitrariness of coordinates, even if coordinates are used to do explicit calculations. From a mathematical perspective, these two steps were crucial for the geometrization of modern physics; from a physical one, they introduced special relativity (the laws of transformation formulas to relate results by different observers in space-time) and general relativity (a framework to describe space-time independently of the arbitrariness of coordinates).

As in these examples, the personal fate and success of an individual can be inspirational; stories of people, in lively contrast to the impersonal results of science, can motivate (or sometimes deter). However, in the inevitable selection of stories to tell, there lies a danger. The importance and influence of those named may seem exaggerated; those unnamed one cannot see. Like ants exploring a jungle to scavenge for food, scientists start at a firm base and spread out to reach new territory. To stay secure, one should not lose ties to others; ties may sometimes be thin, but are strengthened by success: more are attracted to the same spot, to help or envy. Or ties die down when a source is depleted, most scavengers moving elsewhere. It is not a genius ant which finds new food and brings home most of it, but one that is hard-working and keeps its eyes open, to be ready at the right moment and place. It is difficult to do justice to the many contributors to science; referring to the usual stories about those already standing in bright light only makes things worse by further dimming all others. There is not much more annoying than having justice attempted in an unjust way.

What is more, focusing on the bright side ignores, downplays or denies the darkness that can befall scientific developments, like all human activities. It goes without saying that scientists sometimes make mistakes, errors which may be worth pointing out, not to slander but because one can learn much even from mistakes. The best scientists have not hesitated to admit mistakes; some even highlighted them to warn or teach others. A significant number of important theories, including general relativity and quantum mechanics, started with versions that were, in strict terms, wrong. And yet, they were of great influence; the "correct" theories we now work with would not have come to existence without their predecessors. Initial errors were results of some kind of incompleteness which could rather easily be overcome once the groundwork was set down. Nobody should be ashamed of such errors, and no one should fault anybody for them. Only mediocre individuals need be afraid of talking about their mistakes, a true scientist can rest assured in the validity of Theodore Roosevelt's ode to the hero: "It is not the critic who counts: not the man who points out how the strong man stumbles or where the doer of deeds could have done them better. The credit belongs to the man who is actually in the arena, whose face is marred by dust and sweat and blood, who strives valiantly, who errs and comes up short again and again... who spends himself in a worthy cause; who, at the best, knows, in the end, the triumph of high achievement, and who at the worst, if he fails, at least he fails while daring greatly." For physicists trying to translate the book of nature, it holds that[6] "Translation, almost by definition, is imperfect; there is always 'room for improvement,' and it is only too easy for the late-comer to assume the *beau rôle*."

Darkness in science can reach personal levels. Scientists must sometimes face attacks from outside, hindering progress.[7] Although science does occasionally follow the mythical idea of a siblinghood with unselfish interest in understanding nature, many moves, including those by some prominent scientists, can only be understood as attempts to keep competition in check. Science, much like politics and so much else, is to a large degree powerplay. Not only has modern science with its large collaborations and often high stakes (not to speak of tight job and funding markets) made acquaintance with the role of power; examples can be found in the

6) Terence Kilmartin: *A note on the translation* [of *À la reserche du temps perdu* by Marcel Proust].

7) In a letter to Marin Mersenne, Descartes expressed hesitation to publish his new results, in fear of sharing Galilei's fate brought about by the Inquisition. The letter is dated November 1633, not long after Galilei's sentencing on June 22nd, 1633.

days when it was pushed only by a few. Galileo Galilei, for one, attacked his contemporary, and great ally in heliocentric pursuits, Johannes Kepler, with whom he otherwise corresponded in a friendly manner, on the theory of the tides, developed in differing ways by both scientists. (Kepler's version turned out to be closer to the correct explanation.) In retrospect, history seems to be made by the powerful, but in a more Tolstoyan view, progress is driven by the often unseen masses, by the many contributions from individuals.

Some scientists, of course, deserve more credit than others. When contributions made by someone are especially important and attract many followers, even more so when new contributions are made by the same person several times in a lifetime, the person, not the contributions, is often placed in the foreground. However, even if a person is granted multiple discoveries, personal acclaim may not always be justified. Power gained from the first success becomes a self-enhancing tool to facilitate more success. More funding and colleagues can be attracted with more power, and new influential work is a consequence gained more easily than by someone starting from scratch. Power is attracted by the powerful; only during revolutions does transfer of power happen in the opposite way. (Einstein, for instance, joined the powerful after his first works, but his role is special: he continued to stay outside the mainstream, preferring to follow his own ideas as far as he could without the need to buttress old successes. He did not follow and he did not force. One might even go as far as saying that this was his greatest achievement.) Science is often seen as open to diverse opinions, but it is much less susceptible to revolutions or true democracy. The powerful stand in bright light, while others may campaign in vain. In addition to the self-enhancement caused, there is, sadly, a form of corruption in science. Publications, citations and the credit one receives are the currency of science. Power increases one's visibility, and can be used to gain more credit by self-enrichment. Sometimes, the research field occupied by a powerful scientist appears as a fiefdom, as a piece owned by the scientist. Opposing opinions are suppressed, and credit is given to close followers or within a small citation circle of scientists primarily citing one another's work. In the end, the competition stifled, it is not always the correct view that emerges.

When examples of scientists are given, one should see them not as owners, in general, not even as originators of what they contributed. The admirable way of doing science is not to conquer, but to explore; of being creative, not to sire, but to conceive. The new arises not in brief ecstatic breakthrough moments, but by long, painful nurturing. In the main story lines of science, discoveries have been made by long contemplations of existing knowledge and of what was dissatisfying or inconsistent about it. Often with luck and hard work, one converged to a better version. A successful scientist does not own contributions any more than a construction worker owns the house he helps to build. If a piece of the truth is owned by someone, or by a group of people, it ceases to be true.[8)]

8) "GALILEI: The truth is a child of time, not of authority." Bertold Brecht: *Life of Galilei.*

Mathematics The hallmark of science is its principle that, in trying to understand nature, the ultimate judgment about the correctness of our ideas can only come from nature herself. We observe phenomena, form ideas about their origin and build theories about their workings, then perform experiments to test our theories, comparing theoretical predictions with observations. Science begins and ends with nature and our role in it.

In the intermediate stages of this process, one sees another virtue of science: mathematics used to organize observational data, to model phenomena, and to derive new predictions. The true and unique virtues of science can be found in this interplay of noting nature and minding mathematics, the two incorruptible pillars in the world of the mind. Without mathematics, an understanding of the universe as profound as we have it today would not have been possible. Mathematics in science holds a role both of power and of artistic[9] value, a role we will come back to throughout this book.

True to this role of mathematics, equations will be used to strengthen the arguments given in the discussion of phenomena. As already seen, these formulas, most of them at least, appear in boxes, that is, quarantined, to be glanced over by hesitating eyes, lest they spread out like ants and multiply, driven by inquisitive minds, reaching out to questions unanswered or unasked. To witness this nonmathematical multiplication of mathematics, one may look at any theoretical-physics monograph or research article.

Gravity The best-known force is gravity, a force which always acts to attract masses no matter how far apart they are. The magnitude of the force gets weakened as the masses are more distant, but remains. The force is inversely proportional to the square of the distance. It also increases proportionally when either of the masses is enlarged. The precise value of the force, multiplying the values of the two masses between which the force acts, is determined by a constant factor called Newton's constant. It can be determined by measuring the force between two masses, and then takes the same value for all other masses.

At least in principle, one measurement would suffice. Since measurements are always subject to uncertainties and errors, for instance, from imprecise readings of the instruments, one should perform a large number of measurements and then average over their outcomes, so as to cancel out random errors which go one way (leading to a larger value) as well as the other (a smaller value). Even so, the great achievement of finding a force formula is apparent from the fact that a finite number of measurements can explain the potential infinity of phenomena realized for the gravitational force acting between differently positioned masses. In particular, we can measure the force and Newton's constant on Earth, in a controlled laboratory, and then use the same value for the masses we see moving in the sky, the Sun, the Moon and the planets. The first systematic measurements of this type were

9) Beauty in mathematics may not be easy to spot. "The author claims to have discussed a mathematical hypothesis. But he endows it with a tangible visualization, something mathematicians never do," wrote one of the censors of Galilei's inquisitorial proceedings in 1633.

performed by Robert Cavendish, establishing the universality of Newton's law of gravity.

> **Newton's law** We write Newton's gravitational force as $F = -GM_1M_2/r^2$ where G is Newton's constant, M_1 and M_2 are the two masses, and r the distance between them. The minus sign is included to indicate that the force is attractive, pulling the second mass toward the first, against the direction in which we would walk from the first mass.

As with acceleration before, we have two formulas for the same object, F. We can view Newton's law as a definition of the values the force takes at different places if mass M_2 is moving around mass M_1. Newton provided not much more than such a definition based on the observations available to him. As physics progressed, the need for a derivation of the force formula from other principles became apparent, requiring new insights into nature. A related fact has led us to use different symbols, namely, uppercase letters M_1 and M_2, for the masses instead of lowercase ones as before in $F = ma$. There is no obvious reason why the value of a mass that causes the gravitational force should be the same as the value that tells us how an object accelerates reacting to a force.

Vectors When we use force, it matters which way we push. Like all other forces, the gravitational one is characterized not only by its magnitude but also by its direction. The force exerted by one object like the Sun on another object such as the Earth points from the second mass to the first one. The force always points along a line connecting the two objects (as long as their extensions are small compared to the distance, otherwise the force points between the centers of mass). If the objects move, the direction of the force seen in space, not with respect to the line connecting the masses, changes. To keep track of the force, we must consider not just one number as assumed so far, but a number for the magnitude together with additional information to characterize the direction. The additional piece could be the values of two angles because a direction in space is determined by two numbers instead of a single one as on the plane, just as we determine points on the surface of Earth by two numbers, latitude and longitude. It is often more convenient to keep track of the direction of the force by visualizing it as an arrow at every point, the length of which corresponds to the magnitude of the force and the direction to the direction in which the force is acting. Such an arrow, or the set of numbers representing it, is called a vector.

> As Cartesian coordinates label points in space, we use coordinates to represent vectors, then called vector components. We pretend that the base point of the arrow visualizing the vector is the origin of space, and read off the vector components as coordinates of the endpoint. Denoting a vector by an arrow to distin-

guish it from a number, we write $\vec{F} = (F_1, F_2, F_3)$ as the collection of its components F_1, F_2 and F_3.

We arrange directional information by vectors, even though a collection of all the vector components as a set of numbers would have the same content. Vectors may represent actual directions in space, like that of a force, or they may represent more abstract information. Well-known examples for the latter are dials such as those on an analog clock or a speedometer. The hands of an analog clock indicate time by the way they point, even though time itself does not have a direction in space. There is no direct correspondence between time and the directions of the hands other than one based on convention. Still, the spatial representation has several advantages; for instance, it is easier to follow the progress of time, and the cyclic nature of most processes is mimicked by the closed dial repeating itself after every turn.

Another convenient feature of vectors is the possibility to do calculations with the entire object. We add two vectors $\vec{F} = (F_1, F_2, F_3)$ and $\vec{G} = (G_1, G_2, G_3)$ by adding their components, which corresponds to the concatenation of the two directions: $\vec{F} + \vec{G} = (F_1 + G_1, F_2 + G_2, F_3 + G_3)$. We compute the magnitude of a vector $\vec{F}$ from its components, which we denote as $|\vec{F}| = \sqrt{F_1^2 + F_2^2 + F_3^2}$ and which amounts to measuring the length of the arrow. The equality of the length and the quadratic expression follows from the Pythagorean theorem, the most fundamental statement of (Euclidean) geometry.

Pythagorean theorem Names don't matter. When we use coordinates to algebraize geometry, we must make sure that the results our calculations provide are insensitive to some arbitrary choices we make in choosing the coordinate grid. Just as we can rotate the sheet of paper we draw geometric figures on, or move it around on the desktop, we are free to rotate or translate the coordinate system we use. Individual labels of coordinates are devoid of direct geometrical meaning, but there are quantities, which then give rise to the unquestionable rules of geometry, that do not change when we rotate or translate the coordinate grid. The simplest such quantity is the length of a straight line, or the distance between two points, for the distance can be measured without using any coordinates; its value must be independent of the choice of coordinates we make.

An important question in algebraized geometry is how to calculate coordinate-independent quantities from the coordinate expressions we use for instance to represent individual points. To continue with the example of the length of a straight line, we read off the coordinate differences of its endpoints by drawing lines perpendicular to the coordinate axes. Computing the length from the coordinates then amounts to relating the side lengths in a right-angled triangle (Figure 1.7), a task taken care of by the Pythagorean theorem: if a and b are the lengths of the two

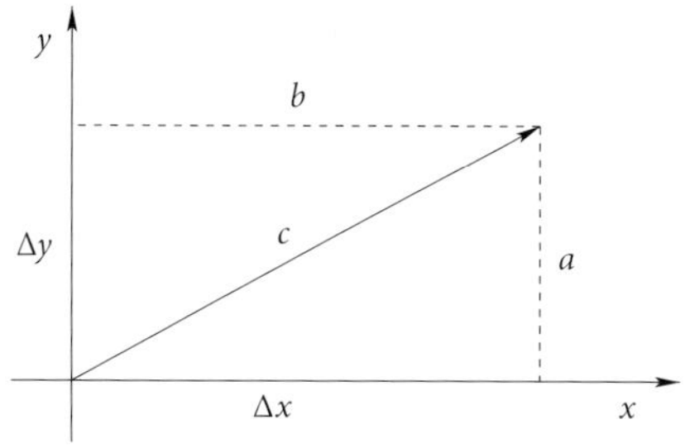

Figure 1.7 The length c of a straight line (arrow) is computed from the coordinate intervals $b = \Delta x$ and $a = \Delta y$ along its extension. As the side lengths of a right-angled triangle, the values of a, b and c are related to one another by the Pythagorean theorem.

shorter sides of a right-angled triangle, by necessity the two sides cornering the right angle, then the third side has length c related to a and b by $a^2 + b^2 = c^2$.

> **Line element** We write this equation as $\Delta s^2 = (\Delta x)^2 + (\Delta y)^2$, rebranding $a = \Delta y$, $b = \Delta x$ and $c = \Delta s$. An expression for the distance Δs in terms of coordinate differences Δx and Δy is called a line element, the most fundamental law of geometry.

Squared numbers appearing in the Pythagorean theorem suggest that, in geometrical terms, the theorem could express a relationship between areas of squares erected upon the sides of the triangle, of side lengths a, b and c. This observation indeed points the way to a proof of the theorem, but recognizing the relationship between the squares is far from obvious. In the special case of $a = b$, the value of $c = \sqrt{2}a$ can be interpreted as the length of the diagonal in a square of side length a. The square in question is obtained by combining the right-angled triangle, in this case also being isosceles, with its own reflection along its long side. Further reflections and translations of the triangle, as in Figure 1.8, provide the squares erected on the sides of areas a^2 (obtained twice) and c^2.

We complete the proof of the Pythagorean theorem for an isosceles right-angled triangle, $a = b$, by looking at the reflections indicated in Figure 1.8. We obtain three new triangles by reflecting the original one on its three sides. As perfect reflections, they share all properties with the original triangle, such as relationships between the side lengths or angles; they also have the same areas. This is one of the basic laws of geometry, for the geometrical relationships we see in a perfect mirror cannot differ from what we see without the reflection.

As indicated in the figure, we reflect the new triangles once more until we obtain all the eight reflected triangles denoted as Δ. By the properties of all Δs as right-angled triangles, the three separate regions covered by the reflected triangles are squares, two of side length $a = b$ and one of side length c; we have two squares of area a^2 and one square of area c^2. Counting the triangles, we see that the squares of side length a are made of two copies of the original triangle, while the square of

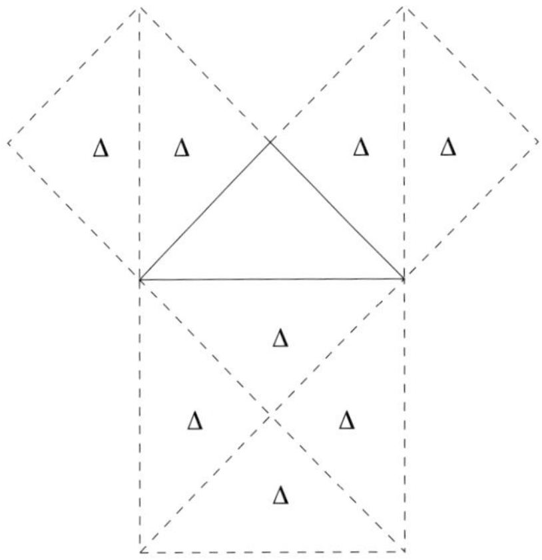

Figure 1.8 Proving the Pythagorean theorem for equal short side lengths by reflecting the original triangle several times. Counting the number of triangles shows that the two squares erected on the short sides of the triangle occupy the same area as the square erected on the long side. All triangles labeled with Δ have the same size and proportions as the original one.

side length c is made of four copies. The reflected triangles do not overlap, and we conclude that the area covered by the large square, made of four triangles Δ, must be the same as the two areas covered by the small squares together, each made of two triangles Δ. We have proven that $c^2 = a^2 + b^2$ in the special case of $a = b$.

Proofs of special cases do not always lend themselves to obvious generalizations; sometimes, considering a more general version than one is interested in allows one to recognize hidden patterns which might even simplify special cases. For the question at hand, the figure we obtain by reflecting our triangle, now with $a \neq b$, is not a simple square (but rather a kite, or deltoid). Judiciously placed reflections nevertheless help to prove the Pythagorean theorem in its full beauty; the new difficulty is that some of the reflections should also be moved in order to show the proper relationships of surfaces.

Attempting to repeat the previous constructions, but now applied to $a \neq b$, we arrange four reflections of our triangle as shown in Figure 1.9. The outermost sides of the four reflected triangles enclose an area of size c^2, that is, they form a square of side lengths given by the longest side of the triangle, but they do not fill this area; the black square in the figure, of size $(b-a)^2$, remains uncovered. (One side of the black triangle runs between the endpoint of the side of length b of the top gray triangle and the endpoint of the side of length a of the right triangle; thus the side length $b - a$.)
We could complete the proof by combining the geometrical considerations presented so far with some algebra: The area covered by the four gray triangles is $c^2 - (b - a)^2 = c^2 - b^2 - a^2 + 2ab$, as given by the difference of the areas of

two squares. The area of a single triangle is then $\frac{1}{4}(c^2 - b^2 - a^2 + 2ab)$. On the other hand, one of these triangles is half a rectangle of side lengths a and b, so its area must be $\frac{1}{2}ab$. Combining these two equations, we have $c^2 - b^2 - a^2 = 0$, the Pythagorean theorem.

Although the algebraization of geometry is a powerful tool, mixing geometrical with algebraic considerations is often considered bad style by the purists. Either stay geometrical all the way through an argument, or turn to algebra from the very beginning. Nevertheless, alternative viewpoints can often provide added insights, even if a calculation or proof has already been completed by some means.

Figure 1.10 shows a rearrangement of the reflected gray triangles of Figure 1.9. The gray area is of the same size as before, since one can recognize in it the same four triangles without overlap. Also included in the figure is the black square not overlapping with any of the triangles. As indicated by the solid lines, the four gray triangles together with the black square occupy an area which equals the size of two larger squares, one of side length a and one of side length b. This area is of the size $a^2 + b^2$, and by Figure 1.9, it equals the area c^2 of the square seen there. Since both areas are made from the same pieces, four gray triangles and the black square, they are equal: $a^2 + b^2 = c^2$.

A geometrical proof, in contrast to churning out a list of algebraic formulas, is often more intuitive (though tricky) because it appeals to our visual sense, not to the nonexistent sense of numbers. If the result is surprising and yet becomes evident by the geometrical considerations suggested, the result is often perceived as beautiful not just (hopefully) by mathematicians.

Dimensions With geometry let loose from the confines of intuition, dimensions can reach unexpected heights. As an example of Descartes' algebraization of geometry by the introduction of coordinates to label points, vectors algebraize directions in space. There is an unexpected pay-off which in algebraic terms seems simple

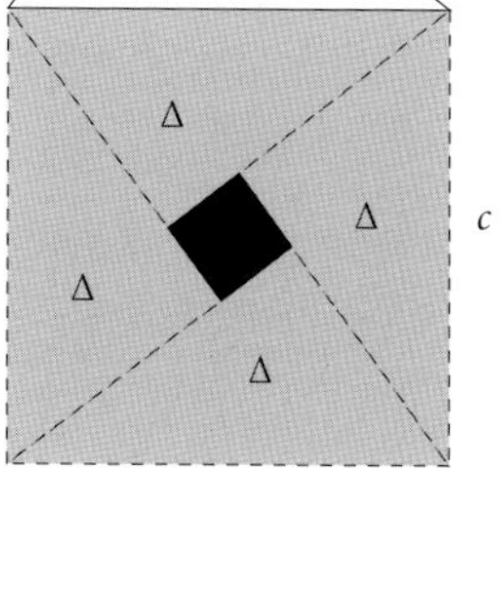

Figure 1.9 Reflected triangles do not fill a whole square of area c^2.

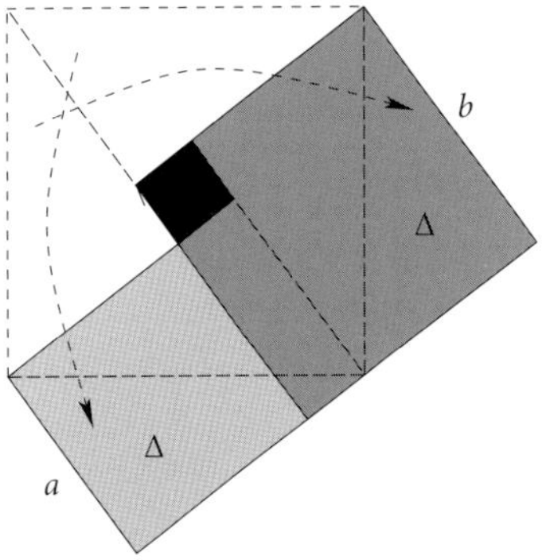

Figure 1.10 Rearranging reflected triangles. Dropping two triangles and adding two more reflected ones does not change the enclosed area, but it shows that the square obtained by the initial reflections, of area c^2, has an area equal to the shaded region, made of one square of side length a (light shade) and one of side length b (darker shade and black).

but has far-reaching consequences. If we use coordinates or vector components to label points or directions in space, nothing prevents us from enlarging the set of coordinates. We can decide that we want to consider a collection of points labeled by four or five or even more coordinates, and directions pointing between those points would be labeled by the same number of components. In the space, we know there are only three independent directions in which we can move, and there is no room for additional coordinates or components. There is no visualization of sets of points with more than three independent coordinates, but their mathematics makes perfect sense. For instance, even though we cannot imagine how vectors may point as directions in such a set and how to concatenate two of them, we can extend the mathematical operation of the addition of vectors by including a few more coordinates.

In this way, we define new mathematical spaces, unrelated to the actual space we inhabit. Since the number of coordinates (or vector components) plays a key role in distinguishing such spaces, we assign a new term to them: it is called the dimension of space. Ordinary space is three-dimensional, while a surface like Earth's is two-dimensional and a curve is one-dimensional. A single point can be assigned the dimension zero because we do not need any coordinates to characterize the point as an element of itself; it is already known once the space of this one point is given. All these are familiar geometrical objects, while higher-dimensional spaces of four, five or more independent coordinates are alien to our imagination.

Fractal dimensions Even more alien is the notion of a space whose dimension is not an integer number but a rational one. The dimension is a fraction, the space called a fractal. Introduced by the mathematician Benoît Mandelbrot, fractal dimensions describe spaces that may appear of changing dimensions when viewed on different scales. For instance, a snaky coastline, as the borderline between two areas, should be one-dimensional, but may occupy a region close to an area of its own. Based on the relationship of geometrical sizes of objects to cover it, one assigns to it a fractal dimension.

Even if we stay in the familiar three dimensions and return to the question of motion, vectors have the advantage of allowing us to keep track of the direction of an object's velocity, not just the magnitude. Adding up velocity vectors separated by small time durations, or integrating the velocity vector, provides the trajectory $\vec{x}(t)$ in three-dimensional space. With vectors, we have full control of the entire dimensional freedom of an object's motion.

Differential equations Motivation leads to motion. If we know the acting force, such as the gravitational one, we know the acceleration of an object of some mass m. The acceleration, in turn, tells us how the object's velocity changes, which shows how its position changes. Starting from some position where we let the particle go with some initial velocity, we can find the positions it will take at all later times by keeping track of all those changes and adding them up. The set of the positions of the object encountered at all times forms a one-dimensional curve, the object's trajectory.

When we determine a curve by keeping track of all the changes of positions encountered along it, we are solving a differential equation. A differential equation, like Newton's second law evaluated for some known force, tells us how the position changes at all times, or how the velocity changes which then tells us again how the position changes. One way to find an approximate solution to a differential equation, used for instance in some computer programs, is to implement the change dictated by the equation not at all times but only after some small intervals. Between those intervals, one lets the object move undisturbed according to the velocity it has at the beginning of the interval. The approximation amounts to a chain of short straight lines which lies close to the curve of the actual solution.

In the approximation, solving the differential equation is much like driving with GPS instructions. After certain intervals, the GPS device tells us which way to turn, or how we are supposed to change direction, the direction of our velocity. If we follow the instructions, we drive along a curve which at all times obeys the required turns; by driving, we solve the equation that lays out the changes of direction. A differential equation prescribes changes of direction (or other quantities) not just after certain intervals, but at all times. It amounts to a frantic GPS shouting out orders at every moment. ("Le-le-le-le-left. Riiiiiiiight. Riiiiight. Straight. Straight! Keep straight!") No one would want to follow such instructions, but with mathematics the solution of a differential equation is quite manageable, even though it can sometimes be exasperating.

There are efficient methods by which one can solve a differential equation, or find the curve along which one drives, with meticulous obedience to the directions of the frantic GPS. Some of them allow one to sidestep driving (and listening) and can produce the whole trajectory at once. These methods are not always applicable, and even if they are, they require much experience (or a good program of computer algebra). No matter how the solution is found, it presents a curve which at every one of its points follows the direction given by the differential equation. Once the solution is available, a mathematical expression providing the position in space as a function of a parameter along the curve, one can evaluate it and compute any desired piece of the curve. Such a solution, called one of closed form, is powerful; in a sense, it amounts to time travel: it allows us to arrive at the destination much faster, without driving and having to follow every single turn.

We have already encountered some differential equations and their solutions. The exponential function is constructed by the requirement that it equals its own derivative; it solves the differential equation $\mathrm{d}f/\mathrm{d}x = f(x)$, with directions and solutions shown in Figure 1.11. It is much easier to compute $\exp(x)$ at some value of x, than starting at the known value $\exp(0) = 1$ and following the directions of vectors until one reaches x. Another differential equation encountered is $\mathrm{d}^2 f/\mathrm{d}x^2 = -f(x)$ (of second order, computing the derivative twice) which we solve by either the sine or the cosine function. We can also find sine and cosine as the two solutions of a coupled set of linear equations $\mathrm{d}S/\mathrm{d}x = C(x)$, $\mathrm{d}C/\mathrm{d}x = -S(x)$; see the previous boxes *Exponential* and *Euler's identity*.
These differential equations have the property that they are linear: they do not contain products of the unknown function $f(x)$ with itself or its derivatives. Such differential equations are easier to solve than nonlinear ones which do not obey this property, but in some cases solutions can be difficult to come by nonetheless.

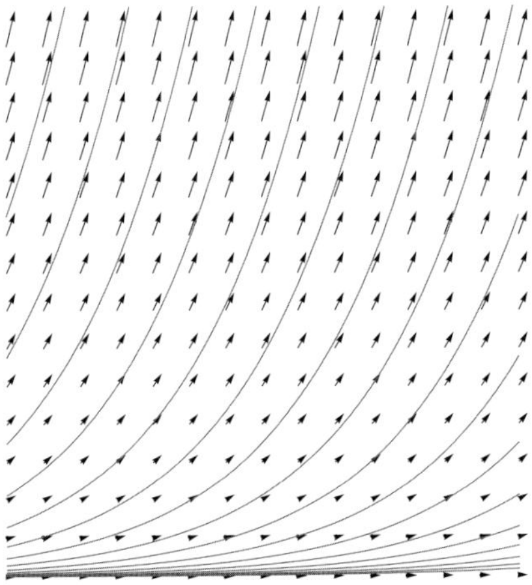

Figure 1.11 Directions provided by a differential equation $\mathrm{d}f/\mathrm{d}x = f$: arrows $(\Delta x, \Delta y)$ with vertical length $\Delta y = y\Delta x$ at point (x, y). All curves following the directions are of the form $f(x) = C\exp(x)$ with a constant $C = f(0)$.

Conserved quantities Constancy in spite of change is often a desirable quality, so too in the case of differential equations. Sometimes, fortunately quite often in theoretical physics, one can find quantities, combinations of the unknown $f(x)$ and coefficients in the differential equation, which must remain constant for any solution, no matter what its form may be. By computing such numbers, called constants of motion or conserved quantities, one can find a great deal of information about solutions to the differential equation; sometimes one can even solve it completely. Conserved quantities also play a large role in physics because they give rise to conservation laws, among them those of energy, momentum and angular momentum.

The role of momentum, defined by mass times velocity $p = mv$, as a conserved quantity is easiest to see. We are aware of this phenomenon when two billiard balls bounce off each other along a straight line, and we could even do simple experiments to quantify the law. Some mechanisms of propulsion are based on momentum conservation: if a rocket ejects fuel particles back out its engines, conservation of momentum requires that the main body of the rocket flies off in the opposite direction. The rocket is much heavier than the fuel particles it ejects in brief periods of time, and so its velocity increase is smaller than the velocity of the ejected particles: larger m means smaller v for the product $p = mv$ to be the same.

On the other hand, we are used to changes of velocity in many cases in which no bouncing-off or ejection of other objects is apparent. Any force, such as the gravitational one, causes a change of velocity and momentum without the need for other objects to be involved, carrying the momentum change. Momentum, indeed, is conserved only if no exterior force is acting on the system (or does not cause sizable acceleration during the times considered). A rocket is subject to the gravitational force, which if the rocket climbs up drains some of its momentum. But in brief instances of time, the increase in velocity, or the boost of the rocket, is provided by the ejecta, and the whole system obeys momentum conservation.

> We write Newton's second law as $\vec{F} = \Delta\vec{p}/\Delta t$, using vectors to show the direction of force and momentum in space. If no force is acting, or all forces can be ignored for some period of time, we have $\Delta\vec{p}/\Delta t = 0$. With a vanishing derivative, $\vec{p}(t)$ never changes and momentum is conserved.

The example of the rocket shows why we consider momentum as the conserved quantity, not velocity. As fuel is ejected, the mass of the rocket decreases; the velocity increases. It is only the product mv that remains constant, not the individual factors.

Energy Energy conservation is an important concept, not just in modern attempts at responsible stewardship of humankind's resources, navigating the changing demands and resources of an evolving society. It is used also to account for evolution in time as dictated by the differential equations of theoretical physics. For physics, energy is always conserved, turned into a law of nature rather than an ethical one.

Are all the efforts put into energy conservation pointless, not because they are undone by human greed (another law of nature, it seems) but because physics ensures that energy is conserved, anyway?

When we talk of energy in these different contexts, physics and society, we do speak about the same concept. A slight difference does exist, however, because economies are interested only in certain forms of energy that can be turned easily in ordered motion, for instance, or structured materials. Electric (or magnetic) energy can be used for many purposes, while the energy associated with a hot material is applied only with difficulty for anything other than heating something else. Heat, while welcome in winter times, is often considered a waste product of technical processes. (The physical notion to distinguish different kinds of energy according to their exploitability is called entropy, a measure for the order that comes with the energy content. Molecules of a substance that can be burned have more order than the random motion produced by heat after burning.) Energy is conserved once all its different forms are accounted for, but this observation is not helpful for practical purposes encountered in modern economies. For theoretical physics, on the other hand, keeping good accounts is one of the secrets of its success, and so the complete accounting of energy conservation is a crucial tool.

Potential Strength is presaged by potential. In physics, we often assume that the force acting on an object is given as the (negative) derivative of some other function called the potential: $F = -\Delta V(x)/\Delta x$. The potential $V(x)$ is the profile of a mountainscape such that F is the force experienced when one hangs on to its slopes. The steeper the mountain, the larger the derivative, and the larger the force. The minus sign is included because a mountainslope towering up at one's right-hand side (toward larger values of x) implies a force to the left (toward smaller values of x). Although the analogy is based on gravity, the mathematical tool of a potential can be used for a much larger class of forces.

Gradient If a force can act only along one fixed direction, described by a single coordinate x, it is always possible to find a function $V(x)$ such that it is a potential of the force. We have to integrate the function $F(x)$, which may be complicated, but is always possible. (The potential is not unique, for adding a constant to it does not change the derivative and so keeps the force unchanged.) If the direction of the force varies in space, however, it has several independent components $\vec{F} = (F_x, F_y, F_z)$ determined by a potential via the limits of ratios of changes in separate directions: $F_x = -\Delta_x V/\Delta x$, $F_y = -\Delta_y V/\Delta y$ and $F_z = -\Delta_z V/\Delta z$. If we move along x by Δx, along y by Δy and along z by Δz, the potential changes by independent amounts: $V(x+\Delta x, y, z) = V(x, y, z) + \Delta_x V$, $V(x, y+\Delta y, z) = V(x, y, z) + \Delta_y V$ and $V(x, y, z+\Delta z) = V(x, y, z) + \Delta_z V$. Instead of just one derivative, there are three independent ones along the coordinate directions; they are called partial derivatives because they take into account only the change along one of the directions keeping the other coordinates fixed, and are denoted as $\partial V/\partial x = \lim_{\Delta x \to 0}(\Delta_x V/\Delta x)$. The vector of all partial derivatives is called the gradient of the potential, $\mathrm{grad} V = (\partial V/\partial x, \partial V/\partial y, \partial V/\partial z)$. Finding a potential for a given force then amounts to solving three differential equations, $F_x = -\partial V/\partial x$, $F_y = -\partial V/\partial y$ and $F_z = -\partial V/\partial z$, for one and the same function $V(x, y, z)$. Now, a solution does not always exist and there are forces not determined by a potential. (An example is a force $\vec{F} = (y, -x, 0)$.)

The value of the potential (our altitude in the mountainscape) is one contribution to the energy; the other one is the kinetic energy, defined as half the product of mass with the velocity squared. Such a combination should have a chance of being conserved, for as one falls down a mountainslope, decreasing $V(x)$, one's velocity and kinetic energy increase. As one rolls into a wall at the bottom of the slope, the damage done (to the wall and oneself) is not proportional to the velocity attained, just as the damage in a car collision (for instance the deformations caused) is much larger than double when the velocity has been twice as large. The quadratic dependence of kinetic energy on velocity takes into account the increased calamity at high speed. Kinetic energy also depends on the mass, a heavy object rolling down more slowly than a lighter one: it is more difficult to pull a heavy object than a light one with the same force. Although one's work increases the kinetic energy by the same amount, a larger mass means a smaller velocity change.

Energy conservation When it comes to energy conservation, good accounting is key. The energy of a moving object of mass m, $E = \frac{1}{2}mv^2 + V(x)$, takes the same value at all times even as v and x change. Its conserved nature is achieved only if the factor of $\frac{1}{2}$ is included in the expression for kinetic energy.

To show that energy is conserved, we consider a small step Δt in time, during which the object moves by an amount Δx. The potential changes from $V(x)$ to $V(x + \Delta x) = V(x) + \Delta V$. The velocity changes, according to Newton's second law, by $\Delta v = \frac{1}{m} F \Delta t$ or, in terms of the potential, $\Delta v = -\frac{1}{m}\Delta t \Delta V/\Delta x$. The velocity squared, in the kinetic energy, changes to $(v + \Delta v)^2 = v^2 + 2v\Delta v + (\Delta v)^2$. Since we assumed Δt to be small (we may think of a limit of Δt approaching zero, turning differences into derivatives) all changes caused are small and a square such as $(\Delta v)^2$ is much smaller than an unsquared quantity such as $2v\Delta v$. The change in kinetic energy is close to $\Delta(\frac{1}{2}mv^2) = mv\Delta v = -v\Delta t \Delta V/\Delta x = -\Delta V$, using the definition of $v = \Delta x/\Delta t$. The change in kinetic energy cancels the change in potential energy, making the total energy E a conserved quantity.

If an object is constrained to move along a line, or in a single dimension, the conserved energy provides almost all the information to understand the motion, as visualized by the mountainscape. The energy is conserved also if there are no constraints on the direction in which the object can move, in which case we need to determine three independent functions for the coordinates that give us the position. We must solve not one, but three differential equations at once, and often the direction provided by one of the equations depends on the solutions of the other two; the differential equations are said to be coupled. In such situations, conserved quantities are important for disentangling the different motions along independent directions, but the energy as just one conserved quantity is no longer enough to visualize all the motion.

> To recognize the direction in space of position $\vec{x} = (x, y, z)$ and velocity $\vec{v}$, we have to include three independent changes Δx, Δy and Δz during a change in time Δt. The individual changes of V are related to the independent components of the force $\vec{F} = (F_x, F_y, F_z)$ by $F_x = -\Delta_x V/\Delta x$, $F_y = -\Delta_y V/\Delta y$ and $F_z = -\Delta_z V/\Delta z$. A change Δt in time, however, comes with changes of all coordinates, and so the changes of potential all contribute: $V(x + \Delta x, y + \Delta y, z + \Delta z) = V(x, y, z) + \Delta_x V + \Delta_y V + \Delta_z V$. The velocity, too, has three independent components $\vec{v} = (v_x, v_y, v_z)$, and its magnitude, like the length of a vector, has the value $|\vec{v}| = \sqrt{v_x^2 + v_y^2 + v_z^2}$. The square of the velocity, as it appears in the kinetic energy, changes to $|\vec{v} + \Delta\vec{v}|^2 = (v_x + \Delta v_x)^2 + (v_y + \Delta v_y)^2 + (v_z + \Delta v_z)^2 = |\vec{v}|^2 + 2(v_x \Delta v_x + v_y \Delta v_y + v_z \Delta v_z) + |\Delta\vec{v}|^2$. Newton's second law $\vec{F} = m\Delta\vec{v}/\Delta t$, using the same arguments as before, shows that $m v_x \Delta v_x = F_x \Delta x = -\Delta_x V$. All changes caused in the expression of the energy $E = \frac{1}{2}m|\vec{v}|^2 + V(\vec{x})$ cancel and the energy is conserved.

Energy is conserved provided we account for all its forms, namely, in the mechanics examples used here, the kinetic and potential energy. This trick helps to make energy conservation in mathematical terms more successful than in practical terms. Physicists are such good accountants that whenever measurements indicate an energy drain in certain situations, they invent a new form or a new carrier of energy to make up for the loss. The most famous example happened when physicists had looked in detail at a certain form of particle reaction, called β-decay, and found that decay products did not seem to carry the same amount of energy contained in the decaying particle. Worried about energy conservation, the theorist Wolfgang Pauli suggested that there might be another particle, one difficult to see, stealing away the missing energy. The surprising lesson of this story is not that physicists would seek refuge in accounting tricks, but that Pauli was right. There is such a particle, long since identified and christened the neutrino, which is not only produced in β-decay but also contains just the right amount of energy to guarantee conservation in the process. Unfortunately, not all problems associated with energy conservation can be solved by accounting tricks.

The fact that particle physicists were able to identify the neutrino by independent means, not by the energy it takes, but by weak reactions it sometimes deigns to undertake with other particles, shows that energy conservation is much deeper than calling whatever energy seems lost a new form of energy (or particle). Energy conservation is a statement about the completeness of processes considered. If we observe a system, but meddle with it by putting in energy from the outside or taking some away, the energy can change by whatever amount we like. We model this situation by using a potential $V(x, y, z, t)$ which depends not only on the position in space but also on time, mimicking some external agent who manipulates the system by turning some knobs. Even if the particle does not move and so neither its kinetic energy nor its position change, the energy can be increased or decreased by simply changing the potential. (Or we could make the mass time dependent, and so change the kinetic energy without compensation from the potential energy.) Energy is no longer conserved because some agents have not been included in the equations. Systems in which energy is conserved, on the other hand, can be considered closed: All factors relevant for how it behaves have been taken into account.

By referring to time-dependent potentials, we hint at another deep property of energy conservation: it is a consequence of what is called time-translation invariance. If parameters such as the mass of an object or the form of the potential do not change in time, the setup of our system is the same at all times; no point in time is distinguished over others. While the object moves and changes position, it is always subject to the same general conditions. If we consider situations in which the potential, say, is time dependent, this dependence, absent an unknown external agent, is caused by another physical process described by its own equations, and carrying its own energy.

Irrespective of its conceptual importance, energy as a conserved quantity is of great mathematical use because it helps us solve the differential equations governing motion. For most potentials or forces, knowing the conserved energy does not suffice to solve completely for the motion. However, there is a class of forces, which happen to play an important role in physics problems, that allow another set of conserved quantities, called angular momentum. Angular momentum is conserved provided that the force everywhere in space points toward (or away from) some center. The potential then depends only on the distance to the center, given in Cartesian coordinates by $r = \sqrt{x^2 + y^2 + z^2}$, not on x, y and z separately. One example is the gravitational force exerted by one heavy and thus barely moving mass on a smaller mass. The heavy mass, such as the Sun, plays the role of the central, fixed point, while a planet would be the lighter, moving mass.

Angular momentum Rotation, the systematic change of angles, is controlled by angular momentum. Its conservation, like the conservation of linear momentum, is a rather familiar experience. Just as moving objects have a tendency to keep moving if no force (or friction, which is also a force) stops them, rotating objects such as a wheel keep rotating. Figure skaters speed themselves up during a pirouette: pulling in or pushing out their limbs, they change the way the different contributions to their mass are distributed. When some parts of their bodies are farther from the rotation axis, they must move longer distances to follow the rotation; they rotate more, as it were, but only a certain amount of rotation (quantified by angular momentum) has been made available when the figure skater started the pirouette. As a consequence, the velocity by which the whole body pirouettes must change, slowing down when the limbs are stretched out and growing when they are held close to the axis.

The conservation of linear momentum means that an object, such as a decaying elementary particle or a rocket spewing out burned fuel, can speed up by shedding mass; angular momentum conservation means that a rotating object can speed up by becoming more compact. (The quantity analogous to the mass, which provides a measure for how an object's constituents are distributed with relation to the rotation axis, is called the moment of inertia.) The speeding-up of a rotating object made more compact is of a different nature than the acceleration experienced for instance by objects falling toward Earth. Ignoring the effect of friction (of the skates on ice or of the body through air), a figure skater can restore the original speed by bringing the limbs back to where they started; the pirouette automatically speeds up with limbs pulled in. There is a conserved quantity at play if a state of motion can be brought back to its exact form after having been changed by the skater.

During a pirouette, a figure skater is not subject to any relevant force; there are only internal forces exerted to move the limbs. An object not subject to an external force is a special case of an object subject to a central force, pointing always toward some fixed point (for an attractive central force). For all such forces, angular momentum is conserved. A first consequence is that any object acted on by a central force must move within a plane through the central point, a familiar phenomenon: A spinning top keeps rotating around a fixed axis unless a force makes it precess. We do not have to balance on a bicycle in motion because the angular momentum of the wheels keeps their axes horizontal. The figure skater rotates around a vertical axis, not having to worry about the limbs twirling around, moving up and down.

If we consider an object in space, able to move in any direction at a given time, the problem of motion is three-dimensional, with three independent coordinates changing in time. As initial values to start the object's motion, we can choose its position and velocity, six numbers that determine where we let the object loose and how much we push it on. After some time, the central force will have changed the objects' velocity according to Newton's second law. Two vectors, the initial velocity and its change along the central force, always lie in a plane; the initial velocity $\vec{v}_1$ and the new velocity $\vec{v}_2 = \vec{v}_1 + \Delta\vec{v}$ must therefore lie in a plane through the central point (Figure 1.12a), and so on at any later time. The plane can lie in space in any way, but it is determined by the initial velocity the object has at one point, together with the line connecting the force center with the object's initial position. Once the object moves, subject to a central force, there is a constant, unchanging and unwobbling axis of rotation normal to the plane, one part of the conservation of angular momentum. From now on, therefore, we can restrict attention to motion within the plane, eliminating one direction orthogonal to the plane which does not change while the object moves.

Motion subject to a central force can be constrained to a plane, explaining why a figure skater's limbs do not twirl about uncontrolled. This observation, however, does not show why the skater speeds up with limbs pulled in. There is another component to angular momentum, a quantitative measure that relates the rotation

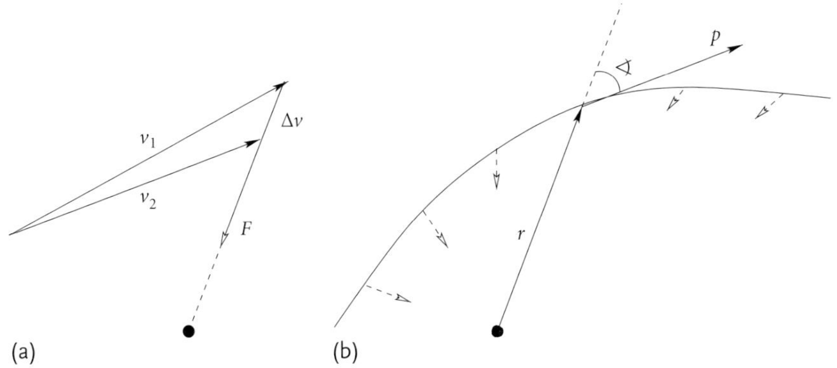

Figure 1.12 (a) With a central force, the velocity and its change always lie within a plane through the center. (b) The quantities used for the definition of angular momentum $L = rp\sin(\sphericalangle)$. The central force is indicated by open arrowheads pointing toward the central circle.

speed to the distance from the rotation axis, or from the central point to which the force is directed. With the quantities illustrated in Figure 1.12b, the distance r of the object from the center, its momentum p and the angle $\sphericalangle$ between these two directions, the value of angular momentum is defined as $L = rp\sin(\sphericalangle)$.

The formula for angular momentum tells us that the object must speed up as it gets closer to the center, for a shrinking r must be compensated for in order to keep L constant. The role of the angle $\sphericalangle$, however, is not so obvious to understand, nor is it clear why the combination of factors as used in L should be conserved, or what the role of the central force is (as opposed to some other force, for which L would not be conserved). To shed some light on these questions, using geometry, one can interpret the value of angular momentum as the area enclosed by a rectangle of side lengths r and $p\sin(\sphericalangle)$, the latter being equal to the projection of p orthogonal to the ray from the center to the object. Equivalently, the value of angular momentum is the area enclosed by the parallelogram with sides $\vec{r}$ and $\vec{p}$, which equals the area of the mentioned rectangle because it is obtained from it by shearing; see Figure 1.13. A central force, then, changes the momentum of the moving object only along the ray from the center, which, for brief time intervals, amounts to a shearing motion of the parallelogram leaving the area unchanged. In this way, one can grasp why angular momentum is conserved by a central force, and why the force must be central for this to be true. However, these considerations are not fully convincing because the fact that also the distance to the center changes, and not just the momentum, is left undiscussed.

Further properties of angular momentum can be illustrated by the problem of a free pointlike object moving through space, with no force acting. There is no obvious central point, but if there is no force, we might as well view it as a vanishing force directed at an arbitrary point we are free to pick. The quantities relevant for angular momentum, at two different times, are depicted in Figure 1.14. We choose the first time, at distance r_1, when the position vector from the chosen cen-

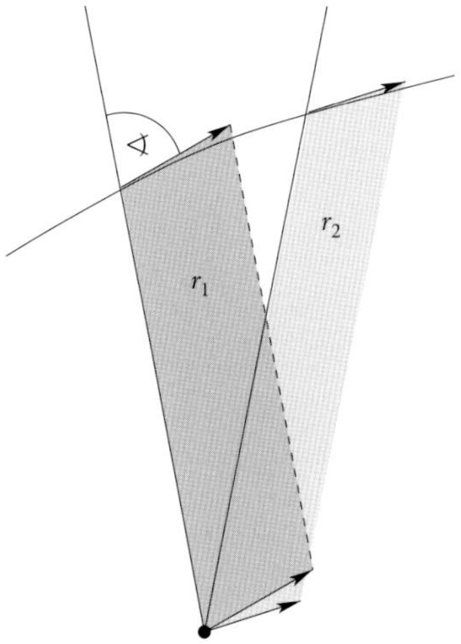

Figure 1.13 The action of a central force amounts to shearing the parallelogram enclosed by the position and momentum vectors, leaving the area constant.

ter and the straight line are orthogonal to each other. With an angle of 90°, the sine function takes the value one and we have $L = r_1 p$. Since no force is acting, the momentum is conserved and takes the value p at all points along the straight line. But the position changes, and at some later time the product $r_2 p$ does not equal $r_1 p$. However, trigonometry in the triangles of Figure 1.14 shows that $r_2 \sin(\sphericalangle)$ is the length of the projection of the vector $\vec{r}_2$ orthogonal to the straight line of motion. This projection equals r_1, for these are the lengths of opposite sides of a rectangle, and so $r_2 p \sin(\sphericalangle) = r_1 p = L$ takes the same value along the whole line.

For general motion subject to a central force, all three quantities making up the angular momentum, that is, position, momentum and the angle between them, change. The changes, illustrated in Figure 1.15, can be related to one another using geometrical rules, and be seen to cancel on general grounds, even if neither the force nor the actual form of the motion is known.

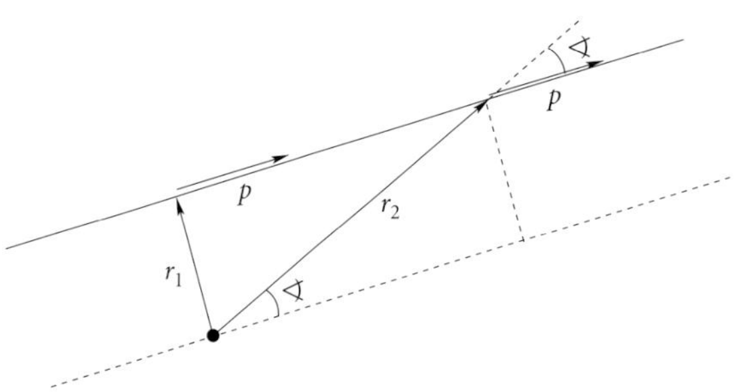

Figure 1.14 Motion along a straight line (solid) when no force is acting. Vectors indicate position and momentum at two different times, and the dashed lines are for auxiliary constructions. The long dashed line is parallel to the solid line.

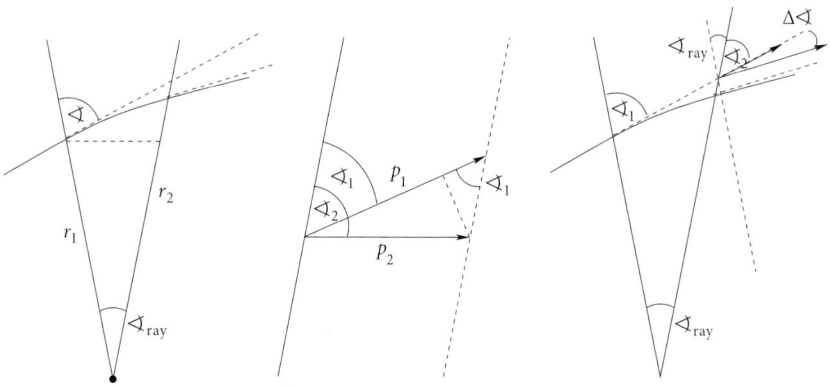

Figure 1.15 (a) The change of distance from the center as an object moves. (b) The change of momentum as an object moves around a center. (c) The change of angle between distance vector and momentum has three sources.

An object in orbit around a center need not move on a circle like a part of a rotating wheel; it may move away from or toward the center, thereby changing its distance to it. In analogy with rigid rotation, we call the distance to the center the radial distance the object has at a given time, or the radius of its trajectory. If the angle between the ray from the center to the object and the object's velocity or momentum is always 90°, the object moves on a circle; otherwise, its radius changes.

As per Figure 1.15a, we consider the change of position of an object moving by an angle $\sphericalangle_{\text{ray}}$ between the rays toward the center, assumed to be small for the following calculations. During that time, the radius, or the distance from the center, changes from r_1 to r_2. The segment connecting the two rays at radius r_1 (dashed horizontal line in the figure) has length close to $r_1 \sin(\sphericalangle_{\text{ray}})$. The triangle in the figure is nearly right-angled (for small $\sphericalangle_{\text{ray}}$), and one of its sides is the radius difference $\Delta r = r_2 - r_1$ while another one is the segment. With the angle $\sphericalangle$ between radius and momentum, we have $\Delta r = r_1 \sin(\sphericalangle_{\text{ray}}) \cos(\sphericalangle)/\sin(\sphericalangle)$.

The angle $\sphericalangle$ of the momentum relative to the radius is used to determine the change of the radius, but it is itself subject to change. According to Newton's second law, the change of momentum during some time interval divided by the interval, provided it is small, equals the force acting on the object. It is here that we make use of the property of the force as a central one, telling us that the change of momentum happens only along rays toward the center. No central force would slow down or speed up an object moving on a circle, but the central force ensures that the object stays on the circle. Even if the object is not on a circle and its momentum might change, it is only the component along a ray that is affected by the force. There is a strict relationship between the change of momentum and the change of angle between momentum and the ray.

> The change of momentum is caused by the force, whose central-force nature we encounter when we calculate this contribution to angular-momentum change. Figure 1.15b places the two momentum vectors at the same point so that they span a triangle enclosing the angle difference $\Delta\sphericalangle$ caused by the force. The end-points of the momentum vectors lie on a line parallel to the ray toward the center, for the force is a central one. This property allows us to relate the angle and momentum changes by geometrical properties of the triangles shown: $p_1\Delta\sphericalangle = -\Delta p \sin(\sphericalangle_1)/\cos(\sphericalangle_1)$ with $\Delta p = p_2 - p_1$.

The change of angle has another contribution, caused by the rotation of the object. If it were moving along a straight line, not subject to any force, the angle between its momentum and the ray changes just because the ray is rotating to keep up with the object. As the ray rotates by some angle $\sphericalangle_{\text{ray}}$, this value must be subtracted from the original angle between momentum and the ray at some earlier time.

> The angle change has two contributions, one already computed from the change of momentum ($\Delta\sphericalangle$), and another one just because the ray connecting the object to the center is rotating; see Figure 1.15c. If the ray rotates by an angle $\sphericalangle_{\text{ray}}$, the angle changes to $\sphericalangle_2 = \sphericalangle_1 - \sphericalangle_{\text{ray}}$ even if the momentum does not change. The total angle change is from $\sphericalangle_1$ at the initial position to $\sphericalangle_2 = \sphericalangle_1 - \sphericalangle_{\text{ray}} + \Delta\sphericalangle$.

All these changes must be combined to see how angular momentum might change from the value $L_1 = r_1 p_1 \sin(\sphericalangle_1)$ it has at some time to the value $L_2 = r_2 p_2 \sin(\sphericalangle_2)$ it acquires later, after the ray has moved by an angle $\sphericalangle_{\text{ray}}$. One can assume the angle $\sphericalangle_{\text{ray}}$ to be small; if L does not change at all in any interval, however brief, it never changes. All changes of r, p, and $\sphericalangle$ depend on the progress of $\sphericalangle_{\text{ray}}$, and they all cancel, making angular momentum conserved.

> Using the relationships computed so far between the changes of r, p and $\sphericalangle$, we calculate the change $\Delta L = L_2 - L_1$ of angular momentum. If we divide by the change of time Δt that causes the change of angle $\sphericalangle_{\text{ray}}$ and let Δt be small, we obtain the time derivative of $L(t)$. If the derivative vanishes, L is constant.
> We combine all the changing quantities to write $r_2 p_2 \sin(\sphericalangle_2) = r_1[1 + \sphericalangle_{\text{ray}} \cos(\sphericalangle_1)/\sin(\sphericalangle_1)](p_1 + \Delta p)\sin(\sphericalangle_1 - \sphericalangle_{\text{ray}} + \Delta\sphericalangle)$. The sine function can be simplified using that $\sphericalangle_{\text{ray}} - \Delta\sphericalangle$ is small: $\sin(\sphericalangle_1 - \sphericalangle_{\text{ray}} + \Delta\sphericalangle) \approx \sin(\sphericalangle_1) - (\sphericalangle_{\text{ray}} - \Delta\sphericalangle)\cos(\sphericalangle_1)$, an equation which is not exact but differs from the correct one only by terms of higher powers of the small quantity $\sphericalangle_{\text{ray}} - \Delta\sphericalangle$. With this simplification, we multiply out all products and write, ignoring products of all small differ-

ences and $\sphericalangle_{\rm ray}$, $r_2 p_2 \sin(\sphericalangle_2) \approx r_1 p_1[\sin(\sphericalangle_1) + \sphericalangle_{\rm ray}\cos(\sphericalangle_1) + (\Delta p/p_1)\sin(\sphericalangle_1) - (\sphericalangle_{\rm ray} - \Delta\sphericalangle)\cos(\sphericalangle_1)]$. The latter expression equals $r_1 p_1 \sin(\sphericalangle_1)$ upon using the relation between $\Delta p/p_1$ and $\Delta\sphericalangle$. We have shown that ΔL is as small as the square of Δt or even smaller, such that $\Delta L/\Delta t$ is zero in the limit of Δt approaching zero.

Angular-momentum conservation can be shown by geometrical considerations, combined with algebraic manipulations of the resulting equations. The boxed statements are general, applying to any central force, but also rather tedious. There is an algebraic line of arguments which is much faster, based on additional notions of vector calculus.

Vector product To see algebraic patterns between vector components, it is sometimes convenient to make vanish as many vector components as possible, while still retaining generality. Components depend on the directions of coordinate axes chosen, and so some of the components can be made to vanish by aligning an axis with some vector. For angular momentum, we can align the x-axis with the position vector, which then has only its x-component nonzero: $\vec{r} = (x, 0, 0)$ of length $r = x$. The momentum vector, in general, cannot be aligned with another axis once the x-axis has been fixed, but we can rotate our coordinate system so that the momentum vector lies in the plane spanned by the x- and y-axes; it then has only its x and y components nonzero: $\vec{p} = (p_x, p_y, 0)$. The factors that define angular momentum are split in the form $L = r[p\sin(\sphericalangle)] = x p_y$, and we may assign to it the direction of the z-axis around which an object with the given momentum would be rotating. Since both the value of L and the direction of the axis are conserved, they define a conserved vector $\vec{L}$. For $\vec{r}$ and $\vec{p}$ of the given form, we write $\vec{L} = (0, 0, x p_y)$ in the z-direction, and denote it as $\vec{L} = \vec{r} \times \vec{p}$, the result of a mathematical operation called the vector product. Without the alignment, the vector product is defined as

$$\vec{r} \times \vec{p} = (y p_z - z p_y, z p_x - x p_z, x p_y - y p_x)\ . \tag{1.3}$$

One of the gratifying aspects of mathematics is that an effort made to understand its concepts and statements pays off in the effort made in understanding or deriving new insights. The not-so-intuitive notion of the vector product, for instance, can be enlisted for a short proof of angular-momentum conservation.

The change of the vector product follows from $\Delta(x p_y) = (x + \Delta x)(p_y + \Delta p_y) - x p_y = \Delta x p_y + x\Delta p_y + \Delta x \Delta p_y$: ignoring the quadratic terms of differences, $\Delta(\vec{r} \times \vec{p}) = \Delta\vec{r} \times \vec{p} + \vec{r} \times \Delta p$. Here, $\Delta\vec{r} = \vec{p}\Delta t/m$ by the definition of the momentum through velocity, and $\Delta\vec{p} = \vec{F}\Delta t$ by Newton's second law. The force

$\vec{F}$, by the assumption of it being central, is proportional to the position vector $\vec{r}$. Each term in $\Delta(\vec{r} \times \vec{p})$ is thus proportional to the vector product of a vector with itself, and any such product vanishes: $\sin(\sphericalangle)$, for the vanishing angle between the vectors, is zero, making the whole product zero. (If the force is not central, we can use these considerations to show that the angular momentum changes by $\Delta\vec{L}/\Delta t = \vec{r} \times \vec{F}$ per time interval Δt. As the analog of a force changing the momentum, the vector product $\vec{r} \times \vec{F}$, called the torque, changes the angular momentum.)

Central-force problem The economy of conservation is evident in mathematics. With energy and angular momentum as two conserved quantities for a system governed by a central force, the equations of motion can be reduced to simpler forms. Even though finding closed mathematical expressions by integrations can still be complicated, graphical methods to visualize properties of the solutions then become powerful.

An object orbiting around a center, staying in a fixed plane orthogonal to the axis of rotation, can be described by two components of its velocity: the change of radius per time interval, $v_r = \Delta r/\Delta t$, and the rotational velocity by the angular change along the orbit, $v_{\sphericalangle} = (2\pi/\bigcirc) r\Delta\sphericalangle_{\mathrm{ray}}/\Delta t$. (The length of the segment stretched out during rotation by an angle $\sphericalangle_{\mathrm{ray}}$ at radius r is $2\pi r\sphericalangle_{\mathrm{ray}}/\bigcirc$.) These are the only nonzero components of velocity because there is no component normal to the plane in which the object orbits. The energy is thus $E = \frac{1}{2}m\{(\Delta r/\Delta t)^2 + [(2\pi/\bigcirc)r\Delta\sphericalangle_{\mathrm{ray}}/\Delta t]^2\} + V(r)$, where we use the fact that the potential of a central force depends only on the radius. In the case of a central mass of size M, the gravitational potential energy is $V(r) = -GmM/r$.

The conserved nature of energy always provides a quadratic expression for the velocity components. For a central-force problem, there are two changing functions, in the expression for energy the radius r and the angle $\sphericalangle_{\mathrm{ray}}$. Such an equation for two independent variables has many solutions, and therefore energy conservation alone does not tell us how an object moves. At this stage, the conservation of angular momentum comes in handy, for we can eliminate the angular change by expressing it in terms of angular momentum, another constant. Only one variable rate of change, that of the radius r, is left in the equation.

Angular momentum is defined as the product of the radius and the component of momentum orthogonal to it. We have already computed the orthogonal component, for it is nothing but the mass of the object times the angular part of its velocity, $(2\pi/\bigcirc)r\Delta\sphericalangle_{\mathrm{ray}}/\Delta t$. We have $L = (2\pi/\bigcirc)mr^2\Delta\sphericalangle_{\mathrm{ray}}/\Delta t$, or $(2\pi/\bigcirc)r\Delta\sphericalangle_{\mathrm{ray}}/\Delta t = L/mr$. Inserting the last equation into the expression for E, we find $E = \frac{1}{2}m(\Delta r/\Delta t)^2 + L^2/2mr^2 + V(r)$, depending only on r and its rate of change. We have decoupled the radial and angular motion making use of conserved quantities: the radius changes according to $\Delta r/\Delta t = \pm\sqrt{\frac{2}{m}[E - L^2/2mr^2 - V(r)]}$ with two separate equations (the two signs) for objects falling in or flying out, and the angle changes according to $(2\pi/\bigcirc)r\Delta\sphericalangle_{\text{ray}}/\Delta t = L/mr$.

Knowing how the radial velocity depends on the radius, for fixed energy and angular momentum and with a given potential, allows us to compute how an object moves. A class of examples is planetary motion in the central-force problem of the Sun's gravity.

1.2 Planets and Stars

The planets and stars move in the sky; some of them have been used for ages as keepers of time. As one tangible application of his discoveries, Galilei suggested to use the phases of the Jupiter moons he saw first to measure time, and solve the long-standing longitude problem in navigation.[10] The stars are a clockwork in the sky but have a much deeper relation to time. They influence time by their mass. Durations in time and distances in space are important notions not just to navigate on Earth, but also to organize the sky. Geometry is too modest a term; its laws constitute much more: cosmometry. However, we should not get ahead of ourselves, for time never does that either.

Although they all appear as sparkles in the sky, planets and stars are different from each other. Stars are the power plants of the cosmos, while planets are residential suburbs (or, much more often, barren terrains). To the untrained eye, stars appear simpler and of smaller variety; it is not important what forms on their surface but what happens inside. There can be no planets without stars, but there are stars without planets. Stars don't seem to move much along the sky, while the planets' movements are much more active and idiosyncratic than what we see of the constant, homely stars.

Planets and comets Newton's law of gravity unifies phenomena on Earth and in the sky. Planets and comets have been perceived in rather opposite ways, the former as steady and regular, although not quite like the stars because they happen to move more; and the latter as sudden, threatening appearances, almost unpredictable because they reappear after more than a whole generation of humans has

10) The House of Representatives of the newly independent (and protestant) Netherlands, a great sea-faring nation, had offered a gold chain as a reward, turned down by Galilei in 1638 in fear of the Inquisition, breathing down his neck.

passed. Viewed from the perspective of Newton's law, they all follow the same principles, just with different values of energy and angular momentum.

An equation that describes the motion of an object under the action of a force does not lead to a unique solution. It is a differential equation, allowing one to choose where and at what speed the object starts. Different solutions can be distinguished by the values of conserved quantities they correspond to, and some ranges of these variables may give rise to solutions of different forms. For instance, since the energy depends quadratically on the velocity, always contributing a positive term, a small value of the energy does not allow large velocities, and some regions in space may be inaccessible for such an object. Planetary motion illustrates this behavior: the planets are bound to the Sun and cannot move farther than a certain maximal radius characteristic of the planet's trajectory. Newton's potential energy $V(r) = -GmM/r$ is increasing toward larger r (where it approaches zero), and cannot be overcome by the planets. Most comets come from much farther away, but they, too, are bound to the Sun and reappear periodically. Only some comets can gain so much energy that they disappear forever, leaving the solar system.

For given energy and angular momentum, we compute the radial velocity $\Delta r/\Delta t$ at every value of r, using the energy equation: $\Delta r/\Delta t = \pm\sqrt{\frac{2}{m}(E - L^2/2mr^2 + GMm/r)}$. The square root requires a positive argument for $\Delta r/\Delta t$ to be a real number. If $E < 0$, r cannot grow to large values. For a given value of angular momentum L, we define the effective potential $V_{\text{eff}}(r) = -GMm/r + L^2/2mr^2$. At all times and for all r, we must have $V_{\text{eff}}(r) \leq E$, with equality when $\Delta r/\Delta t = 0$. Vanishing derivatives mean minima or maxima, according to Fermat's insight. At $V_{\text{eff}}(r) = E$, we have turning points where a planet moving away from the Sun turns around to reapproach it, or a planet getting close to the Sun turns to move farther away. As a plot of the effective potential demonstrates, Figure 1.16, there is always a minimum radius (unless $L = 0$), and a maximum radius if $E < 0$. In the latter case, the object is bound by gravity.

Even without knowing Newton's law, the behavior of the planets was uncovered by the astronomer Johannes Kepler, using precise observations made by his contemporary Tycho Brahe. Planets move along ellipses, with the Sun in one focal point (Figure 1.17). There is one point of minimal and one of maximal distance to the Sun, called the perihelion and the aphelion.[11] Most planets have near-circular orbits, with only a minor difference between maximal and minimal radius. In a plot such as Figure 1.16, the energy is near the minimum of the effective potential

11) Although the law of gravity with a central mass remains unchanged under rotations, that is, the force depends only on the distance between the masses, its mathematical solutions, the orbits, in general do not. Still, the symmetry is realized because the rotated version of any orbit would be another solution to the same laws (as can be seen from Figure 1.17). If the laws were not invariant under rotations, the rotated version of a solution would no longer be a solution. Invariant laws can give rise to nonsymmetric special solutions, just as all humans are equal even though individuals may look different.

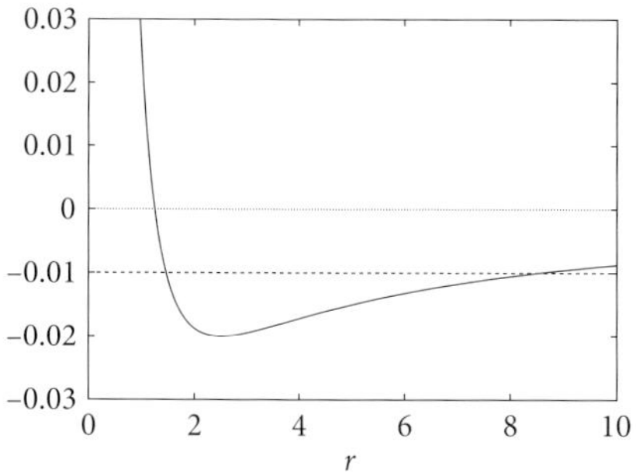

Figure 1.16 The effective potential $V_{\text{eff}}(r) = -GMm/r + L^2/2mr^2$ for $L \neq 0$ shows which regions at a given value of the energy E can be reached: $V_{\text{eff}}(r) \leq E$ restricts the radius r to the left and (for negative E) to the right. (The dashed line shows an example of constant energy -0.01.)

for the given radius. Such a value is rather special; its realization for most planets is explained by their formation process out of an orbiting dust cloud from which they (and the Sun in the center) condensed. The planet closest to the Sun, Mercury, shows an interesting, if small, deviation from the elliptic form. Its perihelion shifts over several orbits, a feature noted by Urbain Le Verrier in 1855 and partially explained by the influence of Venus and deviations of the Sun's surface from spherical shape, to which the closest planet is most sensitive. But there are other causes as well, to which we will come back in Chapter 3.

Another example of objects with $E < 0$ are protuberances, or solar eruptions. As lumps of solar matter ejected by turbulent processes in the Sun, they follow trajectories almost radially outward, with small angular momentum L that does not allow them to follow an orbit. Instead, they reach a maximum radius when $V_{\text{eff}}(r) = E$ at the right, and then fall back to the Sun. The minimum radius at which also $V_{\text{eff}}(r) = E$ is not reached in this case because the lumps hit the solar surface first.

For $E > 0$, an object moving around the Sun is deflected but not bound. It will reach a minimal radius (or fall into the Sun if the minimal orbit radius happens to be smaller than the Sun's radius) and then move out to places far away. High-energy particles that reach us from other stars are examples, and some comets are macroscopic objects of this type. The trajectories are hyperbolas (Figure 1.17), or parabolas in the limiting case of $E = 0$.

Kepler found additional laws characterizing the motion of the planets. His first law states the elliptical nature. The second one states that the line connecting the planet to the Sun sweeps out equal areas in equal times as the planet moves. We have already encountered this law, for any central force, as the conservation of angular momentum; see Figure 1.13. The third law allows us to compare different planets: The square of the orbital period T, the time in which a whole orbit is completed, divided by the cube of the orbit radius r, takes the same value for all planets around the Sun: $T^2/r^3 = 4\pi^2/(GM)$. (We will later derive a similar law in general relativity.) The larger the radius, the longer the planet's year. For instance, Neptune,

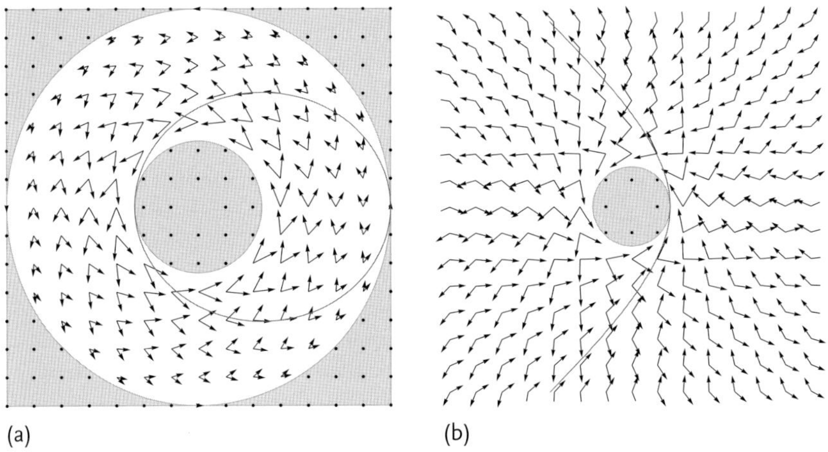

Figure 1.17 Vector fields for the differential equation of Newton's gravitational force, with two arrows at each point for the positive and negative solutions of $\Delta r/\Delta t$ according to the energy equation. Conservation of energy and angular momentum first determines the increments Δr and $2\pi r\Delta\sphericalangle/\bigcirc$ during any interval Δt, which are then used to compute the vector field as the sum of $\Delta r(\cos(\sphericalangle), \sin(\sphericalangle))$ and $(2\pi r\Delta\sphericalangle/\bigcirc)(-\sin(\sphericalangle), \cos(\sphericalangle))$ at position $(r\cos(\sphericalangle), r\sin(\sphericalangle))$. (a) The total energy is negative, so that solutions cannot enter the gray regions, with turning points at the bounding circles. One Kepler ellipse is shown. (b) The total energy is positive, and solutions are hyperbolas.

first identified on September 23rd, 1846 by Johann Galle, did not complete an orbit until July 11th, 2011.[12] Using the law, one can measure T and r for the planets and compute a value for the central mass, the mass of the Sun. For Neptune, the period of 165 years and its orbit radius of 30 times the distance from the Sun to Earth (the latter being 150 million kilometers), we compute the solar mass of two quadrillion quadrillion kilograms.

Ellipse An ellipse aligned with the x-axis is the set of all points (x, y) solving the equation $(x/a)^2 + (y/b)^2 = 1$, with the semimajor axis a and semiminor axis $b < a$. Its two focal points are $(\pm f, 0)$ with $f = ae$, $e = \sqrt{1-(b/a)^2}$ the eccentricity. Every point on the ellipse has the same distance sum $2a$ from the two focal points. The maximal and minimal distance from the ellipse to one focal point, for instance, the center of the gravitational force, are $r_\pm = a \pm f = a(1 \pm e)$. For a Kepler orbit, we compute $r_\pm$, where $\Delta r/\Delta t = 0$, from conservation of energy: $E - L^2/(2mr_\pm^2) + GmM/r_\pm = 0$ implies $r_\pm = -\frac{1}{2}(GmM/E)\left[1 \pm \sqrt{1 + 2EL^2/(m^3G^2M^2)}\right]$. The square root gives the eccentricity, and the semimajor axis is $a = -\frac{1}{2}GmM/E$.

12) Neptune moves so slowly on the sky that its sightings by Galilei, evident from the charts he drew, did not lead to an earlier discovery. The tardy planet was mistaken for a star.

Moon The Moon orbits around the Earth like a planet orbits around the Sun. It obeys the same laws of gravity and motion; it is bound to the Earth and follows a near-circular orbit. Like Mercury, the Moon shows small deviations from what one would expect according to Newton's law: the perigee, that is, the Earthly analog of the perihelion, shifts over several orbits of the Moon. The period, a time of just under nine years when the Moon's perigee and apogee return to their original positions, is much shorter than for Mercury and was noticed early on, in ancient Greece.

Newton's law can account for perigee precession because the Sun's gravity is a considerable contribution to the force that determines lunar motion. We are no longer dealing with a strict central-force problem because there are two competing centers, the Earth and the Sun. Orbits are not elliptic, although still close because the main player in the Earth-Moon system remains Earth. Newton, in his *Principia* published in 1687, already provided this explanation of precession, but a quantitative match of observations and calculations remained elusive until the eighteenth century. For some time, astronomers thought they were witnessing an imperfection of Newton's law, as it did not seem a good model of lunar motion. However, in 1749, the French mathematician Alexis Claude Clairaut noticed that the approximations used to solve Newton's equations with two gravitational centers were not correct. When he included additional terms ignored before, he found a precise match. Newton's law does apply to the Moon's orbit; just nature is not as simple as one often assumes in calculations.

The Moon orbits around the Earth *almost* like a planet orbits around the Sun. There is a slight but important difference: The masses of Moon and Earth are much closer to each other (with a ratio of about 80) than the planetary masses are to the solar mass. It would be more correct to say that Moon and Earth orbit around their common center of mass, close to the Earth's center but not right at it. (It is about 2000 km below the Earth's surface, closer to the surface than to the center.) In other words, the Moon does not just move around Earth, it also pulls along the Earth as it orbits.

A familiar consequence is the tides. When the Earth orbits around the center of mass of the Earth-Moon system, the centrifugal force compensates the gravitational force at the center of the Earth. At the surface, the two forces are not equal everywhere; the gravitational force is stronger than the centrifugal force at parts of the surface pointing toward the common center of mass of Earth and Moon, and it is weaker at parts pointing away from the center. The surface parts, or the more flexible oceans, are pulled away from Earth along a line connecting Earth's center to the common center of mass of Earth and Moon, giving rise to two tidal bulges which move along as the Earth rotates.

Lifting water and land to raise the tides costs energy, which must be payed for by the orbiting motion of Earth and Moon. With the specific characteristics realized for the system, it turns out that the Earth slows down its rotation while the Moon gains energy. The primary reason is that the Earth rotates much faster, completing a full cycle within a day, than the Moon orbits around the Earth, for which a whole month is required. As the Earth rotates, it drags along the tidal bulges due to friction, which therefore point not straight toward the Moon but toward a spot ahead of the Moon along its orbit. There is more Earthly mass in the hemisphere ahead of the Moon than behind it, pulling the Moon further along and transferring energy from the Earth to the Moon.

Lunar escape The Moon is gaining energy, leaving Earth behind. It gains energy by tidal effects, the period of a lunar orbit becomes larger, and the Moon slows down, even though the energy increases: Kepler's law requires a constant ratio for T^2/r^3, which can be maintained only if the orbit radius r and the period T increase as energy and angular momentum are transferred. The additional energy is invested more in gravitational potential than kinetic energy. As the Earth slows down, the Moon inches away.

> The Moon's orbit around Earth is near-circular and close to the equatorial plane; if the latter condition were violated, tidal energy transfer would not be strong. (There is a small angle of 5° between the orbital and equatorial plane.) A circular orbit has values of radius r and angular momentum L_M such that the effective potential $-GmM/r + L_M^2/(2mr^2)$ has a minimum, realized by Fermat's considerations when $L_M = m\sqrt{GMr}$. Here, we denote the mass of Earth as $M = 6 \cdot 10^{24}$kg, and the lunar mass as $m = 0.0123M$. If we know how L_M is changed by tidal effects, we can compute the change of r.
> Angular-momentum conservation equates the change $\Delta L_M = -\Delta L_E$ to minus the change of Earth's angular momentum. The angular momentum of a sphere of radius R and mass M rotating with period T, such as Earth to a good approximation, is $L_E = 2\pi I_E/T$ with $I_E = \frac{2}{5}MR^2$, the moment of inertia (a rotational analog of the inertial mass). Tidal effects leave I_E constant, so that the relative change of L_E is the negative relative change of T: $\Delta T/T = -\Delta L_E/L_E$. Putting everything together, we compute $\Delta T = T\Delta L_M/L_E \approx (\Delta r/r)(m/M)\sqrt{GMr}T^2/(2\pi R^2)$ for the change of an Earth day.

Radar measurements show $\Delta r = 3.8$ cm per year for the departure rate of the Moon, implying a lengthening ΔT of the Earth day by about 5 h in a billion years.

Stars For a long time, stars seemed to embody the constancy of a higher world, contrasted with the unsteady life we see and lead on Earth. In medieval mythology, as laid out for instance in Dante's *Divine Comedy*, only the inferno could seem more chaotic and worse than humankind's action on the surface of the planet; it was placed in Earth's center, as far removed as possible from the elevated sphere of crystalline and innocent stars. The center of the world, or even the surface of the center, was not deemed a good place to be at.

Stars provide orientation. Far out at sea or in unknown lands, the constant stars are welcome skymarks, for centuries the only reliable tools of navigation. Stars embody our sense of abstraction. What meets our eyes is just bright points in the sky. We have no direct sense for their distance; in fact, establishing a reliable distance scale was an age-long process full of birthpains for modern cosmology. Imprecisions and sometimes embarrassing mismatches between different measurements were overcome only in the first few decades of the twentieth century, mainly by Edwin Hubble's work. For science, the road to seeing in stars a multitude of distant worlds, perhaps not so unlike our own, was long and tedious, but it has been taken by abstracting and speculating minds. As such, the stars symbolize our desire to extend the limits placed on our world as well as our knowledge. What we find in the stars is often subject to contingency, a pattern we see but only perceive; a phenomenon assigned too much importance. The constellations, steeped in mythology in diverse cultures, are artificial and have no meaning for the cosmos. (Nor does the position of a planet within or without a certain constellation.) Stars and what we have seen in them are a metaphor for the scientific endeavor; sometimes, also science indulges in the analysis of interesting-looking facts, only to find their contingency, but on the long term it produces persistent intellectual sparkles.

Nowadays, much of what we know about the stars is rather different from ancient or medieval lore. With new insights about them, the stars have lost their innocence in our eyes. They are more violent than could have been imagined, routinely using processes unleashed by humankind only in the most devastating of all weapons. The stars are far from being constant and steady; their changes just happen on vastly different timescales than our own moods and lifespans. The Sun's household seems sustainable, but will expire. Its light is not perfect: When we dissect the spectrum of colors it emits, using a prism or a grating, we see dark spots in it, lines analyzed in 1814 by Joseph von Fraunhofer. The stars are not pristine, ethereal sparkles and they are not all equal: their colors vary, some are brighter than others. Betelgeuse, in Orion's arm, appears red to the naked eye. It is a red supergiant with a radius as large as Jupiter's orbit. Although its mass amounts to "only" 18 times the solar mass, its volume could contain a billion suns. Betelgeuse is bloated. The family name shared by the pair of Sirius A and B belies their difference. Although Sirius B is about half as massive as Sirius A, B's radius is less than half a hundredth of A's, its brightness just a thousandth. Sirius B is a white dwarf, a world collapsed after the depletion of its stellar resources, while Sirius A remains the brightest star in our sky. Stars are born and age. Some fade away, others go out in supernova explosions, or are devoured by their neighbors. Many a star disappears in black holes that swell to masses of millions of suns. Not even the stage of the universe, space and time, is steady; it moves, shakes, and can disappear. When a star becomes too heavy and burns through its fuel, it collapses to a black hole, in whose center space and time disappear. The whole universe could find such a fate. Much in the universe is self-destructive, and also on these grand scales, all one's actions have consequences for oneself. Science, by opening our eyes, has humanized the skies (one might have wished the opposite). Despite all excesses, Earth, for the limited time of its existence, is rather peaceful.

Mathematics The stars have lost their role as exemplars of constancy, by a centuries' worth of discoveries about them. Along the way, a new titan was established:

mathematics. Mathematics takes over many of the roles seen before in the stars. It shows us sparkles in our minds, not in our eyes. It inspires poetry.[13] Mathematics is a metaphor; describing and explaining complex circumstances in compact ways. It can evoke new, unexpected relations and consequences. There is beauty in mathematics, characterized by the same standards used in literature and poetry: necessity outshining constructiveness. For instance, there are only a few number systems with the rules we are used to: the real and complex numbers and, with even fewer rules obeyed, quaternions and octonions. Necessity eliminates arbitrariness; it saves us from the temptation to ask "why" if there is only one way things can be.

Mathematics extends the abstractions inspired by looking at the stars, mere points in the sky whose nonevident distance hides entire new worlds. Mathematics guides the desire to extend the limits of our world and knowledge; it is a metaphor for the whole scientific endeavor. It provides orientation and navigation in an abstract world, but also contributes to the way we orient ourselves in our world. Like stars and their constellations, mathematics, in its relation to nature, is sometimes subject to contingency. Not all the patterns we see in it, often after intense efforts of uncovering them, match with nature's properties. When we model nature with mathematics, we pick the piece of math, such as the methods of differential equations, that is most useful. All the while, mathematics continues to develop on its own, undisturbed by nature.

Mathematics has allowed us to navigate much farther than orientation by the stars could have done. Mathematics is often described as a language, the language spoken by Nature herself as declared by Galileo Galilei in his *Saggiatore*.[14] A contemporary of Descartes', Galilei had not yet been affected much by the algebraization of geometry: he identifies geometrical figures as Nature's letters. (His student, Bonaventura Cavalieri, was one of the pioneers of infinitesimal calculus.)

If mathematics is a language, it is a compact one, and much more. It heeds logic to the strongest extent, with built-in rules to fight falsehoods, or unclarity. An example for its strict and rigorous way is the humble parenthesis, a powerful weapon against mock turtles.[15] In language, this role is played, without perfection by hyphens, or by concatenating words. Alas, as the example of mock turtles demonstrates, maths' rigor is also its loss when it comes to playfulness. (Theoretical physicists, of all people, are often admired by mathematicians, if not so much by the rest of humankind, for being able to maintain a healthy combination of rigor and playfulness. Not only the understanding of nature by physics but also suggesting new mathematical questions and structures often benefits from such a

13) "'O Captain! My Captain,' Charlie asked, 'is there poetry in math?' Several boys in the class chuckled. 'Absolutely, Mr. Dalton, there is... elegance in mathematics.'" N.H. Kleinbaum: *Dead Poets Society* (based on the motion picture written by Tom Schulman).

14) "Philosophy is written in this greatest book, which is open to our glances above all others, I mean the universe, but one cannot understand it if one does not first learn how to understand its language and to know the letters in which it is written. It is written in mathematical language, and its letters are triangles, circles and other geometrical figures. Without these means it is impossible for any human to understand a single word; without them, one errs in a dark maze."

15) The main ingredient of mock turtle soup; see the footnote in Section 1.1.

balanced perspective, for it allows one to bend the rules respectfully, transcending limits to reach new ground. In the rare cases in which this is realized to a strong degree, it plays a large part in what is often described as "genius.")

By its built-in rules, mathematics allows the derivation of new knowledge from established one, together with a test of its correctness. Mathematics thus grows, at a slow but reliable rate. While all active languages evolve, they merely react to new developments for instance in culture, or between cultures when they collide. To a large degree, language evolves due to plain laziness. Mathematics, by contrast, never loses its rigor; it grows out of itself by applying its own rules. This productive discipline is especially powerful when applied to nature, a subfield called mathematical physics. It allows us not only to describe what we observe and measure in efficient ways, we can also use the rules to predict new phenomena. Examples are new planets found by calculating their position based on observed disturbances in the orbits of other planets, or the counter-intuitive phenomena uncovered by applying the mathematics of relativity. Calculations used to explain observed irregularities in the motion of objects in the sky have often predicted new phenomena. In 1845, Urbain LeVerrier studied deviations of Uranus' orbit from a Kepler ellipse and, trusting Newton's law, concluded that a new, unidentified planet should exist. He estimated its likely position and forwarded the results to the astronomer Johann Galle, who trained a telescope on the predicted spot and saw light not claimed by any known star. The planet Neptune was discovered. Ten years later, LeVerrier again noticed irregularities, this time in Mercury's orbit. Astronomers looked for another planet close to and perhaps outshined by the Sun. However, Mercury's orbit did not show a new planet; it helped to confirm inaccuracies in Newton's gravitational force, corrected by general relativity.

The role of mathematics also provides puzzles, the most important one: Why does it work so well? Several other questions have often been asked, for instance: Are mathematical statements discovered (from something preexisting) or created? Is nature, or at least the nature of nature, mathematical? Does mathematics work so well because it is adapted to nature, or because it is adapted to our thinking about nature? Are we blinded by success? There is, after all, much math without (known) applications to nature. And, to a mathematician, the world, namely, the sum of all our observations, Nature herself, is just an approximation. If mathematics is meant for Nature, as her language or essence, it appears to be highly redundant. Sometimes, in particular at the frontiers of modern physics, mathematics seems too rigorous for nature, for instance, when infinity is obtained as a result of calculations, a mathematical faux pas brushed over only with difficulty. Or has the right piece of mathematics just not been found yet? When we think that nature is mathematical, we may be projecting our own logic into nature. Or worse, we might be projecting our own ideals, like some latter-day stoics.[16] Mathematics, when combined with physics, does not operate on the grounds of logic only. We make choices when we follow particular theories. Experiments help us to support some theories and rule out others, but on their own they will never point to one and only one theory. We make selections of theories and mathematics that we can handle and comprehend, mathematics that we like and find elegant.[17] Is what we comprehend mathematical only because we need to formulate it by mathematics in order to make it comprehensible? If so, whatever else there might be remains unseen and ignored.

1.3 Cosmology

When we look at our star-studded sky, we see light not distributed uniformly, but in patterns, apparent or real. There are two kinds of patterns, both filled with myths since the early days of human culture, but of almost opposite degrees of relevance today: We organize stars in constellations, and we are presented with a dense band of stars called the Milky Way. Constellations are useful for orientation and the navigation of old, as skymarks at night times or at places out at sea where no landmarks are visible. The precise patterns, however, are not relevant: what seems close on the sphere of the sky, projected along our line of sight, can be far away in the cosmos; neighboring stars that form the constellations may not be cosmic neighbors at all. Neither are the patterns we see important: There is a set of mathematical statements, called Ramsey theory, saying that any desired pattern can be found in a large sample of spread-out points. Nevertheless, there seems to be a human desire in naming the stars, for it has occurred in different cultures and at different times. To see the cultural influence one need only look at the sources used to identify constellations in the Northern and Southern hemispheres. By the time the Southern stars were seen by Westerners, Greek mythology had long lost its predominance.

16) Nietzsche charged the stoics with imposing their own morality onto nature when they claimed to live according to nature: "You desire to *live* 'according to Nature'? Oh, you noble Stoics, what fraud of words! Imagine to yourselves a being like Nature, boundlessly extravagant, boundlessly indifferent, without purpose or consideration, without pity or justice, at once fruitful and barren and uncertain: imagine to yourselves *indifference* as a power – how *could* you live in accordance with such indifference? [...] In reality, however, it is quite otherwise with you: while you pretend to read with rapture the canon of your law in Nature, you want something quite the contrary, you extraordinary stageplayers and self-deluders! In your pride you wish to dictate your morals and ideals to Nature, to Nature herself, and to incorporate them therein; you insist that it shall be Nature 'according to the Stoa,' and would like everything to be made after your own image, as a vast, eternal glorification and generalism of Stoicism!" Friedrich Nietzsche: *Beyond Good and Evil*.

17) "My formula for finding truth ringed in my ear like a Newtonian principle: 'Of two truths that serve to enlighten the same topic, the yellow wristband is to be awarded to the one that requires fewer words and syllables to be expressed...' " Franz Werfel: *Star of the Unborn*.

It is perhaps not surprising that the stars, during clear nights, were not only seen with joy, but also all too happily endowed with meaning and condolence: what would the night-sky be without stars? (Not to speak of the day-sky without that one star we live from.) Engulfed by darkness, in a world just barely understood, the ancients must have received from the stars not only rays of light but also beams of hope. To their credit, the Greeks, for one, did recognize that higher importance is not always attached to the most fancy and flamboyant. In one inconspicuous corner of the sky, there is Polaris, a lone star surrounded by nothing close enough to grant it a role in a major constellation.[18] Polaris is a constellation of its own, and thereby the only constellation not contingent on Earth's position in the cosmos. Its importance is derived not from a neighborly pattern but from another, intrinsic property: the fact that it does not appear to move. Polaris happens to be situated along the Earth's axis of rotation, a fact which in itself is contingent, though no foresight has placed it there, but all the same makes its location important for us. The direction toward Polaris always points due North.

Milky Way The second pattern we see in the stars is more impressive: the Milky Way, a stream of stars spilled across the sky. Also here, the ancient Greeks show the way from myth to deeper understanding. An old story asserts that this stream once originated from the goddess Hera, breastfeeding the baby Heracles. Democritus already recognized in the Milky Way the cross-section of a system of stars, we might ourselves be part of. Now, we know that the galaxy is a vast disc of stars in rotation, bound by the same force, gravity, that acts on Earth and holds the planets of the solar system together. The center of the disc, Sagittarius A*, has been found, playing the same role for the stars of the Milky Way as the Sun, one of those stars, plays for the planets. One might expect this center to be an even grander object, a superstar.

To some degree, this is true. With sufficient resolution and patience, one can follow single stars orbiting around the center, and determine the shape of their trajectories. Several of those stars have been observed for years. They all follow ellipses around the same point, just as the planets do around the Sun. According to Kepler's laws, which apply as in the solar system because the same force is acting, properties of the orbits can be used to estimate the size of the central mass. The result is indeed as grand as befits the grandeur of the Milky Way it holds around it: the central mass is as large as millions of suns. However, we cannot see anything bright or shiny in the spot where this mass must be. All this mass is concentrated in such a small region that our current telescopes are unable to resolve it, that is, so far at least, for there is hope that the right resolution will soon be obtained with interferometric observations, combining the images seen with several telescopes recording electromagnetic waves of millimeter wavelength. The only reliable theoretical explanation for this phenomenon is not a star or a superstar but a black hole, a mass so dense that it forces all galactic stars to swarm around it, not giving

18) Polaris is the final star in the shaft of Ursa Minor, a constellation so weak that to find it, or Polaris, one is advised to extend one of the sidelines of Ursa Major (the Big Dipper) by about five times its length.

itself away by its own light. A humble superstar, the center of the Milky Way covets privacy.

Distances Another rather old observation has often given rise to speculations. Some parts of the sky contain diffuse regions, called nebulae, of an appearance similar to the Milky Way but much smaller in size and brightness. Throughout history, it has been speculated that these regions are indeed star systems of their own, not of smaller stars but of stars farther away from us than the stars of the Milky Way.

As with many astronomical observations, Galileo Galilei was the first to resolve individual stars in nebulae and to conclude their status as worlds of their own, separate from our Milky Way. In his *Sidereus Nuncius*, published in 1610, he writes: "The *galaxis* is nothing but a collection of numberless stars heaped together; for whatever one of its regions the telescope is aimed at, it immediately offers to the eye a huge set of stars, of which some appear rather large and noticeable, while the multitude of the small ones is beyond investigation ... What is more, the stars astronomers call *nebulae* are piles of small stars squeezed to enormous density. While every single one of them escapes the eye because it is small or far away from us, it is the mixture of all their beams that creates the sparkle that has hitherto been understood as a denser part of the sky which can reflect back the rays of the stars or the sun." Being able to resolve the stars in nebulae with his powerful telescope, Galileo draws conclusions about their nature and distance from us. He had no means to determine the actual distance; most likely he could not have imagined how far they are. Instead, he turns from the galaxies back to our solar system and notes another discovery, which he deems even more important: "We have briefly communicated the observations about the Moon, the stars and the galaxy. Now, it remains to note what we deem the most important of the current endeavor, and to inform the public."

This new insight, thought to be even more important than the discovery of new worlds separate from our Milky Way, are the four moons Galileo saw around the planet Jupiter, new worlds as well. Nowadays, we would tend to disagree with Galileo's sense of what is more important, distant galaxies or moons. But in Galileo's time, moons around planets other than the Earth were unheard of. The prevailing (and church-sponsored) world view was a geocentric one, in which the Moon and the Sun orbited with the other planets around a static Earth. Nikolaus Copernicus had already published his new heliocentric model, and found an instant admirer in Galileo who began to accumulate empirical evidence for it. The moons of Jupiter were one of the pieces of his puzzle. If there are moons orbiting around the moving Jupiter, there is no reason to use the Earth's moon as part of the proof that the Earth must be static. We can see Jupiter moving around with its moons as a model of the Earth moving around with its single moon. It is not possible that both remain static, for we see Jupiter move across the sky. Therefore, we may as well assume both as moving around the Sun. Galileo produced many more arguments, for instance the phases of the planet Venus as it orbits the Sun, but the moons of Jupiter were an important piece, to him more important than the discovery of new worlds of stars, of which many were already known. At a time when much land on Earth remained to be discovered by Western culture, Galileo's favoring of the Jupiter moons may also have been influenced by the promise of actual new territory, however far it may be. He recognized the lucrative potential by naming the moons the Medicean ones, after his main financial supporter.

For centuries after Galileo, thinkers like the philosopher Immanuel Kant and the writer Edgar Allen Poe have given voice to the idea of nebulae as distant worlds separate from the Milky Way. However, the interpretation remained difficult to prove: one would have to determine the distance of those stars from us, a difficult task. For stars nearby, one can make use of the parallax, a slight difference in the position of a star on the sky as seen from different places on Earth's orbit. Positions on the sphere tell us angle differences of the line of sight at different times. Using the known distance between opposite places along Earth's orbit (twice the radius of Earth's orbit), geometry provides the length of the line connecting Earth to the displaced star.

The question of how far the stars, even those in the Milky Way, are from Earth has been of immense interest for centuries, but it was answered only in the first half of the nineteenth century, mainly by work of the mathematician Friedrich Wilhelm Bessel. This success marked the climax of the long and daring adventure to turn geometry, literally, rules for measuring distances on Earth, into what could be called cosmometry. Just as we experience the gravitational force on Earth but use the same laws that govern it to describe the motion of planets and stars, we first establish distance measures and relations between them on Earth (or rather, on small parts of its surface) and boldly extend them to find the distances of stars.

Geometry Greek mythology attempted to extend the human range on Earth to the skies by projecting its thoughts into constellations or in the nature of the Milky Way. In astronomy and cosmology, we are extending the mathematical laws we have found on the human scale to much grander ones. There is no guarantee that this process leads to correct results: The laws might change depending on the scale. If we do geometry on a desk, drawing triangles and determining relations between their side lengths or angles, the laws are different from those we would find if we drew triangles stretching between the continents on Earth. On our flat desktop (or a sheet of paper), we learn how to draw a parallel for any straight line, another line not intersecting the first line. And, given one straight line, there is always a unique parallel through a given point; there is no choice involved in drawing the parallel. Try to apply these rules on Earth's surface for continent-long lines (or on a globe) and you will find them to have lost their meaning. How do we even draw a straight line on Earth's surface, if the surface keeps bending away from the straight direction? For triangles on the spherical surface of Earth, we must use spherical geometry rather than the planar geometry we learn on a planar sheet of paper.

When we extend geometry to cosmometry, we cannot be sure that no other gap in the rules exists. Perhaps we will have to change our geometrical laws when the distances we compute become large, much larger by far than the distances on which we can test our geometry by direct view. The way to seeing whether this is the case is the same one we use to find that planar geometry no longer suffices on the whole of Earth's surface. We first try the known rules, notice some inconsistencies, and attempt to improve them. Improving or changing rules is always a dangerous process, for we must suspend some of them in order to adapt the system. If this is done too crudely, we might reach a situation in which no consistent set of rules exists, that is, complete anarchy. Finding consistent systems of rules is a delicate task in mathematics, which itself is not guided by any rules. One would drop some of the known rules and see if the remaining ones would leave the situation too uncontrolled, a development akin to the historical formation of political systems. (One can do without a king, but then one better has something like a parliament to represent the people.)

In geometry, this process was completed by Carl Friedrich Gauss. However, the result is a mathematical classification of different possibilities; it does not tell us which of them is realized for the cosmos we measure. With knowledge of the possibilities allowed by mathematics, a second tedious phase begins: One attempts to apply one of the known systems of rules, measuring a multitude of distances between different objects. Geometry tells us how these distances should behave in relation to one another, so that we are provided with means to test the consistency of our set of distances with the system of rules we used. With the information available now about distances between different stars and galaxies, it is clear that the rules of geometry change as we look at larger distances, an observation that gives even deeper meaning to the distinction of cosmometry.

Standard candles The parallax method can be used only for stars close to Earth; for stars farther away, the apparent motion becomes smaller and smaller and soon falls below the resolution of telescopes. Another way to determine the distance of stars is based on the observation that the brightness of a star as seen from Earth decreases with the distance of the star. Seen from a distant star, Earth occupies only a small portion of a sphere around the star. If the star shines with equal brightness in all directions, the amount of light sent from a distant star to the area occupied by Earth is small. Some early measurements of the distances of stars were based on this method, but they could not yield the correct distance because no information on how much light the star sends out was available. (Often, it was posited that a star considered had the same brightness as the Sun, an assumption which is almost never correct.) When we observe a star, there is only one piece of information, the apparent brightness we see, and without further input we cannot disentangle the two separate pieces of information given by the actual brightness and the distance.

To map the cosmos in spite of these difficulties, astronomers have developed the method of standard candles, based on the observation that certain types of stars have regular relationships between some intrinsic properties, for example, their color spectrum or variation rates, and their actual brightness. By meticulous observations of all kinds of stars in our neighborhood, which can still be mapped by the parallax method, those relationships were uncovered. The first important case of such standard candles, that is, stars shining with a well-defined and, as it were, standardized brightness, were the cepheids, variable stars with a clear relationship between brightness and their pulsation rate. First found by Henrietta Leavitt, the standard feature was exploited by Edwin Hubble in order to map the sky out to much larger distances than the parallax method would allow. With painstaking observations, Hubble was able to identify many new cepheids in different regions of the sky, including some of the nebulae, and to measure their properties. He confirmed for the first time that nebulae are star systems of their own, far removed from the outer limits of our own Milky Way. In this way, Hubble extended our worldview beyond the constraints of a single galaxy.

Nowadays, dedicated searches for new galaxies have revealed millions of them, organized in clusters of galaxies attracting one another by the gravitational force: the largest systems bound by gravity. Beyond the size of galaxy clusters, the universe becomes more homogeneous and less structured, but some unexplained (and perhaps contingent) irregularities remain. There are voids, underdense regions one of which happens to surround us.

Expansion There is much information in the simple sparkle of a star. Some stars are variable, as exploited when using cepheids as standard candles. And even though it is difficult to notice with the naked eye adapted to the darkness of nighttime observations, stars come in different colors. Just using a camera, mounted on a tripod, one can reveal the colors by aiming the lens at the sky with several minutes exposure. The stars will appear as lines, small segments of circles drawn out by the rotation of the Earth, and in different colors from blue to red (with hardly any green). Stars send out their light in colors, and with characteristic distributions of the frequencies of light they emit. Stars consist of hot gases, mainly hydrogen, which emit and absorb light in characteristic ways, preferring certain colors or frequencies. This feature of the stars provides us with another method to compare Earthly properties with heavenly ones: we can analyze hot hydrogen in the laboratory to determine which colors of light emitted or absorbed it prefers. Stars are eloquent about the colors hydrogen prefers in them, and we can compare those preferences at the vastly different spots, on Earth compared with stars.

It turns out that the preferred colors do not match. As noticed by the astronomer Vesto Slipher, following this method of comparison, the preferred colors of hydrogen in stars appear systematically shifted to the red end of the spectrum of visible light. Moreover, the shift is more pronounced the farther away the star is. Should the laws of physics, which we can use to compute the preferred colors on Earth, be different in stars? If so, we should give up all hope of a scientific understanding of the cosmos, for we can only try to explain it using the laws of nature we uncover by our experiments on Earth. Fortunately, shifts of color can be explained by a simple property of the star, instead of changing the complicated building of physical laws.

Color is the sensual quality associated with the frequency of light, a wavelike excitation traveling through space. The frequency of an emitted wave depends on the velocity of the emitter: if the emitter is approaching us while sending out a wave, its own motion makes it catch up with the wave. After one hill of the wave has been emitted, it is traveling away with the characteristic speed of the wave. The emitter, catching up, will send out the next wave hill at a position closer to the previous one than an emitter at rest would do. The wave length, as the distance between two successive hills, is smaller than for an emitter at rest, and so an observer, in a given observation time, will receive more hills, with larger frequency. Similarly, an emitter moving away emits the wave with smaller frequency as seen by us. For light, higher frequency is associated with blue colors, while lower frequency with red. The systematic shift toward the red part of the visible spectrum can be explained by the assumption that stars, most of the time, move away from us.

Slipher, in this way, discovered an escape motion of distant stars from us. The reason for this flight remained unclear, as did the relationship between the escape velocity and the distances of the stars, insufficiently known. Hubble, by improving the distance measurements for many stars, including distant ones, noticed that the escape velocity is proportional to the distance. If a star is twice as far away than another one, its escape velocity is doubled. The proportionality suggests an explanation based not on properties of individual stars and their motivation to escape, but on properties of all of space. There must be uniform expansion of everything, even where there are no stars but only empty space, that is, an expansion of space itself. The more distant a star is, the more space there is between us and the star. If all of space is expanding uniformly, more space gives rise to more velocity at the position of the star. Hubble, by his observations made around 1925, did not only extend our worldview beyond the Milky Way, he uncovered a new property of the cosmos: the fact that even space which forms the universe is not constant but changing, expanding uniformly.

Origin The notion of nonconstant, expanding space is so radical that it would have been difficult to accept, had it not been hinted at by observations as well as theoretical considerations at almost the same time. New studies in relativity theory, as revolutionary as Hubble's insights, had, by the 1920's, given rise to several mathematical models, most of which seemed to be consistent only if space is not constant. While the majority of physicists remained hesitant, some adventurous theorists had embraced such models, suggesting that a question long deemed out of the grasp of science could be tackled. If space is expanding, it must have been ever smaller the farther we imagine it back in time. If, in thoughts, we turn back the expansion, we reach a phase of immense density in which all there is now was concentrated nearly in a single point. It is impossible to go back beyond a single point, and so one might hope to derive, by mathematical tools applied to a model of an expanding cosmos, information about the origin of the whole universe. A single point is structureless, unlike the complex distribution of matter we nowadays observe in the universe, justifying hopes that a theory of the beginning of the universe may be simple enough for us mere humans to grasp.

Again, there is a daring extension of the known rules of physics to much more extreme regimes than we explore on Earth. Matter in these early, nearly pointlike stages must have been much denser than anything we can probe, much denser even than the matter we can observe in stars. With high density comes along large temperature, further indicating the extremes of the regime to be considered. Nothing probed so far can help us imagine what the laws of physics in such extreme phases of the universe are; we might try to apply Earthly laws unquestioned, but utmost care should be observed. We know, now, that many changes in the laws do happen as the density of matter increases.

While we have to leave the question of the origin unanswered, for now, the expansion of the universe itself is an unquestioned conclusion not only of astronomical observations but also of theoretical inquiry. Edwin Hubble's discovery had been previewed by theory, even while it had not always been predicted with firmness because its consequence is perhaps too grand and revolutionary. However, it was contained in theories which already existed at the time of Hubble's observations: the expansion of space is a consequence of general relativity, a theory formulated in 1915 and evaluated for cosmological implications soon thereafter. The main motivation of developing this theory was to extend Newton's law of gravity beyond the realm of solar-system physics, removing some of the inconsistencies seen in Newton's formula. General relativity provides us with a more precise and (quite literally) more universal description of gravity, a topic which brings us back to the role of mathematics in the exploration of the cosmos.

When the titans had beaten the gods, it fell to them to govern the universe. Its growth had to be sustained, keeping inflation in check. Energy supplies had to be secured. Primitive life forms that were to sprout from wafting ponds had to be nurtured and inspired – even the mightiest titan sometimes desires a pet. Standards of measures and communications had to be maintained and synchronized.

As the titans gathered in a loud, boisterous round to split the tasks between them, they quickly grabbed their most favorite ones. Unsteady Chronos, always late for meetings but the first to decide, picked the task of ensuring standards. Economastos took control of universal growth. Lively Biothyma, lounging in Chronos' lap, cared for all the life, Physikos for energy supplies, and stern Sophia chose general law. Never has the universe been seen in a more excited state.

2
Relativity

At a basic level, there are two classes of objects in classical physics, sources of different phenomena and targets of almost incomparable physical descriptions. On the one hand, we have objects of certain masses and shapes, whose motion or deformations under forces are the subject of physical investigation. On the other, we have the more ethereal waves of light or heat radiation. We often use light as a tool to see and study the motion of massive objects, but light is an interesting object in its own right. It has no clear shape and no obvious mass; instead, it comes with different frequencies (the colors) and has other characteristic qualities. Less prosaically, one could say that massive bodies build the world, but light and heat make it worth living in. In classical physics, the behavior of massive objects is described by mechanics, while the massless light is described by electrodynamics. (In quantum physics, the demarcation line between the two classes of fundamental objects will be drawn rather differently, not according to their mass but according to a more subtle particle property of intrinsic rotation, the spin.)

2.1
Classical Mechanics and Electrodynamics

Mechanics Mechanics is the branch of physics that deals with the motion of objects or their parts. These objects may be extended like a whole planet or a fluid flowing through a pipe, but one often simplifies the description by ignoring all properties of the constituents relative to one another and looking only at the motion of the whole object in a certain direction. The main intrinsic characteristic affecting the motion is then the object's mass, which one often thinks of as located in a single point (the center of mass) or in a small homogeneous and structureless sphere; this way, there is no extension and no decomposition into different parts of the object to be taken into account. The object's motion is described only by the center's position at different times, and derived notions such as velocity or acceleration. With this simplification, no additional numbers need be taken into account for the orientation of a complex body in space or relative positions of its parts.

If the object is extended and made from different parts, there are additional sets of variable parameters to be considered. First, the mass of a single object may

The Universe: A View from Classical and Quantum Gravity, First Edition. Martin Bojowald.

change if it breaks up or collides and merges with another object, or decays like an unstable radioactive particle. In such situations, the mass can no longer be seen as an unchanging parameter constant in time, which complicates considerations. Even though the mass may change, there is still the constant of motion derived before: the momentum $p = mv$ is constant in time unless a force is acting. If the mass is allowed to change, the velocity of whole objects or their constituents is no longer constant; for instance, a particle at rest may decay by emitting two particles flying away in opposite directions. All objects involved have different masses and velocities, but conservation of momentum implies that the two decay particles must move such that their combined momentum $m_1v_1 - m_2v_2$ (taking into account the opposite directions by a minus sign) equals the initial momentum of the particle at rest, which is zero. Such "kinematical" relationships are often used in the analysis of particle decay.

Another set of qualities of extended objects is their shape, or the mass distribution that defines an objects's density throughout its interior. Instead of specifying the density at all points, one often uses a convenient description that details deviations of the shape from spherical form. For a homogeneous, structureless sphere, the main characteristic is the total mass contained within the objects's confines. However, real objects are not perfect spheres, even though they often come close in astrophysics. Deviations from spherical form are encoded by additional parameters, of which in general an infinite number is necessary. However, for near-spherical objects, only a small finite number of those quantities is substantially different from zero, and those numbers are the ones most important for a mathematical description.

The first deviation from a sphere is the dipole moment; it describes how the shape of an object can be seen as obtained by squishing a sphere along one axis. In this process, the object acquires two poles where the axis intersects the surface, hence the name "dipole" moment. A large class of examples is given by objects rotating around an axis, such as the Earth. The two poles here are the actual geographical poles, and the surface deviates from spherical form by Earth's rotation whirling out the equatorial regions compared to polar regions. Measured at the equator, the distance to the center is about 20 km longer than at the poles, just 0.03% of the radius of more than 6300 km. Earth's surface, by the planet's rotation, becomes ellipsoidal rather than spherical. There are also deviations from the ellipsoidal form, such as mountain ranges or marine trenches. They can be described by additional "multipole" moments, measuring further distortions of the shape associated with squishing in several directions at the same time, with different strength in different directions. An infinite number of parameters is necessary to describe the complete shape by multidirectional compressions, of which the dipole moment is the foremost one to determine variations of the gravitational force on Earth's surface by the changing distance to the center. For precise measurements, such as the definition of the height of a mountain above sea level, additional variations of the mass distribution and density of Earth must be taken into account. This deformation of the rotational ellipsoid, called the geoid, requires more than the dipole moment to be specified.

When an object rotates around itself or orbits around a second one, or some other imaginary center, another conserved quantity is its angular momentum provided that any acting force points toward the center. Then, there is the body's energy, composed of kinetic and potential contributions, which sum, too, remains constant as the body moves. These conserved quantities, namely, momentum, angular momentum, energy, are sufficient to determine the motion of the body, constituting a powerful way of characterizing change and evolution by the constants in the object's life. Mathematical expressions for the conserved quantities provide equations to be solved for. This process may be difficult, but the equations themselves, based on simple principles of conservation, are unquestionable as long as the underlying setup, that is, moving masses as objects changing their position as time goes on, remains valid.

Light Cosmology deals with mass distributions in rotation or more complicated motion; mechanics is thus an important part of the theoretical formulation. Properties of light also play a role, used as a messenger to probe the distant past. At some times in the history of the universe, light can even be so intense that it contributes the predominant part of the total energy in the universe. Its motion and other properties are then important to find consequences for everything that evolves in the universe; even light can be so dense and intense that it dominates whole eons.

Light shares many properties with massive objects as they play a role in mechanics. Light has energy: it can heat a room. It has momentum, too, as one may infer from the laws of reflection: A light ray hitting a mirror orthogonally bounces straight back, while a ray hitting the mirror at an angle bounces back casting the same angle with the mirror beyond the reflection point. While the vertical component of the ray's momentum bounces back at the mirror surface, the component parallel to the mirror moves on with constant momentum. In the absence of a force, which in the example acts only once, when the mirror is reached, light thus has its momentum conserved. There is even an angular momentum associated with light, a quality that is more difficult to visualize, but has become important for modern observations, both microscopic and macroscopic.

If one thinks of light as a traveling wave, its angular momentum is encoded in distortions that give the wave fronts a staircase rather than planar pattern. In addition to these properties shared with the massive objects of classical mechanics, light has qualities more closely associated with waves, such as its frequency made visible by the different colors, and its polarization. Most massive objects have colors, but the origin lies in complex properties of their constituents and in the way they interact with light; there are no colors in a dark room. Light, on the other hand, cannot exist when its quality of colors is removed. There is no black light. In a sense that will become precise as we delve deeper into relativity theory, black holes make light disappear by taking away its color (a phenomenon called infinite redshift). Polarization, the second quality not present for massive objects, is a directional aspect of light. With polarization filters, one can block light in direction-dependent ways, at different rates when the filter is oriented.

Not all of these properties are familiar to us from regular observations, but they give rise to interesting effects. Perhaps even more surprising is the fact that light is a close companion of a force. In classical mechanics, we describe the motion of objects subject to forces acting on them. The nature of the forces we use must be determined by independent means, by new theories that are not part of mechanics. One example is the gravitational force, whose properties and laws play a large role for the history of the universe. However, even though the gravitational force is the most familiar one experienced every day by us massive bodies, it is not the only one in nature. It does not become active in those manipulations of other objects in which we are exerting a force in the literal sense. The forces we use are electric or magnetic in nature, independent and rather different from the gravitational force. And light, as has been shown in the nineteenth century as one of the great successes of physics, is nothing but electric and magnetic forces having acquired a life of their own. Light is those forces on a spree, liberated, freed from the domination by massive bodies.

Electric force Matter is a constant balance of charged particles, pulling and pushing one another. For standard matter as we know it, only two different types of charged particles are relevant, the negatively charged electrons and the two thousand times heavier protons of the opposite charge. A charged particle exerts an electric force on another charged particle, just as a massive object exerts the gravitational force on another massive object. Charge or mass are nothing but the qualities that we assign to objects according to the forces they experience. In the case of gravity, the mass cannot be negative and the force is always attractive, pulling two bodies together. The charge, by contrast, can be positive or negative, and two objects charged both positive or both negative repel each other, while two oppositely charged objects attract each other. (What we then call the positive charge is a convention; one could as well have defined protons to have negative charge and electrons to have positive charge.)

Coulomb's law The force formula, named after Charles-Augustin de Coulomb, resembles Newton's gravitational force in Section 1.1: $F = Cq_1q_2/r^2$, where C is a positive constant, q_1 and q_2 are the two charges, and r the distance between them. The minus sign of Newton's law has turned into a plus sign: two charges of equal sign repel each other, while two masses are always positive but attract each other.

Since protons and electrons attract each other, they can build complicated bound structures, the atoms. In a single atom, the light electrons fly around a heavy nucleus made from protons, together with uncharged neutrons that bind together with the protons. Here, we see a third fundamental force at play, for protons left alone in a small region could not stand but would rather repel one another. Accompanied by a healthy dose of neutrons, whose number can differ for different isotopes of the same element, the protons are held together by a strong nuclear force active only on so short distances that we do not experience it in day-to-day life. It is the force (and associated energy) exploited by nuclear technology.

If there are as many protons as there are electrons in an atom, the combined charge obtained by adding up all the individual charges is zero. However, when two objects approach and begin to encroach, there are spill-over forces that can allow atoms to interact with other ones and form larger systems such as molecules, or to hang out in large crowds of liquids or solids. The electrons around the atomic nucleus are not always distributed uniformly; they may sometimes come closer to one side as opposed to the other. While the total charge remains zero, we then have one positive charge on one side and a more diffuse negative one on the other. For another atom nearby, the positive charge, say, is closer and implies a force stronger than the one exerted by the displaced negative charge. Although all charges present still add up to zero, the attractive and repulsive forces, centered at two separate points, do not compensate each other. The combined force, after partial combination of two electric forces, is much weaker than the force exerted by a single isolated charge, but it remains strong enough to hold other atoms close, leading to bound objects. Cohesion felt between the different molecules in a liquid, allowing it to flow without falling apart into little pieces, or the strict order in solids such as crystals arises in this way.

Magnetic force A charge reacts not only to other charges, but also to electric currents. Deflecting an electron with this force is a common method of imaging with (now somewhat old-fashioned) TV tubes, aiming the beam at different spots to be illuminated. Conversely, charged particles cause not only their electric force but also an independent one when they are moving, that is when they participate in an electric current. Two parallel wires with a current flowing through each of them attract or repel each other depending on the directions of the two currents, even though the wires remain electrically neutral.[1] In the circuits used normally, the resulting forces are small and remain unnoticed most of the time. However, there are mechanisms by which forces from many small contributors, laid on top of one another by coiling a wire, can add up to produce sizable ones, much like the concentration of many massive particles in huge objects such as the planet Earth can enlarge the gravitational force. The Earth is an example also for the magnified forces resulting from electric currents. The planet is rotating, and its molten interior is hot enough for some atoms to be ionized, the electrons stripped off the nuclei and no longer bound to them. The overall electric charge remains zero because the electrons are still present in the interior, but they no longer occupy places tied to the nuclei. Some parts of the Earth's rotating interior constitute strong circular currents, exerting forces on other currents.

All other planets, or the Sun and the Moon, are too far removed from Earth to notice these forces. However, since Earth's current is so strong, it can be detected by

1) Electrocardiography was developed by Willem Einthoven in 1903 based on this phenomenon. Weak currents, caused by the muscles of a beating heart and picked up by wires attached to a patient's chest and extremities, can be measured by the motion of a light conducting piece in the circuit.

a much smaller current on the surface. Our usual electric gadgets don't operate at large-enough currents and are too heavy to notice these forces; one need not worry that one's cell phone will be ripped out of one's hand when switched on. However, thin circular wires or light coils of them could notice the force, as do other forms of circular currents provided by nature. Atoms have electrons in motion around the nuclei, forming circular currents. These currents are weak and normally distributed randomly in a sample of atoms, so that the forces experienced by individual atoms all cancel one another. However, some materials, called ferromagnets, are ordered enough in their solid state for the force to remain; these materials, such as iron, interact with Earth's current and align themselves with Earth's rotation. The effect is well-known and much used in magnets; the force in question is called the magnetic force.

Electric forces arise from any charged object, magnetic forces from moving ones. Except for the different dependence on motion, the two forces are related. Indeed, the motion dependence is made irrelevant by a unified treatment of both forces, in which they are seen as just two contributions to a single force, the electromagnetic one. For now, however, we will dwell a little more on the magnetic force in isolation because its properties are less familiar than those of the electric one, which latter simply points from one charged particle to another one, or away from it.

In order to describe the magnetic force in more detail, let us look at a strong circular current flowing around some axis denoted by a vector $\vec{B}$. If the current flows through a large wire loop, or an extended region such as the Earth's interior, the force does not depend much on the position of a small test wire (or magnet) placed in the interior, a case we use for simplicity. (For Earth, we cannot put a magnet in the interior, but rather detect the force outside of the region in which the current flows. The force then depends more strongly on the position of the magnet.)

The weak current in the small test wire or the magnet reacts to the force caused by the strong current; by comparison, it is too weak to produce its own force. From handling compass needles, we know that the force does not make the magnet move along the axis provided by the strong current – a compass is not pulled North – it only serves to reorient the magnet. There is no force along the direction $\vec{B}$, but a force is required for reorientation. As illustrated in Figure 2.1, this force must be perpendicular to $\vec{B}$ and to the direction of the current (or the velocity of its moving charged particles), for any other way would make the compass needle move even when it is aligned with the axis. After the needle has been aligned, even the twice-perpendicular nature does not make the force vanish, but it means that the needle is just stretched and deformed. These weak forces are absorbed by the magnet's internal cohesion (which, as with cohesion in liquids, is based on spill-over electric forces). This twice-perpendicular force between electric currents or moving charges is called the Lorentz force.

Electric and magnetic field Unneutralized and uncompensated, all objects engage in constant pushing and pulling, exerting electric forces at all times and magnetic forces when they move. In cosmology and astrophysics, strong magnetic forces can

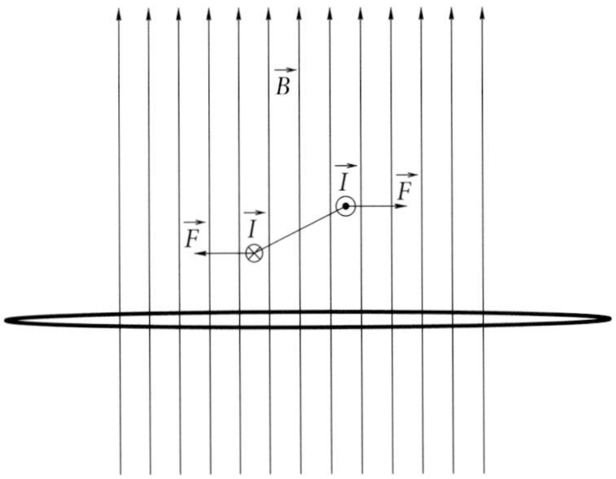

Figure 2.1 A large current in the wide wire around the axis $\vec{B}$ exerts a force $\vec{F}$ on a smaller current $\vec{I}$ through the slanted wire. (The direction of the current is perpendicular to the plane in which the drawing is placed. A current moving toward the reader is indicated by ⊗, as if one were to look straight at the tip of an arrow head. A current moving away from the reader is indicated by an open circle ⊙, as if looking at the back of an arrow.) The direction of the force is perpendicular to the axis $\vec{B}$ and to the current $\vec{I}$ flowing through the small wire. In this way, the position of the small wire does not change, but its orientation is affected in such a way that its axis aligns with $\vec{B}$.

be generated, while electric forces play only minor roles. Once an object is charged, it attracts other charges of the opposite sign, inducing its own neutralization. Nevertheless, one can see the consequences of the force on our day-to-day human scale. For instance, light particles are often attracted to a plastic rod charged up by rubbing it with a towel. When our shoes rub the carpet during a dry day, the charge on our body can be so large that it discharges in a flash when we are about to touch a metallic object (or sometimes a fellow person).

In these situations, although we touch and rub an object to charge it up, the resulting electric force acts at a distance. The discharging flash bridges a gap of air between our hand and a metallic door knob, and all our hairs repel one another without touching when they are charged. The action at a distance is what makes charged-up hair annoying, spreading it out more than we like. However, the same property also makes the electric force so ubiquitous and useful in many different applications.

In order to describe the effect of a charged object on another, distant charge, one makes use of the notion of the electric field. The charged object, for instance from a charged-up rod, generates a field that does not stay on the surface but spreads out into the surrounding region, weakening like the inverse square of the distance. This field is invisible and, by itself, does nothing at all, but it comes to action when it is encountered by another charged object. Then, the combination of the charge q and the electric field $\vec{E}$ produces an electric force on the charged object. The mathematical product $\vec{F}_{\text{electric}} = q\vec{E}$ shows that the electric force is, literally, the product of two contributions, the charge of the object acted on and the electric field generated by another, distant charge.

The advantage of using the field instead of formulating the force in terms of the two charges is that both physical processes involved are visible and separate: We can first charge up the first object, generating the electric field but not yet a force. A force comes to action only when another charge moves into the region where the electric field is strong, resulting in the force as the product.

Similarly, we describe the magnetic force by a magnetic field $\vec{B}$ generated by a moving charged particle or an electric current. The direction of the field is given by the axis of the current, as used earlier and shown in Figure 2.1, and its magnitude depends on the size of the current. The magnetic force exerted on a moving charged particle is then determined by three quantities: the charge q of the particle and its velocity $\vec{v}$, both specifying the current $\vec{J} = q\vec{v}$, and also the magnetic field $\vec{B}$. It must be a product of all three quantities, literally and mathematically, but one that takes into account the three directions of the velocity, the magnetic field and the twice-perpendicular magnetic force on the moving particle. As the behavior of magnets showed, the force is perpendicular to both the velocity and the magnetic field, a result which mathematics denotes as the vector product $\vec{F}_{\text{magnetic}} = \vec{J} \times \vec{B} = q\vec{v} \times \vec{B}$.

Lorentz force While the Lorentz force $\vec{F}_{\text{magnetic}} = q\vec{v} \times \vec{B}$ is twice-perpendicular, to both $\vec{v}$ and $\vec{B}$, irrespective of how those two vectors are aligned, its magnitude depends on the angle between the two vectors. If they point in the same direction, the force vanishes; if $\vec{v}$ and $\vec{B}$ are perpendicular, the resulting force is maximal. The magnitude of the force is $|\vec{F}_{\text{magnetic}}| = q|\vec{v}||\vec{B}|\sin(\sphericalangle)$ with the angle $\sphericalangle$ between $\vec{v}$ and $\vec{B}$.

The vector product used in the equation for the Lorentz force has occurred before in the final formula for angular momentum in Section 1.3 in the box *Vector Product*. One can confirm that the formula (1.3) provided there ensures that $\vec{v} \times \vec{B}$ is always perpendicular to both $\vec{v}$ and $\vec{B}$, and that the magnitude depends on the sine of the angle between the two vectors.

Fields The two long-distance forces feel different for us, but at first mathematical encounter, they appear in almost identical formulas. Coulomb's law for the electric force looks just like Newton's law for the gravitational force, and it shares one of its main problems. It has the form of a nonlocal law, manifesting action at a distance: the force between two charges depends on the distance between them, and even though the force is small for distant charges, it does not vanish. However, there is no time dependence in the formula which would tell us how the force builds up at distant locations; the force on all charged particles in some large region seems to jump up in an instant when we move in a single new charged particle.

Physicists, starting with Newton himself, are always suspicious about instant action at a distance, and so the force laws, similar but also similarly flawed as they are for electricity and gravity, must be reformulated using new fundamental insights. At the end of the process, completed in James Clark Maxwell's electrodynamics and

Albert Einstein's general theory of relativity along strenuous mathematical routes, the two brotherly forces show rather different faces.

Although the notion of the electric field as an intermediary has allowed us to separate the force from the charge of particles it acts on, it does not solve the problem of action at a distance. We may view the process of charging up an object or moving in a new charged particle as building, over a certain period of time, an electric field to which other charged objects react. We can imagine the increasing electric field as a wave seeping out of the charged object into its surrounding space; the charge acts like a fertilizer that makes the field grow, stronger at nearby places than at distant ones. However, the actual build-up of the field remains to be formulated in a detailed mathematical version which guarantees that the field, as it spreads out of the object being charged up and then reaches the places of other charged particles, does indeed take its time. Such a process seems much more complicated, and by its nature, more time dependent than what the simple distance dependence of Coulomb's law would indicate.

A slight modification of the view of the electric field as something spreading out of objects being charged up leads to the electric field as a space-filling entity, always present but not always active. Just as in agriculture, the field itself does not move or spread out; it is always there, at all points in space, but it may be barren, with vanishing growth, too small to be noticed. If a large charged object moves into a region, like a refreshing shower in the desert, the electric field starts blooming; it becomes excited, tuned up at all points depending on the distance from the nurturing charge. A moving charged object leads to growth of an electric field which behaves much like a wave. It rises up around the charged object, stays alive as the charge passes by, then dies down in the wake. Just as the desert ground is always there even if no rain nurtures growth, the electric field is always present but is near zero at most places far from charged objects.

A water wave has another characteristic: it can move on its own once it has been excited. Cohesion pulls up nearby water at a wave crest, and gravity pulls it down. The interplay of these two counteracting forces, both serving to hold the level shape of the surface but overshooting slightly, leads to the well-known phenomenon of traveling waves. The electric field alone cannot accomplish travel. It must cooperate with another force, a different phenomenon that produces its own rain as the field flourishes. The fresh growth must quickly develop into a rain forest that can sustain the newfound life and carry it elsewhere, a traveling oasis. This is how light comes to life.

Electromagnetic field Electric and magnetic fields show constant strife. A growing electric field, like one generated by a moving charge or a current, generates a magnetic field around it, the rain that can sustain it further. Vice versa, a changing magnetic field, new rain, reinvigorates the electric field. In fact, it first shrinks the electric field it owes its existence to, keeping the overall force in check, but remains active and overshoots when the electric field reaches zero. The direction of the electric field is turned around by the changing magnetic field, and once the electric field has turned, it continues to change the magnetic field in the opposite direction, making it decrease. Some time later, the electric field will have made the magnetic field vanish, but only at one instant because the electric field still changes. The magnetic field now turns around, again making the electric field decrease in magnitude and so on. The resulting process is self-sustaining and able to move. A changing electric field generates a magnetic field not just at its own position but, like the current in a wire, in a region around it, pushing ahead into barren low-field territory. And behind the full-grown electric field, the desert returns to unexcited fields.

Electric and magnetic fields give rise to traveling waves, self-sustained. They do not need matter for their existence; they can be nonzero in vacuum where the same process of their interplay occurs. When these fields travel through vacuum regions, there is no analog of waste, decay or friction; the fields just keep on moving. The distances traveled can be enormous, truly astronomical, without any end to the interplay. Since no matter is required for these waves to propagate, it is only fair to grant them the notion of an independent entity of their own, recognizing the collaborative nature: we speak of the combined, or unified electromagnetic field, a physical entity interacting with but not relying on matter. It can be interpreted as a potentiality of electric or magnetic forces, which may exist even if no charged particles are around to manifest the forces. However, there is a much more familiar interpretation, for one incarnation of these waves of electric and magnetic fields is light. It shares the property of being able to travel unabated through empty space. Moreover, light is generated by the charged constituents of atoms, sparking a change in the electric field to set off the sustained interplay process.

Electromagnetic waves Unity ensures advance. Without magnetic fields, an electric field could not travel, and without electric fields, a magnetic field would stand still. The union of both fields is important for their excitations to propagate as waves. With this intimate link, we view the fields as one, as two contributions to a single electromagnetic field. As an independent entity, the electromagnetic field, our self-sustained oasis in the desert, has a full-fledged domestic economy with interactions between its parts, as well as its own ambassadors: electromagnetic waves.

Although the electric and magnetic forces seem different, they are two equal contributors to a unified whole. Diplomatic relationships, namely, the existence of messenger waves, is regarded by physicists as an important condition to grant independence as a physical object. The colors we perceive may originate in the properties of matter, but they are a quality of light itself, or of the electromagnetic waves that reach our eyes after they have been reflected or scattered from material objects. We see and measure electromagnetic waves, granting them the same empirical status as matter.

Well-maintained relationships with the surrounding material world also demonstrate the harmonious interplay of the fields amongst themselves; they show that the electromagnetic field can persist on its own, independent of matter. The electromagnetic field is not just a tool to describe forces, invented by efficient physicists; it is a physical degree of freedom of its own. Without wave phenomena, the electric and magnetic fields could only react to the motion of matter and charges; thanks to its emissary waves, the electromagnetic field can act in its own right. The interrelation of the electric field $\vec{E}$ and magnetic field $\vec{B}$ is reflected in the fact that they must obey differential equations relating their changes in space and time.

Maxwell's equations In electrodynamics, there are two laws of a form that goes back to the mathematician Carl Friedrich Gauss,

$$\frac{\Delta_x E^x}{\Delta x} + \frac{\Delta_y E^y}{\Delta y} + \frac{\Delta_z E^z}{\Delta z} = d_{\text{charge}}, \quad \frac{\Delta_x B^x}{\Delta x} + \frac{\Delta_y B^y}{\Delta y} + \frac{\Delta_z B^z}{\Delta z} = 0 \,. \tag{2.1}$$

These two equations determine how the electric field is built around a charge distribution of density d_{charge}, and by having a zero in the second equation, require the absence of a magnetic analog of charge. We refer to rates of change, or derivatives, in different directions, as indicated by subscripts. If a field varies in all three directions of space, $\Delta\vec{E}$ may depend on the direction. We separate the three independent changes by writing $\vec{E}(x + \Delta x, y, z) = \vec{E} + \Delta_x \vec{E}$, and similarly for y and z. (One often writes $\partial\vec{E}/\partial x$ for $\lim_{\Delta x \to 0} \Delta_x \vec{E}/\Delta x$, a partial derivative.) The independent changes of the components of the electric and magnetic fields cannot be arbitrary, but must obey the two Gauss' laws. The interplay of both fields and their reaction to a current $\vec{J}$ is apparent in the two remaining laws of the version completed when James Clerk Maxwell wrote down the fundamental equations of electrodynamics, based on experimental insights by Michael Faraday and André-Marie Ampère:

$$-\frac{\Delta_t \vec{B}}{\Delta t} = \left(\frac{\Delta_y E^z}{\Delta y} - \frac{\Delta_z E^y}{\Delta z}, \frac{\Delta_z E^x}{\Delta z} - \frac{\Delta_x E^z}{\Delta x}, \frac{\Delta_x E^y}{\Delta x} - \frac{\Delta_y E^x}{\Delta y}\right) \tag{2.2}$$

$$\frac{1}{c^2}\frac{\Delta_t \vec{E}}{\Delta t} + \vec{J} = \left(\frac{\Delta_y B^z}{\Delta y} - \frac{\Delta_z B^y}{\Delta z}, \frac{\Delta_z B^x}{\Delta z} - \frac{\Delta_x B^z}{\Delta x}, \frac{\Delta_x B^y}{\Delta x} - \frac{\Delta_y B^x}{\Delta y}\right) \tag{2.3}$$

with a constant $c = 299\,792\,458$ m/s, a velocity.

Maxwell's equations of electrodynamics not only couple changes of the electric and magnetic fields, they also involve changes $\Delta_t \vec{E}$ and $\Delta_t \vec{B}$ in time. A *changing* magnetic field with $\Delta_t \vec{B} \neq 0$ generates an electric field, and a *changing* electric field with $\Delta_t \vec{E} \neq 0$ (or an electric current $\vec{J}$) generates a magnetic field. This mathematical form is the interplay required for waves to propagate.

Wave equation In the absence of charge and current, $\vec{J} = 0$ and $d_{\text{charge}} = 0$, (2.2) and (2.3) combined, using also Gauss's laws, imply a wave equation for $\vec{E}$ and $\vec{B}$: We compute a further change in time of the change in time $\Delta_t \vec{E}$ by inserting $\Delta_t \vec{B}$ from (2.2) for the components of $\vec{B}$ on the right-hand side of (2.3).

With several manipulations, we obtain

$$-\frac{1}{c^2}\frac{\Delta_t(\Delta_t \vec{E})}{(\Delta t)^2}+\frac{\Delta_x(\Delta_x \vec{E})}{(\Delta x)^2}+\frac{\Delta_y(\Delta_y \vec{E})}{(\Delta y)^2}+\frac{\Delta_z(\Delta_z \vec{E})}{(\Delta z)^2}=0 \qquad (2.4)$$

as the quintessential form of the wave equation, to be fulfilled also by $\vec{B}$. For a plane wave moving along the x-direction, the y- and z-components obey $-\Delta_t^2 E/(c\Delta t)^2 + \Delta_x^2 E/(\Delta x)^2 = 0$, a less complex equation solved below.

There is a multitude of phenomena associated with electromagnetic waves. Radio waves are generated by currents in the electric circuits of an antenna. Less obviously, these waves also include those used for X-rays, the radioactive gamma rays and, as the prime example, light in the visible spectrum or as ultraviolet or infrared versions. In addition to the electric nature of the generation of many of these waves, they can act on charged particles as we expect it for electromagnetic phenomena. Strong lasers have fields so intense that they could move large charged bodies, and a light particle such as a single electron in an atom can be moved and excited even by weak light.

The different types of all these electromagnetic waves, as diverse as they may look, are distinguished only by the frequency or the wave length by which the wave changes in time and space, relating $\Delta_t \vec{E}$ to $\Delta_x \vec{E}$ and the other spatial changes. For visible light, the frequency determines the color we see. If the frequency falls outside of the visible part of the spectrum, we do not perceive the wave anymore with our naked eyes, but the waves remain noticeable by other types of detectors. In addition to the frequency, the intensity of the waves, such as the brightness of light or the norm $|\vec{E}| = \sqrt{(E^x)^2 + (E^y)^2 + (E^z)^2}$, is a characteristic that determines how much of a wave we have.

Speed of light Light moves at the speed of light. In addition to its frequency and intensity, the speed by which it moves in space seems an important parameter to specify light. From material objects, we are used to the fact that they can move at different velocities, or not at all. For light, manipulating the velocity is less obvious.

In order to isolate the self-sustained propagation speed of interacting electric and magnetic fields from exterior influences such as a matter medium, we consider an electromagnetic wave traveling through empty space, through a vacuum region in which no matter and no charge is present. The value of the speed of light is then not arbitrary and subject to choice; it is predicted by Maxwell's equations.

The combination of rates of change in different directions in (2.4) seems difficult to satisfy, but if we restrict attention to fields changing only in one spatial direction, say x, the general solution can be expressed in a rather simple form. All components of any such field must look like $E(t,x) = f(x-ct) + g(x+ct)$ with two free functions f and g. The shape of the wave is unrestricted, but its propagation is determined: The f-contribution moves to the right as t increases, that is, if it has a maximum at $x = x_0$ for $t = 0$, it has a maximum at $x = x_0 + ct$ at time t, while the g-contribution moves to the left. If only one spatial direction is considered, we have a combination (or superposition) of left-moving and right-moving waves. Their speed is the constant c that appears in (2.3), the speed of light.

The speed of different kinds of electromagnetic waves in vacuum is no distinguishing feature at all because it must be the same for all waves. The velocity is a general property of all electromagnetic waves, and therefore we speak of the speed of light, rather than the speed of light of a certain color. And although it is not implied by the name, other electromagnetic waves move at the same speed that light does. (There are even more versions of waves which always travel at the speed of light, for instance, a gravitational analog of electromagnetic waves.) The speed of light is traditionally denoted by the letter c in physics, taken from the not-so-common[2] word "celerity" for speed, a word whose root can still be found in "acceleration." We will use "celerity" as a short form to denote the speed of light in vacuum.

If an electromagnetic wave moves through a region occupied by matter transparent in some frequency range, such as visible light in water, a multitude of interaction effects between matter and the electromagnetic field can occur. Matter is built from charged particles, the electrons and protons in atoms, and charges are affected by the electromagnetic field. By resonance and other effects, light of different frequencies interacts differently with the charges, and deviations for instance in the speed of different colors result. Electromagnetic waves are often scattered or deflected at the surfaces of matter distributions, in a way that depends on the frequency. One consequence is the fact that the sky is blue: When it hits molecules in the atmosphere, blue light is scattered toward Earth's surface more strongly than red light.

When they move through matter, waves are absorbed, at least to some degree even in a transparent medium. The charged particles making up atoms and molecules are moved by the changing electromagnetic field of the wave; they are accelerated and made to oscillate by an electric field, or deflected in their motion by a magnetic field; currents of charged particles can be turned around or modified otherwise. All this requires energy to move massive particles, supplied and paid for by the electromagnetic wave causing the motion. The electromagnetic wave loses energy, decreasing its intensity in the absorption process.

The degree of absorption depends on the wave's frequency. Water and glass are transparent to a good degree for visible light, but not for ultraviolet light. The structures and tissues in a human body absorb X-rays differently, as exploited in X-ray imaging. Some of this frequency dependence can be attributed to resonance phenomena: just as a tree, a building or a bridge can be excited to large oscillations

2) For a random example of its use, we quote "Meanwhile, with a celerity that Jules Peterson, as a business man, would have quite understood, the remains were carried into another apartment of one of the most fashionable hotels in the world." from F. Scott Fitzgerald: *Tender is the Night*, 1934.

by pushing it in the special rhythm of a certain frequency, an electromagnetic wave can excite the charged particles of atoms and molecules in matter more easily when its frequency shakes them with the right rhythm. As a consequence, electromagnetic waves of different frequencies lose energy at different rates, and also the velocity of these waves is reduced by frequency-dependent amounts.

When an electromagnetic wave, including light, moves through matter, its velocity is usually less than what it would be in vacuum, the speed referred to as the unqualified speed of light c. In vacuum, however, when an electromagnetic wave travels on its own, unperturbed by matter, it always moves at the speed of light. More surprisingly, and counterintuitively, the speed measured is the same irrespective of how fast the light source or the measurement device is moving with respect to the wave. If we walk or throw a ball on a moving car, for example, a stretch-limo, so as to have enough space, in the same direction in which the limo is moving, a person staying on the sidewalk will see a speed equal to the sum of the limo's and our own or the ball's velocity, or the difference for motion toward the back side of the limo. However, if we hold a flash light in the limo's direction, that is, if we "throw light,"the beam of light still moves at the speed it has if we shine it standing on the sidewalk.

The speed of light is enormous and difficult to measure in comparison with a common limo's speed (even of one not obeying the traffic laws). More refined measurements make it possible to confirm this counterintuitive feature of light. Experiments by Albert Michelson and Edward Morley in 1887 first showed that light rays moving along with or perpendicular to the Earth's motion in space have the same speed. Also, mathematics tells us that there can be only one speed of light in vacuum, irrespective of how a light source is moving: we derived the speed for all electromagnetic waves using Maxwell's laws that govern the interactions of electric and magnetic fields. The result is a simple constant which cannot depend on the motion of a measurement device. Reconciling our traffic experience with the incorruptible mathematics of Maxwell's equations requires a revolution, a new thinking about space and time.

2.2
Special Relativity

In classical mechanics, an object that moves uniformly, that is, with unchanging velocity, on another object moving uniformly moves at the sum of the two speeds. This conclusion is simple, agrees with our day-to-day experience, and cannot be avoided. The experimental and mathematical observation that light in vacuum is so stubborn that it always moves at its own speed, no matter whether we shine our flash light while sitting in a moving limo or standing on the street, violates what we have seen so far about the mechanical laws of motion. We have two independent and oft-confirmed but mutually incompatible theories of important phenomena: on one hand, classical mechanics with its laws of motion for material objects, and electrodynamics with the laws of electricity, magnetism and electromagnetic waves on the other. One of these theories must be wrong.

Velocity Composition is the way to complex understanding. Our laws of motion determine how velocities are to be composed if an object is moving on another one. We should expect composition laws to tell us how combined velocities can be computed, just like we combine numbers to sums or products in other operations. If two objects move in the same direction, one on top of the other, the following should be true for the combined velocity, denoted by an operational symbol $\oplus$:

1. If the first object moves at velocity v_1 on the second object moving at velocity v_2, the combined velocity $v = v_1 \oplus v_2$ should be the same as $v_2 \oplus v_1$, the velocity of the second object moving at velocity v_2 on the first object moving at velocity v_1. Which object moves on the other one is just a matter of viewpoint. We might say that we are moving on a stretch limo, both with respect to the sidewalk, or that the sidewalk moves underneath the stretch limo which moves underneath our feet. In both cases, we obtain the velocity of us on the limo with respect to the sidewalk.
2. If the second object is not moving (or moves at speed $v_2 = 0$), the combined velocity $v = v_1 \oplus 0 = v_1$ equals the first velocity. The second object is simply standing on the first one and follows its motion.
3. If the second object moves against the first object at the same (but oppositely directed) speed $-v_1$, the combined speed $v = v_1 \oplus (-v_1) = 0$ vanishes. We might just separate the two objects, forget about the first object's motion, and say that the second one does not move at all.
4. If there is a third object moving at speed v_3 on the second one which still moves on the first one, it should not matter whether we combine $v_1 \oplus v_2$ with v_3 or v_1 with $v_2 \oplus v_3$. The result in both cases must be the velocity at which the third object is seen moving from the sidewalk: $(v_1 \oplus v_2) \oplus v_3 = v_1 \oplus (v_2 \oplus v_3)$.

In these statements, we have used the symbol $\oplus$ to denote the combined velocity in compact form. One can verify that all requirements are fulfilled for normal addition of numbers, $\oplus = +$. However, normal addition is not the only operation satisfying these requirements. We could use multiplication if we just write 1 instead of 0 and v^{-1} for $-v$, and there are many more possible choices (given by so-called Abelian groups) if these composition rules were the only information available. To determine how to combine velocities, we must consider additional information, or pose some principles that we expect the laws of motion to satisfy. The form of motion is not fully determined by the unbiased laws of mathematics.[3] We must state additional assumptions, and our own expectations, which might very well be wrong, enter the game.

3) Combining different quantities depends on physical properties, such as interactions. If we combine 100 ml of 96%-ethanol with 100 ml water, we only mix 185 ml $\neq$ 100 ml + 100 ml of liquor

Galilei transformations In order to decide how to combine velocities, we must be more specific about our description of motion. To do so, we look at an object moving in some direction, whose distance from some fixed point is denoted by $x(t)$ as a function of time. If the object is moving at a constant speed v_2 without acceleration or slowing down, a process called uniform motion, its position changes by $x(t) = x_0 + v_2 t$, with $x_0 = x(0)$ the position where the object starts at time $t = 0$.

If the object is moving on another object, the frame of our stretch limo, the base point x_0 is not constant but depends on t. On the limo, we measure the distance traveled starting from a fixed point x_0, but like the whole limo, this base point moves by $x_0(t) = x_0(0) + v_1 t$ if v_1 is the speed of the limo. Combining these equations, we obtain $x(t) = x_0(0) + v_1 t + v_2 t = x_0(0) + (v_1 + v_2)t$. The combined velocity is $v = v_1 \oplus v_2 = v_1 + v_2$, the ordinary sum. The transformations involved, rewriting the motion of an object moving on a moving one, in particular making the base point time dependent by $x_0(t) = x_0(0) + v_1 t$, are called Galilei transformations.

In terms of derivatives, we write $v = \Delta(x_1 + x_2)/\Delta t = \Delta x_1/\Delta t + \Delta x_2/\Delta t = v_1 + v_2$ if $x_1(t)$ is the motion of the frame and $x_2(t)$ the motion of the object moving on the frame.

We strengthen the law of combining velocities by relating it to geometry. The laws of geometry as spelled out by Euclid follow from the basic Pythagorean theorem, summarized in the line element $\Delta s^2 = \Delta x^2 + \Delta y^2 + \Delta z^2$ for the distance of a short straight line between two points in space separated by $(\Delta x, \Delta y, \Delta z)$ in their Cartesian coordinates. The line element is invariant under several transformations of space, showing us universal symmetries. Geometrical relationships we calculate do not depend on the position in space; we can shift all points (or the whole sheet of paper on which we draw triangles) by a fixed amount. Nor does the geometry depend on our orientation in space; we can rotate all points or the sheet by some angle around an axis of our choice.

In combination, these are the most general transformations of space that leave Euclidean geometry unchanged. However, there is another transformation if we also allow time to play a role. If the sheet of paper on which we draw triangles to compute relationships between their side lengths and angles is moving, we do not expect those relationships to change. We could do our geometry homework on the bus to school (or on the bike, as one of my teachers liked to joke). Our handwriting might be rather messy, but we need not worry that the results are wrong because we have been moving while doing the math.[4] Transformations to a moving system are accomplished by the Galilei transformations just derived, which as another class of

4) To be sure, the last statement can be made at least when the bus is moving at a constant velocity, with uniform or inertial motion. When the bus accelerates, there are forces acting on the sheet of paper which might deform it together with the geometrical figures drawn. For now, we consider only uniform motion to avoid complications of acceleration, coming back to general effects at a later stage.

transformations leaving Euclidean geometry unchanged, play a fundamental role in physics.

> A shift by a fixed amount is obtained by replacing (x, y, z) with $(x + a_x, y + a_y, z + a_z)$ for some constants (a_x, a_y, a_z), the displacement vector. The line element and even all the individual coordinate differences Δx, Δy and Δz do not change because constant displacements do not matter; for instance, we have $\Delta x = x_2 - x_1 = x_2 + a_x - (x_1 + a_x)$.
> A rotation by a fixed angle $\sphericalangle$, for instance, around the z-axis, is formulated with the sine and cosine functions: (x, y, z) is replaced by $(x \cos(\sphericalangle) + y \sin(\sphericalangle), -x \sin(\sphericalangle) + y \cos(\sphericalangle), z)$. (See, for instance, Figure 1.4.) Here, the calculation is more involved, but from properties of the functions used, namely, $\cos(\sphericalangle)^2 + \sin(\sphericalangle)^2 = 1$ by the definition of the functions on the unit circle, it follows that the part $[\Delta x \cos(\sphericalangle) + \Delta y \sin(\sphericalangle)]^2 + [-\Delta x \sin(\sphericalangle) + \Delta y \cos(\sphericalangle)]^2 = \Delta x^2 + \Delta y^2$ is unchanged.
> A Galilei transformation is a translation by a time-dependent displacement $(a_x, a_y, a_z) = (v_x t, v_y t, v_z t)$. Since no time differences Δt are considered in the Euclidean line element, the dependence on t does not spoil the invariance (while a dependence of the displacement on x, y or z would).

Our considerations show that the combination of velocities is ordinary addition, as intuition told us. However, if the second object is a light ray, its velocity is $v_2 = c$ and if $v_1 \neq 0$, $v = v_1 + c > c$ is larger than the speed of light. The laws of combining velocities in classical mechanics violate the observation that the speed of light is the same no matter how fast the measurement device is moving. Some of the assumptions used must be wrong, but we referred only to basic and intuitive laws of motion. They all seem reasonable, and it is not obvious what we can change in the previous considerations in order to be consistent with properties of the motion of light. Nor is it clear whether we should change the laws of classical mechanics at all, or perhaps correct our understanding of electrodynamics used to derive the speed of light.

Reconciliation Clashes, and reconciliations, of well-established theories often signal significant progress in physics. If tensions can be overcome, a better, more general description of nature is obtained. Reconciliation not only brings existing knowledge in mutual agreement, but can also predict new phenomena that follow from the combined theory. However, overcoming the tension is hard work, and it requires good intuition about how to proceed when the correct modifications of existing equations are not known yet. Just throwing out all that has been found before and starting from scratch would be hopeless, but not all the traditional equations can be correct either. Picking one equation after another and seeing if and how it can be modified to find agreement would be a too lengthy and uninspired process. Only good intuition can tell what equations or principles one should modify and how to do so, guiding the way of working out the consequences. In the end, everybody wins when opposing views are reconciled.

In the situation at hand, we could distrust classical mechanics, or we could distrust electrodynamics (or both). There are good reasons to trust the laws of composing velocities in classical mechanics, as they are simple and intuitive, and agree with our day-to-day experience, not to speak of many more elaborate measurements. Electrodynamics, on the other hand, is not very tangible; its laws must be and have been extracted in a long series of tedious experiments. Its tension with classical mechanics follows from a mathematical statement, the derivation of the speed of light (a universal constant) from Maxwell's equations. Can we really trust mathematics so much that we would risk throwing over board our intuition for its sake?

It must be experiments, bridging the gap between the traditional realm of velocities in classical mechanics and electromagnetism, that decide what parts of the theories retain their validity. It turns out that the power of mathematics is stronger than the force of our intuition. Classical mechanics is to be modified so as to solve the velocity problem, while electrodynamics can remain untouched. Our intuition, thankfully, will not be violated too much because classical mechanics remains valid to a good degree for objects moving at velocities much less than the enormous speed of light, but it must be adapted for objects of large velocities so as to agree with the constancy of c. With hindsight, this outcome is reasonable because our intuition as well as most measurements with macroscopic objects do not involve too-large velocities. The electrodynamical result about the speed of light was one of the first statements about motion at large velocities; this push to new frontiers is what led us to observe tensions with classical mechanics.

Light clock Light lights the way to a reconciliation of mechanics and electrodynamics. If we accept the notion that light in vacuum always travels at the same speed c, no matter how we are measuring the speed and how we are moving with respect to the light beam, we can use light to determine velocities of other objects, by measuring distances in space and durations in time without having to worry about which laws of mechanics we can still trust. The wager of special relativity theory is to put all our trust in one of the players involved, ignoring as much as we can about the remaining chatter from competing theories, and this player is electrodynamics. We bet that the admirable efficiency of Maxwell's equations in deriving a unique velocity is more trustworthy than our intuition of motion gained by observing traffic at rather small velocities. When new or extended theories are to be formulated, the only ingredients we should use are the basic ones, laws that we cannot do without, such as the motion of light and some of the rules of geometry to relate distances. While assumptions and choices are always necessary, the amount should be kept limited.

With light and geometry, we can define and evaluate a minimalistic clock: a light clock. The lack of irrelevant ingredients and fancy features makes it light, and light is its key device. If we let a light ray bounce back and forth between two parallel mirrors and by some (electronic) means count the number of bounces, we measure time intervals as that number multiplied by the time it takes light to cross the gap. The time, in turn, is characteristic of the clock once the mirrors are set up: it is the length of the gap divided by the speed of light c. Our two players have appeared: Electrodynamics to tell us the speed of light, and geometry to measure the distance between the mirrors.

Light is fast. Light clocks are not so handy to measure the time durations we are usually interested in, during which a large number of bounces must be counted. However, the important property is that anything we need to know to operate and evaluate measurements is independent of how the clock (its two mirrors) is moving. We can take time and space to the test and see how they are perceived from a moving viewpoint. Time and space, in turn, are used to compute or measure velocities. Understanding time and space as measured by a light clock will show us how velocities depend on the viewpoint, and how they are composed.

In Figure 2.2, cross-sections of the two mirror surfaces are drawn in a space diagram, its x-axis parallel to the surface along the cross-section and its y-axis perpendicular, pointing from one surface to the other. A bouncing light ray moves along the y-axis, at a fixed value of the coordinate x. If it starts at point A on one mirror, it reaches the second mirror at B and a short while later arrives back at the first mirror at C. The ray, perpendicular to the surface, has not moved along the surface the second time it reaches the first mirror, and point C is identical to point A. For coordinates, this means $\Delta x = x_C - x_A = 0$; there is no displacement between the two points.

Space-time diagram We perceive time only as motion and change, but in thoughts we can take a snapshot of all that happens in time, all at once. We introduce a new axis, an axis along the direction of time. Moving our eyes along the time axis shows the actual motion of things, every instant caught by a cross-section of the diagram. The whole diagram gives us a bird's-eye view on all of space and time, or space-time for short.

In our light clock, the two identical points A and C of Figure 2.2 are reached by the light ray at different times. With time included, A and C are not identical in a space-time diagram, in which not only the positions of points A, B, C, and the

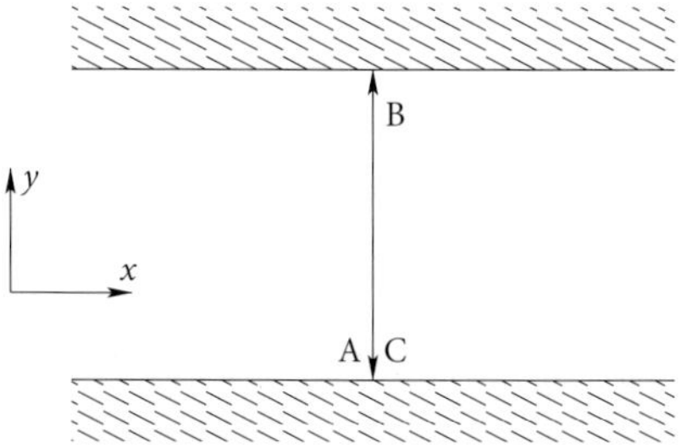

Figure 2.2 A light ray bounces back and forth between two parallel mirror surfaces. Starting at point A on one mirror, the light ray reaches point B on the second mirror, from where it comes back to the first mirror at A = C.

mirror surfaces in space are drawn, but also the times at which those places are occupied. When a point in a diagram is specified by its position in space and time, we call it an event.

If we retained both coordinates x and y and included time t as well, we would need a three-dimensional diagram. The coordinate x is not so relevant yet because it does not change along our light ray, and we ignore it for our diagram and work with just two directions, y for space and t for time. Another question regards the scales of the axes. We cannot compare a time duration with a spatial distance: is one second larger or smaller than one meter? This difficulty can be resolved easily. We are dealing with a moving light ray. Taking the lessons of electrodynamics, we know that light propagates with speed c, no matter how we are moving relative to the ray and the mirrors. A velocity like c, on the other hand, allows us to assign a space distance to any time duration by $c\Delta t$, the distance that light travels in the given amount of time Δt. Accordingly, we use y and ct as the two coordinates for our space-time diagram. The lines along which light moves are then to be drawn at directions of $\pm 45°$: the y-increment will equal the ct-increment or its negative. Together with the mirror surfaces at constant y, we obtain Figure 2.3 for the moving light ray.

Geometry The speed of light tells us time lapses between events, and geometry gives us the distances. Starting at A and coming back to the same mirror at C, the

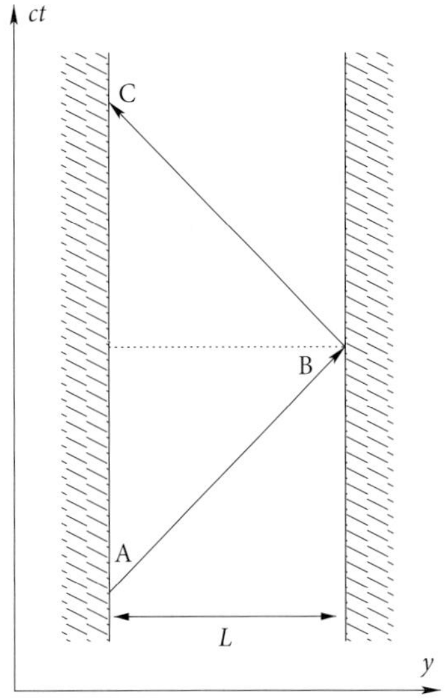

Figure 2.3 A light ray bounces back and forth between two parallel mirror surfaces, shown in a space-time diagram. Starting at point A on one mirror, the light ray reaches point B on the second mirror, from where it comes back to the first mirror at point C. The final point C has the same spatial position as A, but is reached at some later time. The distance between the mirror surfaces is $\Delta y = L$.

light ray must twice traverse the separation L between the mirror surfaces, moving forth and back. It takes a time $\Delta t = 2L/c$ to move from A to B to C.

We write space-time separations between the two events as $(c\Delta t, \Delta x, \Delta y, \Delta z) = (2L, 0, 0, 0)$, a formula that summarizes the motion of light in terms of events in space-time. This latter statement, using coordinates of points, is better suited to uncovering geometrical properties of space and time which, as we saw earlier, are fundamental for classical mechanics. In classical mechanics, it is the Pythagorean theorem that tells us basic geometrical relationships. With points and coordinates, it is contained in the form of the Euclidean line element. The Euclidean line element, in turn, is invariant under certain transformations, that is, translations, rotations, Galilei transformations, showing how to change geometrical and mechanical quantities when we change our viewpoint.

One can recognize geometry in Figure 2.3. The dotted line splits the isosceles triangle ABC into two identical right-angled isosceles triangles. The equidistant sides of these two triangles are of length L, and so the vertical distance from A to C in the diagram is $2L$. On the other hand, it equals c times the time required for light to travel from A to C via B, earlier called Δt. We arrive at the same formula $c\Delta t = 2L$ seen before. This result seems to indicate that we can use the usual laws of geometry and the Euclidean line element, even if one of the directions is time, not a spatial one, a conclusion which seems trivial in the diagram but would be astonishing because it tells us something new about nature, almost in passing: the geometry of space-time rather than just space.

By considering light clocks, we will indeed unravel the geometry of space-time. So far, however, we have used only one Euclidean law in our calculation, a rather elementary one for which we just had to recognize the isoscele nature of our triangles. More general laws such as the Pythagorean theorem have not been used. In order to probe the complete geometry even in the time direction, we must consider a large class of situations in which those laws apply, and test whether they still agree with what we know about the motion of light.

Minkowski line element A moving light clock probes relationships between distances in space and time. If the clock is moving along the direction of the light beams, its rate does not change: Light will have to move longer and the clock experiences a delay on one way (e.g., from A to B), but can enjoy a shortcut on the way back to C. During full cycles from A to C, differences in the time lapses cancel, and the clock keeps ticking at the same rate.

It is more interesting to look at our light clock from a perspective[5] moving parallel to the mirror surface, with some velocity v. With the mirror slipping sideways underneath it, the light ray arrives at a shifted point in space when it returns to the first mirror. We now obtain a nonzero displacement in the x-direction (the direction in which we assume we move). By recalculating the corresponding time lapse,

5) Changes of separations and geometry may be direction-dependent, as illustrated by Galilei transformations in the x-direction, with $x' = x + vt$, but $y' = y$ and $z' = z$ unchanged.

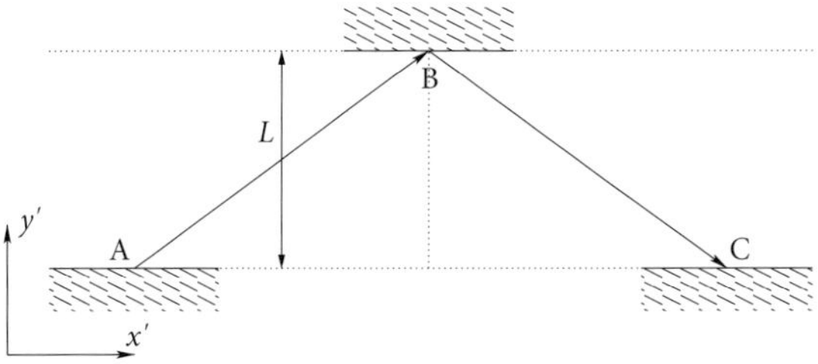

Figure 2.4 A light ray bounces back and forth between two parallel moving mirror surfaces. The three images of the mirrors are shown at different times because the mirrors move as the light ray travels.

we will see how coordinate differences are related and what geometrical laws they obey.

We return to a space diagram because we now have motion along two directions in space, the light ray along y and the mirrors along x. Such a diagram will allow us to use just spatial geometry; we will not need (and not be tempted) to apply the intuitive geometrical laws of space also to time intervals. As shown in Figure 2.4, the two points A and C are now different from each other even in space because the first mirror has traveled a certain distance along the x'-axis while the light ray was traveling. To be on the safe side, we have added a prime to the x-coordinate because it may differ from the coordinate at rest. The mirrors are, after all, moving, and we should expect a transformation (like a Galilean one in classical mechanics) to change the coordinates in the moving system compared to the one at rest.

The distance between points A and C is $\Delta x' = v\Delta t'$: we have assumed that the mirror moves with velocity v while the light ray takes some time $\Delta t'$ to travel from A to B, and then to C. The distance traveled by the light ray, along the arrows in the figure, can be computed in a way similar to what we used before: as indicated by the dotted lines, we split the triangle circumscribed by the light ray and one mirror surface into two right-angled ones. Unlike in Figure 2.3, these are right-angled triangles in space (no time direction is drawn in Figure 2.4) and we may use the Pythagorean theorem. With the vertical distance between the mirrors still of size L and the horizontal distance from A to B (half the distance from A to C) being $\Delta x'/2$, the distance traveled by light is twice the side length $\sqrt{L^2 + (\Delta x'/2)^2}$.

At this moment, we should pause and reconsider our assumptions. We were cautious enough to denote the new coordinates x', y', z' and t' in the moving system with primes to distinguish them from the values they would take if the system were at rest. Indeed, the result we obtain for $\Delta x'$ differs from the value $\Delta x = 0$ with moving mirrors. However, since $\Delta y'$ is the vertical separation between the mirrors along the y'-axis, how can we be sure that it takes the old value $\Delta y' = L = \Delta y$? Well, we cannot be sure at this stage. However, we are considering motion in the x-direction while the y-direction is independent and perpendicular. The assumption we are making is that motion in some direction will only affect distances and coordinate differences in this same direction, not in perpendicular ones. This assumption sounds reasonable. After all, independence of perpendicular directions is necessary to view them as different dimensions of space. However, at the current stage, it remains an assumption along with those we make about other properties of space, for instance, that it is still described by Euclidean geometry. Ultimately, observations must be consulted to test whether all that went into our constructions was correct. We will do so once we have derived some consequences for actual measurements.

Another assumption we made now turns out to be much more crucial, the one motivated by electrodynamics rather than geometry: we posit that light still travels at speed c (not $c' \neq c$) when seen from the moving system. The distance from A to B and on to C, computed as $2\sqrt{L^2 + (\Delta x'/2)^2}$, then equals $c\Delta t'$ where $\Delta t'$ is the time elapsed while the light ray moved. The equality $c\Delta t' = 2\sqrt{L^2 + (\Delta x'/2)^2}$ can be written as $-(c\Delta t')^2 + (\Delta x')^2 = -4L^2$. The vertical separation L has not changed, and so we express its square by coordinate differences found in the system at rest, where $c\Delta t = 2L$, as $-4L^2 = -(c\Delta t)^2$. Eliminating L from the equations, we have $-(c\Delta t')^2 + (\Delta x')^2 = -(c\Delta t)^2$, a relationship not unlike the invariance statement of the Euclidean line element. To make things look more democratic among the coordinates, removing the preferred direction we introduced by assuming we move along x, we add squares of the vanishing coordinate differences $\Delta y' = \Delta z' = \Delta x = \Delta y = \Delta z = 0$ between points A and C and obtain the invariance

$$-(c\Delta t')^2 + (\Delta x')^2 + (\Delta y')^2 + (\Delta z')^2 = -(c\Delta t)^2 + (\Delta x)^2 + (\Delta y)^2 + (\Delta z)^2 \tag{2.5}$$

of $\Delta s^2 = -(c\Delta t)^2 + (\Delta x)^2 + (\Delta y)^2 + (\Delta z)^2$, called the Minkowski line element.

Space-time Geometry is defined by what stays constant even as shapes change. Just as the Euclidean line element, defining geometry of space, is invariant under translations, rotations and Galilei transformations, the Minkowski line element (2.5) does not change if we transform to a moving system according to measurements by a light clock, taking the constancy of the speed of light into account.

The Minkowski line element is the correct one from the point of view of electrodynamics. It strikes a nice compromise because the Euclidean part is still visible in its spatial contributions, but it introduces an important generalization by adding the time lapse Δt. If we set $\Delta t = 0$, considering only spatial displacements, the Minkowski line element is reduced to the Euclidean one. All we know about the geometry of space remains valid, but it is extended by statements about distance measurements in time.

Our derivation of (2.5) shows that not only space differences, but also time differences must be included for complete invariance. Moving in space changes the perception of time durations: $\Delta t' \neq \Delta t$. It is no longer meaningful to think of time and space as separate, as we were allowed in Euclidean geometry and classical mechanics. Space and time together determine geometry, the geometry of a larger stage of space-time in which the three spatial directions and time span four dimensions, all on an equal footing.

Time, spanning the fourth dimension, is one of the wings of space-time, and it makes us fly. In space-time, we cannot always choose the direction in which we move. We are bound to move forward in time, chained to time. "Time flies," as we colloquially (and incorrectly) say:[6)]

Gather ye rosebuds while ye may,
Old time is still a flying:
And this same flower that smiles today,
Tomorrow will be dying.

However, time does not fly on its own. Without space, time has nowhere to fly to, and without time, space would have no time to move. It is only the whole of space-time that makes things fly, in space and time. Motion is the interplay of changing space and time.

The strict directedness of time remains puzzling in physics, philosophy, and sometimes literature. Mathematically, it is contained in the form of the Minkowski line element, whose new time term comes with a minus sign as opposed to the positive spatial terms familiar from the Pythagorean theorem. This innocent-looking dash of sign makes physics at high speeds more savory.

Constant distance Constants are the fundamental notions of geometry, constants that remain unchanged under transformations of space. Geometry, incorporated by a line element, shows under what moves space or space-time remains invariant, or what the fundamental symmetries are. The Euclidean line element of space is invariant under translations and rotations, and so its properties do not depend on where we are and in what direction we look.

The Euclidean line element is also invariant under Galilei transformations provided we treat time as a separate parameter, not as part of space-time. Electrodynamics has shown that this viewpoint is incorrect. We must use the four-dimensional space-time geometry with line element (2.5) so as to be consistent with the observer independence of the speed of light. Spatial translations and rotations remain symmetries because the Minkowski line element still contains the Euclidean part. Also, translations in time are symmetries of (2.5) because none of its coefficients depend on time. However, we cannot rotate a space direction into the time direction because the signs by which coordinate differences squared enter the line element are not equal for space and time. Yet, there are symmetries transforming space in time (and vice versa). In order to derive their form, we step back and see how rotations as symmetries can be seen in the Euclidean line element.

It is sufficient to work with just two of the three spatial dimensions, say x and y. We draw a space diagram with two perpendicular axes for the two coordinates, centered at some arbitrary point p. For symmetries of the line element, we then look at all points that have the same distance, that is, the same line element $\Delta s = \sqrt{(\Delta x)^2 + (\Delta y)^2} = R$ with some constant R, from p. This formula, put in words, means that we are looking at all points on a circle of radius R, centered at the point

6) Robert Herrick

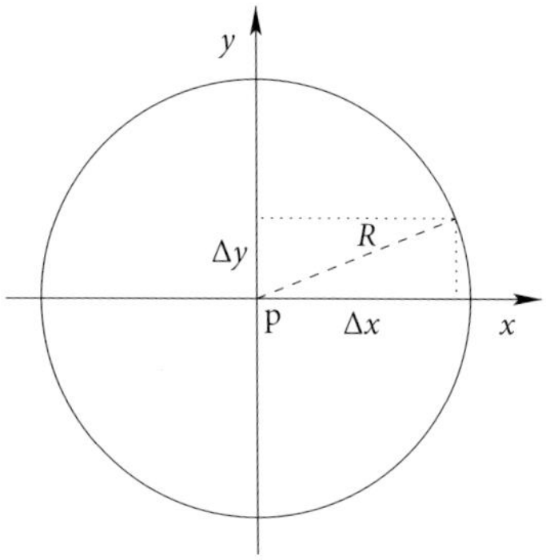

Figure 2.5 A circle as the curve of all points with fixed Euclidean distance $(\Delta x)^2 + (\Delta y)^2 = R^2$ from the center of a two-dimensional space diagram.

p as shown in Figure 2.5. One point that satisfies the condition of $\Delta s = R$ is the point $(R, 0)$ on the x-axis with displacement $\Delta x = R$, $\Delta y = 0$ from the center, but it is not the only solution. If we try to follow the solution curve upwards by adding a positive amount Δy to the y-coordinate, the contribution $(\Delta y)^2$ to the line element increases. In order to keep Δs^2 at the constant value R^2, we have to decrease the x-displacement Δx of our point. Our solution curve moves to the left as we move upwards. Once Δy reaches the value R, we cannot compensate for any additional increase by decreasing Δx because $(\Delta y)^2$ already takes all that is allowed for $(\Delta s)^2$. As we try to decrease Δx further, to negative values, Δy must turn back and decrease from its maximal value R. This whole pattern is realized by the circle shown in Figure 2.5. In particular, the curve remains in a bounded region and cannot move far away from the center.

We have constructed the circle from algebra, as the set of all solutions to the condition that the distance to the center be a constant R. The same curve can be viewed as the set of points obtained by applying all symmetry transformations of Euclidean geometry to some starting point, say $(R, 0)$, under the condition that the center point p be fixed. Only rotations around p satisfy the latter condition, and a circle is, almost by definition, the set of all points obtained by rotating a starting point around some center.

We are familiar with the symmetries of Euclidean space; the algebraic construction of the circle as a solution set seems an unnecessary complication. For Minkowski space-time, the situation is the opposite. We do not have any familiar notion of its symmetries, intuitive or otherwise, but the algebraic solution procedure can be carried out by a rather direct analogy. Once again, algebra is used to generalize familiar notions of geometry and extend its laws to new domains.

We repeat the previous line of arguments for a space-time diagram with two coordinates x and ct. The curves satisfying $\Delta s^2 = -(c\Delta t)^2 + (\Delta x)^2 = R^2$, at constant distance from the center as determined by the Minkowski line element, are no longer circles but hyperbolas as shown in Figure 2.6. A simple solution to this equation is the point $(R, 0)$ on the x-axis, as before, with x-displacement

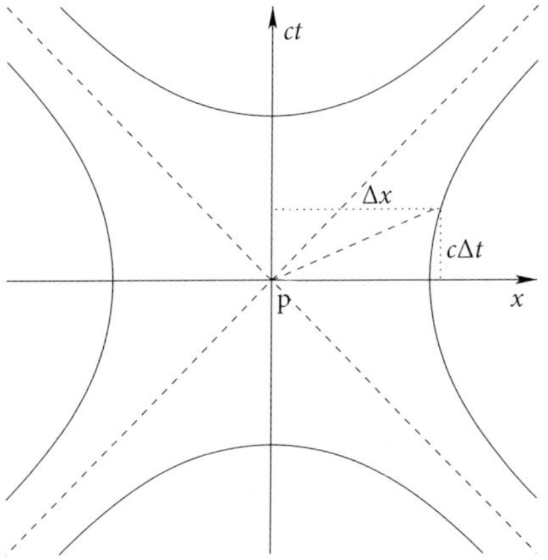

Figure 2.6 Hyperbolas as the curves of all points having fixed Minkowski distance $-(c\Delta t)^2 + (\Delta x)^2 = \pm R^2$ from the center of a two-dimensional space-time diagram.

$\Delta x = R$ from the center and no ct-displacement. If we move upwards by adding some positive $c\Delta t$, to follow the curve of solutions, the negative sign in the time part of the line element means that we must compensate for the larger $c\Delta t$ by increasing also the Δx-displacement, not decreasing it as in the Euclidean case. The two increments of $(c\Delta t)^2$ and $(\Delta x)^2$ in $\Delta s^2 = -(c\Delta t)^2 + (\Delta x)^2$ then cancel each other. With x increasing, the solution curve moves to the right, to larger x-values, and can do so without limit. Any arbitrary amount added to $c\Delta t$ can be canceled by a suitable amount added to Δx. Unlike the circle in Euclidean geometry, the Minkowski hyperbola is not bounded and can reach large values of both Δx and $c\Delta t$. However, for positive $\Delta s^2 = R^2$, the Δx-contribution, which comes with a positive sign in the line element must always outweigh the $c\Delta t$-contribution that counts as negative. In the diagram, the curve stays below the diagonal line where $c\Delta t = \Delta x$, along which $\Delta s^2 = 0$. If we follow the curve downwards, it bends to the right once we cross the x-axis and stays above the other diagonal $c\Delta t = -\Delta x$.

Along the curve just constructed, we never reach the other axis of our space-time diagram, nor do we reach the negative part of the x-axis. The Euclidean circle brought us into those regions and intersected with the axes at $(R, 0)$, that is, our starting point, $(0, R)$, $(-R, 0)$ and $(0, -R)$. The point $(-R, 0)$ is a solution to $\Delta s^2 = R^2$ also for Minkowski geometry, giving us a second branch of the hyperbola not connected to the first one. The other two points, now lying on the ct-axis, do not solve $\Delta s^2 = R^2$ but rather $\Delta s^2 = -R^2$, resulting in two new branches of hyperbolas. Unlike in Euclidean geometry, the line element in Minkowski geometry is not restricted to positive values, and so we should include the hyperbolas of negative Minkowski distance $-R^2$ in our considerations.

Rotations Algebra is the hardheaded sibling of visual geometry. Its notions and operations are not as intuitive, but often more powerful in recognizing new patterns. In our example, although we have a visual understanding of rotations, it is convenient to find an algebraic formula that tells us how to change the coordinates of one point on a constant-distance curve so as to obtain another point on the same curve.

We put our previous constructions of those curves into equations: For the circle, we started at $(R,0)$ and increased Δy, making us decrease Δx to keep $(\Delta x)^2 + (\Delta y)^2$ constant, equal to R^2; the curve had to turn left when we moved upwards. Solving the equation $\Delta s^2 = R^2$ for the new Δx after having changed Δy from zero to its new value, we write $\Delta x = \sqrt{R^2 - (\Delta y)^2}$. This equation expresses the circle as a functional relationship between the two coordinates Δx and Δy of its points, a relationship just as we are looking for. However, our solution is incomplete: it is not applicable to the left half of the circle. Or, it is applicable if we take both positive and negative numbers as two possible values of the square root. For instance, $\sqrt{R^2}$ is equal to R or $-R$ because $R^2 = (-R)^2$. In this way, we do get both halves of the circle, but the formula does not tell us what half we are on if we only specify to be at some value of Δy. The problem is that the circle bulges out to both sides of the y-axis, and a value of y or Δy cannot tell us what side we are on.

Instead of the y-coordinate to vary along the curve, we should better use the angle between the x-axis and a line connecting the center to the point on the curve we are considering (dashed in Figure 2.5). As we move along the circle, the angle takes values in its whole allowed range, from 0 to $\bigcirc$. The coordinate values $(\Delta x, \Delta y)$ of a point on the circle with angle $\sphericalangle$ do not have a simple relationship with $\sphericalangle$ in terms of the familiar functions of polynomials or square roots, but they can be given by trigonometric functions. Indeed, these functions are defined such that $\Delta x = R\cos(\sphericalangle)$ and $\Delta y = R\sin(\sphericalangle)$, and they satisfy an identity that guarantees $\cos(\sphericalangle)^2 + \sin(\sphericalangle)^2 = 1$ for all $\sphericalangle$, so that $\Delta s^2 = R^2$.

With our constant-distance curve written in terms of the parameter $\sphericalangle$, we can find transformation formulas for changing the coordinates of one point on the curve to obtain another such point. Any other point is obtained by an angle increment, mapping $\sphericalangle$ to $\sphericalangle + \Delta\sphericalangle$. Cartesian coordinates of the new point, $\Delta x' = R\cos(\sphericalangle + \Delta\sphericalangle)$ and $\Delta y' = R\sin(\sphericalangle + \Delta\sphericalangle)$ can be computed using the trigonometric identities $\cos(\sphericalangle_1 + \sphericalangle_2) = \cos(\sphericalangle_1)\cos(\sphericalangle_2) - \sin(\sphericalangle_1)\sin(\sphericalangle_2)$ and $\sin(\sphericalangle_1 + \sphericalangle_2) = \cos(\sphericalangle_1)\sin(\sphericalangle_2) + \sin(\sphericalangle_1)\cos(\sphericalangle_2)$:

$$\Delta x' = \Delta x\cos(\sphericalangle) + \Delta y\sin(\sphericalangle)\,, \quad \Delta y' = -\Delta x\sin(\sphericalangle) + \Delta y\cos(\sphericalangle)\,. \tag{2.6}$$

Rapidity We can transform space-time (almost) as we rotate space. To apply the previous arguments to the space-time diagram with Minkowski geometry, we first look for an equation for our constant-distance curves, now given by hyperbolas. As suggested by circles, we attempt to write such a formula using the angle between the x-axis and a straight line connecting the center to a point on one part of the hyperbolas, dashed in Figure 2.6. Since our geometry is of Minkowski type, this angle is not an angle of the form we usually deal with in geometry; to make this clear, we call it "rapidity," denoted by a rocket $\Uparrow$. The name is motivated by the fact that its value determines the ratio of a spatial distance Δx to a time duration $c\Delta t$, of the form of a velocity rather than an angle, that is, the ratio of spatial distances in two directions.

As we define trigonometric functions on a circle, we have "hyperbolic" analogs for hyperbolas. The horizontal coordinate of a point with rapidity $\Uparrow$ is denoted as $\Delta x = R\cosh(\Uparrow)$, the vertical one as $c\Delta t = R\sinh(\Uparrow)$. The hyperbolic origin is indicated by the extra "h" appended to the functions (pronounced "sh" for clarity). In terms of more-familiar exponentials, we write $\cosh(\Uparrow) = \frac{1}{2}[\exp(\Uparrow) + \exp(-\Uparrow)]$ and $\sinh(\Uparrow) = \frac{1}{2}[\exp(\Uparrow) - \exp(-\Uparrow)]$. These functions satisfy the identity $\cosh(\Uparrow)^2 - \sinh(\Uparrow)^2 = 1$ analogous to $\cos(\sphericalangle)^2 + \sin(\sphericalangle)^2 = 1$. The reversed sign ensures that we obtain points on a hyperbola, as required for constant Minkowski distance. If we let $\Uparrow$ run through all real values, we cover the right branch of our hyperbolas: the cosh function always takes positive values. For the others, we have to include the curves $\Delta x = -R\cosh(\Uparrow)$, $c\Delta t = R\sinh(\Uparrow)$ (the left branch) and the flipped versions $\Delta x = R\sinh(\Uparrow)$, $c\Delta t = R\cosh(\Uparrow)$ (the top branch) and $\Delta x = R\sinh(\Uparrow)$, $c\Delta t = -R\cosh(\Uparrow)$ (the bottom branch). (For the circle, those flipped versions would correspond to rotated circles, identical to the original one.)

> Transformations to new points on the hyperbolas are accomplished by analogy with rotations on the circle. We write
>
> $$\Delta x' = \Delta x \cosh(\Uparrow) + c\Delta t \sinh(\Uparrow)$$
> $$c\Delta t' = \Delta x \sinh(\Uparrow) + c\Delta t \cosh(\Uparrow) \qquad (2.7)$$
>
> These equations follow from the relationships $\cosh(\Uparrow_1 + \Uparrow_2) = \cosh(\Uparrow_1)\cosh(\Uparrow_2) + \sinh(\Uparrow_1)\sinh(\Uparrow_2)$, $\sinh(\Uparrow_1 + \Uparrow_2) = \cosh(\Uparrow_1)\sinh(\Uparrow_2) + \sinh(\Uparrow_1)\cosh(\Uparrow_2)$, or the identities of the underlying exponential function.

It turns out that we cannot map a point on the right branch of the hyperbola to a point on the top or bottom branches. The different branches play distinct roles, hinting that space and time retain some of their identity even in a combined space-time treatment.

Lorentz transformations Space and time are not separate from each other, but interchangeable. Space-time transformations (2.7) change Δx in a Δt-dependent way. In Euclidean geometry, we have seen similar-looking equations in the case of Galilei transformations, changing the viewpoint to one of an observer moving at velocity v. To make these equations more similar, we write the rapidity-dependent functions in terms of the velocity so that $\Delta x'$ looks similar to the Galilei version $\Delta x - v\Delta t$ to subtract an observer's velocity. This aim requires $\sinh(\Uparrow) = -(v/c)\cosh(\Uparrow)$ and $\cosh(\Uparrow)$ close to 1. The cosh and sinh functions are related by

$\cosh(\Uparrow)^2 - \sinh(\Uparrow)^2 = 1$ because they specify a hyperbola. They are not independent of each other; both functions contain the information of the single parameter v in different but transmutable form.

> We write the cosh as $\cosh(\Uparrow) = \sqrt{1+\sinh(\Uparrow)^2}$, then solve $\sinh(\Uparrow) = -(v/c)\cosh(\Uparrow) = -(v/c)\sqrt{1+\sinh(\Uparrow)^2}$ for $\sinh(\Uparrow)$ in terms of v/c: $\sinh(\Uparrow) = -(v/c)/\sqrt{1-v^2/c^2}$. The same relationship tells us that $\cosh(\Uparrow) = 1/\sqrt{1-v^2/c^2}$. We write our space-time transformations as
>
> $$\Delta x' = \frac{\Delta x - v\Delta t}{\sqrt{1-v^2/c^2}}, \quad c\Delta t' = \frac{-(v/c)\Delta x + c\Delta t}{\sqrt{1-v^2/c^2}}, \tag{2.8}$$
>
> called Lorentz boosts.

If we compare Lorentz with Galilei transformations, we find that the $\Delta x'$-part is changed only by an additional factor of $1/\sqrt{1-v^2/c^2}$. For small velocities v, much smaller than the speed of light c, this factor does not differ much from 1 and we have almost a Galilei transformation. At large velocities, in particular when we consider light itself, this factor is important and must be taken into account to be consistent with electrodynamics. (For v equal to the speed of light, the factor is infinite. We will come back to this divergence and its consequences at a later stage.)

Moreover, the Δt-part of Lorentz transformations shows a crucial new feature compared to the Galilean counterpart. In classical mechanics and Euclidean geometry, we have bothered only about space and its transformation, for instance, under rotations; time, although it was allowed to appear as a parameter in some space transformations, remained unchanged and untransformed. Time was seen as a rigid, absolute stream of unrelenting ticks which we cannot stop or delay or influence in any other way and by any means, certainly not by changing our position. Lorentz transformations, however, include a formula for changes of time intervals. Time durations are no longer absolute and equally long for all who measure them; their length depends on how fast we are moving. If we consider only velocities much smaller than the speed of light, all terms of v/c in $\Delta t'$ are near zero and $\Delta t'$ almost equals Δt. However, for larger velocities, the assumption of an absolute time is no longer correct. If we change our motion, our perception of both spatial distances and time durations transforms. We must take the notion of combined space-time seriously. Just as we can transform the width of an object into its height by turning our head by $90°$, we can transform space into time (and vice versa) by changing our velocity.

Rotating axes The malleability of time has important and surprising consequences. Just as the transformation of time itself, some implications are unexpected and counterintuitive, or may appear paradoxical. In such situations, it is useful to have geometrical aids to illustrate phenomena not only by the algebraic equations used to derive them but also by diagrams that analyze the same effects as seen by different observers. To that end, we should extend the space-time diagrams we worked with so far by including two sets of axes, one each for two observers moving at different speeds.

To see how the different axes are drawn in one diagram, we once more return to the more familiar example of rotations in Euclidean space. We rotate the axes of one coordinate system (x, y) into a new one (x', y') by literally rotating the axes by some angle $\sphericalangle$ as in Figure 2.7. The result of this familiar transformation need not be derived in detail to see what happens. However, for a generalization of our considerations to Minkowski space-time, it will nevertheless be helpful to go through the construction more systematically.

In order to transform to a new system, we start with the units on both axes, that is, the points $(x, y) = (1, 0)$ and $(x, y) = (0, 1)$, and move them so that they remain at unit distance from the coordinate center p. They move at a curve of constant distance from the center or, with Euclidean geometry, on a circle. Moving the x-unit $(1, 0)$ along a circle of radius 1 by some amount determines the rotation angle $\sphericalangle$. This amount and the rotation angle are free for us to choose, but then the shift of the y-unit $(0, 1)$ is determined. We move it along the circle by the same angle $\sphericalangle$ because we are considering one complete rotation of the whole plane, not two independent rotations of the axes. While the angle should be the same everywhere, there is a second choice to make: If we move the x-unit upwards, we must move the y-unit to the left. That is, not only the amount of the shift but also its direction, or the orientation of the rotation, is determined once we decide how to move the x-unit.

In Euclidean space, we do not really need to think about all these choices because we just know what a rotation does. If the x-unit moves one way along the circle, the y-unit must move the same way, or else the two units might collide for an angle of 45° and the coordinate axes would collapse to just one line. We also know that rotations in space preserve geometrical relationships such as the angle between straight lines. We begin with two perpendicular axes for x and y; a rotation must then preserve the right angle and provide perpendicular axes for x' and y'. This

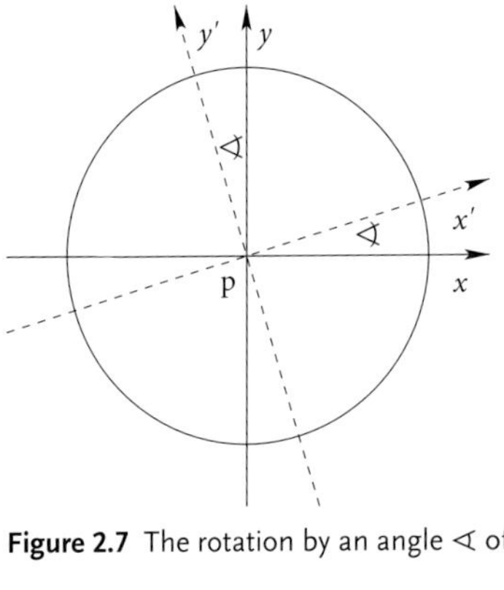

Figure 2.7 The rotation by an angle $\sphericalangle$ of a spatial coordinate system in Euclidean space.

conservation of angles can be accomplished only if the y-unit moves to the left when the x-unit moves upwards.

Coordinate axes If it is still necessary at all, we can convince ourselves of the correct nature of rotations by looking at the formulas (2.6). The new x'-axis is determined by having $\Delta y' = 0$ for all its points because the axis intersects the y'-axis at zero y'-value. By solving the equation $0 = \Delta y' = -\Delta x \sin(\sphericalangle) + \Delta y \cos(\sphericalangle)$ for a relationship between the old coordinates Δx and Δy, we obtain $\Delta y = \Delta x \sin(\sphericalangle)/\cos(\sphericalangle)$. For a given rotation angle $\sphericalangle$, this is a linear relationship between the coordinates. The curve described by it is a straight line, with slope $\sin(\sphericalangle)/\cos(\sphericalangle)$: the new x'-axis. Similarly, we solve the equation $0 = \Delta x' = \Delta x \cos(\sphericalangle) + \Delta y \sin(\sphericalangle)$ which determines the position of the new y'-axis by $\Delta y = -\Delta x \cos(\sphericalangle)/\sin(\sphericalangle)$. The negative sign means that the straight line, that is, the new y'-axis, moves downwards rather than upwards as Δx is increased, a behavior that follows by rotating the y-axis to the left while rotating it to the right would have produced an increasing line as Δx changes. Moreover, the slope of the y'-axis, $-\cos(\sphericalangle)/\sin(\sphericalangle)$ is the inverse of the slope of the x'-axis, which is the correct result for two axes that remain perpendicular after the rotation.

Minkowski diagram We do not have intuition in space-time, but we still have algebra. When we look at the axes of a space-time diagram, they must be changed for a new observer, but how to do so is much less obvious than simply rotating Euclidean axes. We have no geometrical intuition for the action of a Lorentz transformation, especially for a boost to a new velocity as determined by the rapidity parameter $\Uparrow$. Although we know that Lorentz transformations preserve Minkowski geometry just as rotations preserve Euclidean geometry, we cannot be sure at this stage how an angle between two straight lines, as one example for a geometrical notion, is represented.

When we draw a space-time diagram on a sheet of paper, we see angles between the lines we draw. However, even though such a diagram illustrates space-time, not all the geometry contained in it can be taken at face value. We draw a space-time diagram in space, on a sheet of paper on which we have Euclidean geometry. The angles we see are angles of the familiar Euclidean form and size; they are not necessarily angles as determined by the new form of Minkowski geometry we try to illustrate. We have learned how to implement some Minkowski features, by using hyperbolas instead of circles as curves of constant distance to the center. However, we have not yet seen how angles are represented. As one consequence, we are unable to rotate the axes even though we know that the original angle between them, but an angle in Minkowski geometry, must be preserved.

Compared with the Euclidean rotation, only the last, and apparently superfluous method of finding the new positions of coordinate axes is now available. We start with the x-unit and move it up along the hyperbola as a constant-distance curve.

The new position shows the rapidity of the Lorentz boost we are representing. We should then move the ct-unit by the same amount because there is just one rapidity parameter for all of space-time, not two independent ones for the two axes. Again, we must decide in which direction we have to move, that is, to the left or the right, but now there is no intuition to guide us. Euclidean experience, suggesting to move to the left, would not be the correct result: We must move the ct-unit to the right, closer toward the x-unit.

> We look at (2.8) for Lorentz transformations and compute the positions of the new axes. For the x'-axis, we solve $0 = c\Delta t' = [-(v/c)\Delta x + c\Delta t]/\sqrt{1 - v^2/c^2}$ to obtain $c\Delta t = (v/c)\Delta x$, a straight line with slope given by the rapidity ⥣ related to v/c. For the ct'-axis, we solve $0 = \Delta x' = (\Delta x - v\Delta t)/\sqrt{1 - v^2/c^2}$ to obtain $c\Delta t = (c/v)\Delta x$. Also here, the two slopes are inverse of each other, v/c versus c/v, but in contrast to the Euclidean result there is no minus sign in the linear equation for the ct'-axis. As Δx is increased, $c\Delta t$ also increases along this line. The new ct'-unit must move to the right, not the left, as shown in Figure 2.8; the axes in Minkowski geometry turn opposite to what we would have expected by experience with Euclidean geometry.

This surprising and perhaps counterintuitive result must be accepted as the unavoidable consequence of solving the Lorentz-transformation equations, and it is indeed meaningful as further considerations show. For rotations, we were worried about colliding units and collapsing axes if the y-unit were to move to the right as the x-unit moves upwards. Both units, after all, move on the same circle of constant distance from the center and will collide if they move toward each other.

In Minkowski geometry, on the other hand, the constant-distance curve around the center is made of four unconnected parts of hyperbolas. The x-unit and the ct-unit move on different branches, and will never collide no matter how large the

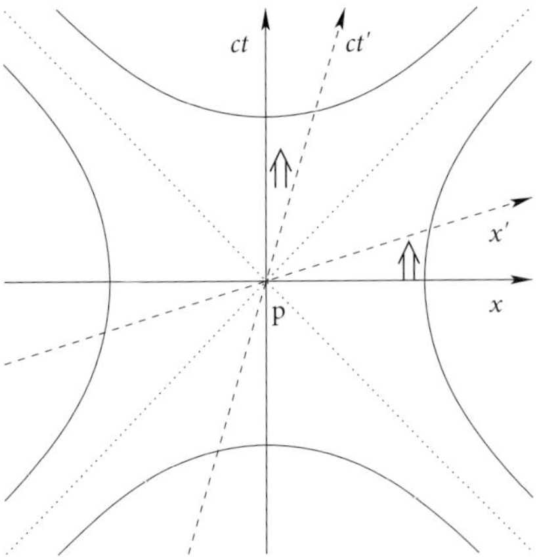

Figure 2.8 Boosting the axes of a Minkowski space-time diagram by rapidity ⥣.

rapidity is. In the extreme case of $|v| = c$, or infinite rapidity, both axes fall on the diagonal and the system does collapse to a single line. However, v in Lorentz transformations is the velocity of an observer doing measurements at a certain speed, and an observer cannot be made of light and move at the speed of light. As long as the speed stays below c, even if it is very large, we obtain a good space-time spanning system of two independent axes. If we do approach the speed of light or even try to surpass it, we would run into trouble with our coordinate system. As we will confirm later, this is a clear warning shot, an indication that the speed of light is special not only in that it is independent of the motion of an observer measuring it, but also in that it presents a limit for all motion.

The shifts of units and axes in our transformation are meaningful in spite of first appearance. The distance between each new unit and the center, on first sight, seems to have increased while we moved along the hyperbola. However, while this appearance is correct for the illustration on a Euclidean sheet of paper, the distance remains constant as determined by the Minkowski geometry we are representing. Hyperbolas, after all, have been determined as constant-distance curves in Minkowski geometry, and so points moving along them, such as our units, stay at constant distance from the center.

The angle between the axes, as another geometrical quantity that is supposed to be preserved, seems to change: the axes no longer look perpendicular after the transformation. Again, it is the angle we intuitively see on our Euclidean sheet of paper that has changed, while the angle computed in Minkowski geometry remains constant. In Minkowski geometry, the new system is still perpendicular. However, we do have to be careful because the illustrations we use do not make the right angle obvious, and so we must pay special attention to how we determine coordinate values of points in the diagram. These values are read off at the intersection of the coordinate axes with lines laid through the point one considers, parallel to the coordinate axes. As seen in Figure 2.9, this process ensures that a periodic coordinate grid is woven through the plane, while readings along lines perpendicular (Euclidean) to the axis would not be part of a uniform plane-filling grid.

Relative simultaneity If time durations change for a moving observer, observers cannot all agree about what is happening at a given time. All observers considered might obey a mutual understanding that they start their clocks at one instant, or retroactively set the zero time to that instant, marked by an unmistakable event, for instance, the time when they first saw a specific supernova. Later events are then assigned a time according to the number of unit intervals passed since the zero moment. However, if time durations Δt differ for observers in motion, in a way that depends on spatial distances Δx, events at different places will be assigned varying times even if there is one observer who experiences all these events at the same time.

As an application of geometrical methods, we look at a diagram similar to Figure 2.9, but with two different events A and B drawn in. As points in a space-time diagram, A and B specify not just a position in space, but also an instant in time such as t', chosen to be equal for both of them in Figure 2.10; with their partial

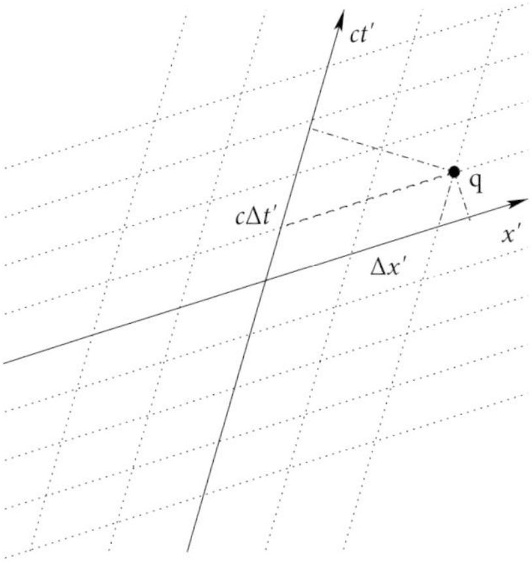

Figure 2.9 In a space-time diagram that does not look right-angled, coordinate values of a point q are determined by lines parallel to the axes (dashed), not lines perpendicular to the axes (dash-dotted). The x'-coordinate $\Delta x'$ is the value along the x'-axis where it is intersected by a line through q parallel to the ct'-axis. The ct'-coordinate $c\Delta t'$ is the value along the ct'-axis where it is intersected by a line through q parallel to the x'-axis. Only this way guarantees that a periodic grid of coordinate increments fills the plane by repeating unit intervals.

timelike nature they mark sharp events. In the diagram, we have two events A and B happening at the same time t' at two different points in space as indicated by their x'-coordinates, with $\Delta x' \neq 0$. The locations and times seen by a different observer moving relative to the primed system are read off from the unprimed axes, where time coordinates t of the two events are not the same; A and B do not happen at the same time of the moving observer. What appears to be simultaneous for one observer may not be simultaneous for another observer in relative motion. This phenomenon, called relative simultaneity, is a clear indication that there is no absolute time, which, if it existed, could be used by both observers to decide whether events are simultaneous.

Relative simultaneity is not all too surprising if we remember that space-time properties are determined by the constancy of the speed of light, the basic principle underlying relativity theory. If observers use light signals and visual observation to see events far away, the light rays have to catch up with a moving observer and may need more (or less) time if they have to travel from two different locations before they can be seen. From the point of view of the unprimed observer in Figure 2.10, where event A happens before event B, light travels farther from A to the primed observer in relative motion than to the unprimed one, and less far from B. Both events were chosen to take place at such times that light signals they emit reach the primed observer at the same time, making the observer conclude that the events do happen at the same time. For the unprimed observer, however, the signals arrive at different times and the events do not appear simultaneous.

Figure 2.11 shows the motion of signals by light rays, as always at 45° in (ct, x) space-time diagrams, no matter how the observer is moving. Light rays are drawn

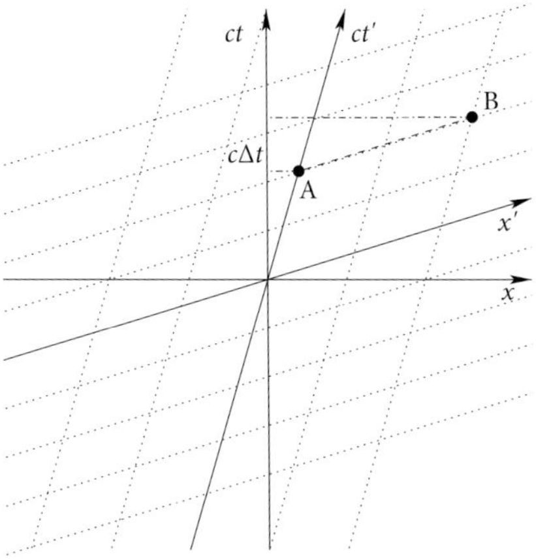

Figure 2.10 Two events A and B happening at the same time t' do not take place at the same time t as seen by a moving observer.

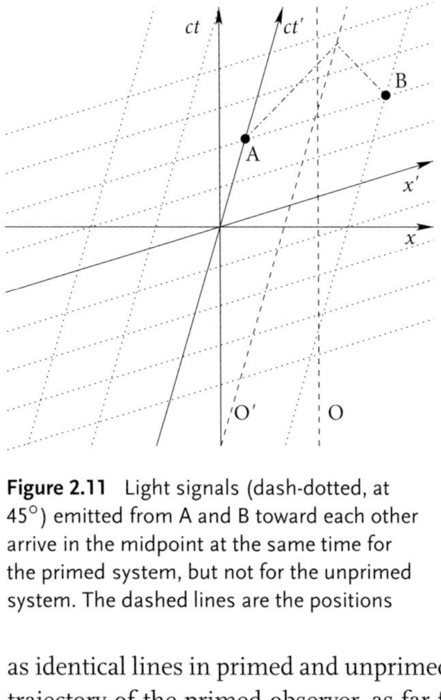

Figure 2.11 Light signals (dash-dotted, at 45°) emitted from A and B toward each other arrive in the midpoint at the same time for the primed system, but not for the unprimed system. The dashed lines are the positions taken by two observers at all times, O moving along with the unprimed system and O′ moving along with the primed system. Both observers have position coordinates in the midpoint between the coordinates of A and B.

as identical lines in primed and unprimed systems. These lines reach the midpoint trajectory of the primed observer, as far from A as from B, at the same time: they intersect right at the observer's line. They appear simultaneous. For the unprimed observer, the signal from A arrives at the midpoint trajectory before the signal from B and is seen first.

Relative simultaneity applies only to events happening at two different places. If two events happen at the same location and at the same time for one observer, they are just one and the same event and will be seen at the same time for all observers.

The relevance of different spatial locations follows from the argument using light signals, but it can also be shown using our transformation equations (2.8). If we have two events happening simultaneously at different points for the primed observer, they are characterized by coordinate separations of $\Delta t' = 0$ with some $\Delta x' \neq 0$. To see what this means for a moving observer, we transform to unprimed values $\Delta t = [\Delta t' + (v/c^2)\Delta x']/\sqrt{1 - v^2/c^2} = (v/c^2)\Delta x'/\sqrt{1 - v^2/c^2} \neq 0$. The time difference between both events in the unprimed system is nonzero if the events happen at different locations, but remains zero if the events are identical.

Space-time The enlightened picture of space-time, obtained from taking properties of light seriously, is different not only from our intuitive understanding, but also from the mathematical formulation used in classical mechanics. We assume nothing but the fact that the speed of light (in vacuum) takes the same value no matter how fast we are moving with respect to it. If this assumption were violated, we would be in conflict with electrodynamics, the astonishing theory that allows us to derive the value of the speed of light from its equations, with no dependence on the velocity of an observer. Using light and mirrors (and some counting device for bouncing light rays) as the most basic building block of a fundamental clock based on physical principles, we were able to derive the geometry of space-time. Space and time are linked to each other. There aren't two separate sets of geometrical laws to determine space relationships on the one hand and time relationships on the other, but only one combined geometry for space and time at the same footing, given by the Minkowski line element.

By keeping the geometry, embodied by the line element, preserved, the form of space-time transformations is determined and must be of Lorentzian form. These transformations, in turn, relate measurements of space distances and time durations made by observers moving at different speeds. As shown by relative simultaneity as the prime example, space and time are not rigid even for physical interpretations. They are transformed into each other in the sense that one observer may see less time separation than another one, but would compensate for that by a different spatial distance between the two events. Space and time, in this way become much less solid than we are used to.

With old foundations shaking, we can find some support by taking account of what can still be trusted. Space and time are no longer rigid and distinct, but there remain some properties that do distinguish space distances from time separations, in a way that all observers would agree on irrespective of their velocities. There *must* be some distinguishing feature between space and time because we perceive space and time differently. We can move at will in space but not in time.

If space and time were completely transmutable, we could transform some time interval into space, move backwards in space, transform back to time and thus travel to the past. No one has ever done so, which does not quite mean that it is impossible; the process might just require velocities so large that it is difficult

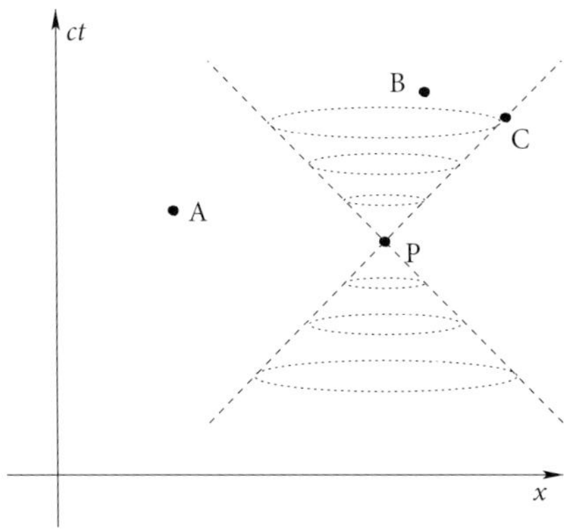

Figure 2.12 A is spacelike separated from P, B timelike, and C lightlike. In a diagram with one additional space coordinate, the cross of light rays would appear as a cone with B in its interior.

to accomplish a complete transmutation for anything but the lightest elementary particles. Even so, there would be severe conceptual problems if time travel were possible.

If we could travel to our own past, we could affect our history so that our future travel becomes impossible, for instance, by preventing, somehow, our own conception or birth from taking place. However, if we prevent our future travel, we cannot have gone back to the past to perform the act of prevention, so we might as well have traveled to the past. Can we even experience travel to the past, or would we, assuming that time travel were possible, just relive a piece of our history without remembering that we had been there all along? Paradoxa and puzzles like these show that space and time, with all their malleability, must still preserve some of their identity and idiosyncrasy, at least so much so that time remains a one-way street.

Comparing the Minkowski line element of space-time with the Euclidean line element of space suggests one crucial distinguishing feature. Time separations enter the line element with a minus sign, and so the total value of Δs^2 may be negative; it is not required to be positive, or perhaps zero, like the Euclidean line element. Taking this cue, we introduce three different classes of separations between events, those with $\Delta s^2 > 0$ (called spacelike separation), $\Delta s^2 < 0$ (timelike separation), and $\Delta s^2 = 0$ (lightlike separation). In Figure 2.12, A is spacelike separated from P, B is timelike separated, and C lightlike. For instance, two events with $\Delta t = 0$ but $\Delta x \neq 0$ are spacelike separated, two events with $\Delta t \neq 0$ and $\Delta x = 0$ are timelike separated, and two events with $\Delta x = c\Delta t$ are lightlike separated. As the last example demonstrates, in space-time, it is possible for two events to have vanishing separation Δs but not be identical, provided they are connected by a light ray.

Light cone Drawn in space-time, a light cone spreads out by 45° to all sides. The regions of spacelike and timelike separated events, as seen from some fixed event P, are delimited by the light cone through P: the set of all light rays that emanate from P to the future or reach P from the past. In a space-time diagram, we draw these light rays as straight lines of 45°, no matter what coordinate system we use. The light cone and space-time features associated with it are invariant under changing an observer's state of motion, just as the value of Δs (which vanishes along all light rays). More-specific distance measures, such as the time delay Δt between two events, on the other hand, depend on which observer is doing the measurement, as seen by relative simultaneity. In particular, the notion of event A in Figure 2.12 happening before or after event P depends on the observer; some will see A happen earlier, others P, and a few will see both events occur at the same time.

There is no observer-independent way of telling what temporal relationship A and P have with each other. There can be no causal relationship between the two events in a consistent formulation; that is, it is not possible for event P to have an influence on event A, or vice versa. For instance, if P is a star exploding in a supernova, a planet at the position of event A cannot be burned by the time of event A. Such an influence, or any causal relationship, would be observer-independent because it would be recognizable just by looking at the state of A, without measuring observer-dependent spatial or temporal distances. For those observers who see A happen before P or at the same time, P can by no means have an influence on A. No observer can ascribe anything happening at A as caused by something happening at P; there cannot be any causal relationship.

Event B, on the other hand, appears to the future of event P for all observers: it is always in the top interior of the light cone of P, an observer-independent notion. Also C, at the boundary of the top interior of the light cone, will be in the future of P for all observers. By the same token, all events in the bottom interior of P's light cone or on its boundary were in the past of P, for all observers.

For any event P, there are clear regions in space-time that mark its future or past, but only a vague notion of the present in which events might be seen as simultaneous with P by some observers, or preceding or following P by others. The best we can do in an observer-independent way is to attribute the same causal meaning to all events, such as A, outside of the light cone of P. For lack of a better word, we still call it the present because the region does contain events simultaneous with P for some observers. Moreover, for any event in this region, there is at least one observer who sees it happening at the same time as P. Compared with the Newtonian view in which time is absolute, with an observer-independent meaning of simultaneity, the present is more vague, but also larger as a space-time region. Relativistic causality is "more present" than Newtonian causality; see Figure 2.13.

Proper time Time is not just measured, but also properly experienced. As we compare measurements with other observers, we find that time durations change. However, many physical processes are characterized by intrinsic time scales, specific times of periodicity or decay (and sometimes growth) that help to define the process, irrespective of who is observing it. A neutron is unstable and decays after less than 15 min, a statement in which it would be awkward to list all possible values that could be measured by moving observers. A single number, as the one quoted, is more proper.

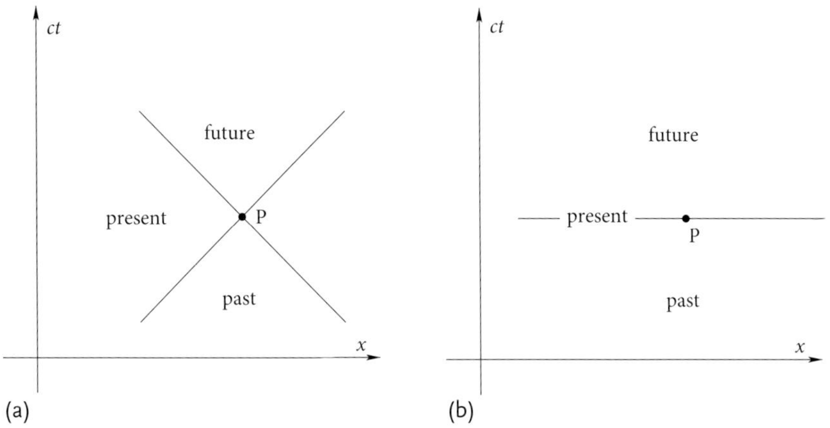

Figure 2.13 The light cone of an event P, as the main observer-independent notion, determines causal relationships with other events happening in the past, future, or present of P (a). In the Newtonian view with an absolute, observer-independent time, the present is the set of events happening at the same time as P (b), a much smaller space-time region than in the relativistic view.

While different observers measure different time lapses on a moving object, there is one notion of time that they can all agree on. As an object is moving at some velocity, perhaps slowing down or speeding up depending on the circumstances, it experiences external influences according to its own progress of time. If the object's velocity is changing, we would constantly have to perform Lorentz transformations in order to keep up with the change of time perception, a tedious process, but one that can be used to determine a notion of time associated only with the moving object, not with any external observers in relative motion. If we compute the corresponding time lapse, we have an observer-independent notion of time, an invariant time also called proper time (the time belonging to the object). While direct time measurements by observers would yield results different from those computed for the object, any observer who keeps track of the object's motion could compute the object's proper time from more direct measurement results.

The computation of proper time is not difficult, if we only realize the power of invariants. We have so far used the invariant line element $\Delta s^2 = -(c\Delta t)^2 + (\Delta x)^2 + (\Delta y)^2 + (\Delta z)^2$, which as the fundamental law of Minkowski geometry is independent of the state of observers, to split space-time into different causal regions as determined by the light cone. Proper time is our second application of invariance.

At every instant of time, proper time is the lapse of the time coordinate Δt as determined by an observer moving along with the object. For such an observer, the object does not move at the instant of time we consider, that is $\Delta x = \Delta y = \Delta z = 0$. With these conditions, the time lapse is $\Delta t = \sqrt{-\Delta s^2}/c$, a real number if $\Delta s^2 < 0$, for timelike separation of events along the objects motion.

Some time later, the object's velocity may have changed; we should transform to a new observer moving along with the object, measuring time lapses $\Delta t'$. Performing explicit Lorentz transformations to keep track with the motion would be the tedious process mentioned above, but we can compute proper time by equating it with the result for Δt obtained by an observer moving along with the object, written in terms of the invariant line element with the same value for all observers. Thus, along an object's motion, we distinguish proper time as the quantity

$$\Delta T = \frac{\sqrt{-\Delta s^2}}{c} . \tag{2.9}$$

One of our main premises is that the speed of light c is constant and observer-independent, and Δs^2 is invariant as the prime statement of geometry. Proper time ΔT obtained from (2.9) is then invariant as well. At every instant of time, it agrees with the time lapse Δt measured by an observer moving along with the object at that instant. It agrees with the invariant time experienced by the object itself.

The Minkowski form of the line element brings (2.9) to $\Delta T = \sqrt{(\Delta t)^2 - [(\Delta x^2) + (\Delta y)^2 + (\Delta z)^2]/c^2}$. Factoring out Δt and identifying $\Delta x/\Delta t$ and so on as the velocity components of the object, we have $\Delta T = \Delta t \sqrt{1 - (v_x^2 + v_y^2 + v_z^2)/c^2}$ or

$$\Delta T = \sqrt{1 - v^2/c^2}\,\Delta t \tag{2.10}$$

with the magnitude $v = \sqrt{v_x^2 + v_y^2 + v_z^2}$ of the velocity.

Proper time ΔT is an invariant notion of time agreed upon by all observers, unlike the specific time lapses Δt measured directly. However, while one can always compute ΔT for a moving object, using its trajectory or the time-dependent velocity in (2.10), it is not possible to trade in the invariant T for the noninvariant t in the line element or other expressions referring to the geometry. A time coordinate of this form would be assigned to a single event, or a single point in space-time. Proper time, on the other hand, is defined only as the time lapse experienced by an object moving between two events. Its value, via the velocity in (2.10), not only depends on the positions of both events rather than on just one event, it also depends on how the object is moving from one event to the other, or on the worldline as in Figure 2.14. These dependencies make sense because proper time is supposed to be experienced by an object moving in a specific way, but they prevent us from interpreting T as a time coordinate assigned to single events. In particular, an invariant proper time is not an absolute time, and it cannot be used to define a notion of absolute simultaneity.

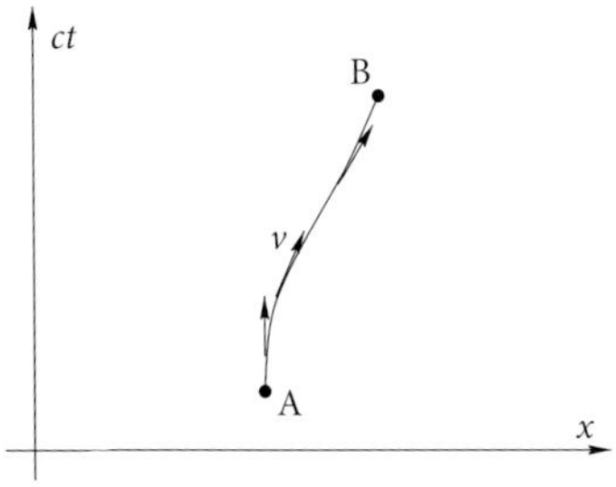

Figure 2.14 An object moving along some (timelike) curve from event A to event B, called a worldline, experiences the progress of time depending on its velocity v. Time lapses as seen by the object are given by proper time (2.9).

Time dilation What we measure is not always what others experience. Proper time experienced by a moving object, related to coordinate time by (2.9), is always slower than any coordinate time measured by observers, or at best, as slow if the object is not moving: The factor $\sqrt{1 - v^2/c^2}$ is a number less than one, or equal to one if (and only if) $v = 0$. The proper-time interval ΔT is always smaller than any coordinate-time interval Δt measured by an observer moving with respect to the object. For a process unfolding on the moving object, an observer not following the motion measures a duration $\Delta t > \Delta T$ larger than the duration seen directly on the object. For the observer, processes on the moving object seem to be drawn out, or dilated.

For a sizable effect of time dilation, the velocity v of an object must be rather close to the speed of light c, or we must have precise measurements of time intervals. The latter is possible with modern atomic clocks,[7] the former if we make use of natural processes occurring with elementary particles at high speeds. Unstable particles provide us with an intrinsic, if imprecise, time unit by their lifetime. If time dilation occurs, unstable particles at high speeds should stick around for longer than they would live at rest.

A classic example is given by muons, unstable particles which decay into electrons (and neutral particles of small mass called antineutrinos) just $\Delta T = 2.2\,\mu s$, or about two millionth of a second, after they have been produced, for instance, in collisions of other energetic particles. Such particle reactions occur in the upper atmosphere above Earth's surface when cosmic rays hit the nuclei of nitrogen or oxygen atoms. The muons then travel towards Earth with a velocity close to the speed of light (at $v = 0.999c$, say). However, even at such high speeds, their lifetime would allow them to travel just $\Delta x = v\Delta T = 660\,\mathrm{m}$ downwards, not far enough to reach the ground. And yet, particle detectors measure those particles in numbers which cannot be explained by rare random events of muons fit enough to outlive the average lifetime.

7) Relative simultaneity and time dilation have been discussed as practical problems for security regulation, for instance, in a study commissioned by the UK government (www.bis.gov.uk/foresight, accessed 1 April 2012). I am grateful to Holger von Juanne-Diedrich for the reference.

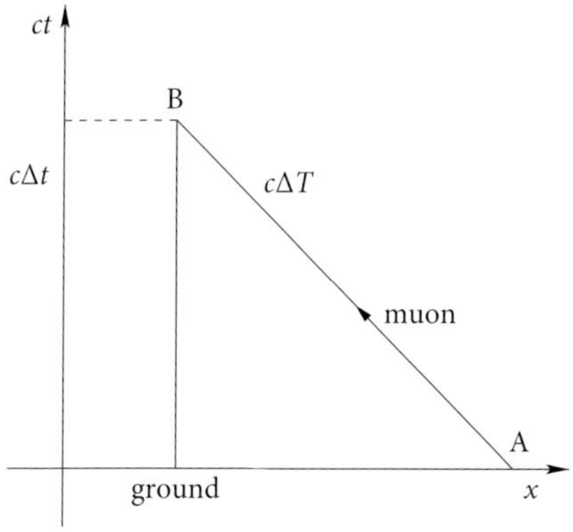

Figure 2.15 A muon is produced in the upper atmosphere (event A) and moves downwards to smaller heights x. At event B, it reaches the ground and meets an observer who stayed at the ground level. The nonmoving observer's worldline is vertical; the muon's worldline has an angle close to 45° because the speed is almost the speed of light. The observer at rest measures the time duration Δt, while the muon's travel time in its own frame is proper time ΔT. Time dilation implies $\Delta t > \Delta T$ such that even the limit of 2.2 μs for ΔT, posed by the muon's lifetime, allows travel distances traversing the main part of the atmosphere.

Time dilation shows that the proper-time interval[8] $\Delta T = 2.2\mu s$ experienced by a muon from birth to death amounts to a dilated duration of $\Delta t = \Delta T/\sqrt{1-0.999^2} = 22\Delta T$ for an observer staying put on Earth's surface. During this time, a muon can travel 14 500 m downwards, more than the meek 660 m. However, the atmosphere is hundreds of kilometers high, still too much for the moving muons to traverse during their dilated life. A closer look reveals that much of the outer atmosphere is too thin to lead to many collisions with cosmic rays, and the production of muons. The densest part of the atmosphere is the troposphere, which contains about 80% of the total mass but is only about 7 km thick (at the poles) and maximally 17 km (at the equator). Muons produced in this layer of the atmosphere can reach the ground if time dilation is taken into account.

In Figure 2.15, the processes are illustrated by a space-time diagram. The time lapse Δt measured by an observer on the ground corresponds to the difference of time coordinates. The proper-time lapse ΔT, to be compared with the muon's lifetime, is computed along the worldline of the muon. As shown by (2.10), we have $\Delta t > \Delta T$ by time dilation. In the diagram, the proper-time distance looks larger than the coordinate-time distance, but just by an apparent effect of our eyes and intuition being used to Euclidean geometry on the plane on which we draw a space-time diagram. Minkowskian distances in space-time, computed with the correct line element, indeed have $\Delta t > \Delta T$.

8) The proper-time interval is the proper time-interval to be used for the calculation, an example in which language appears associative at least in its validity, even though the information conveyed by the two different hyphenations is not the same.

Twin travels The debate of nature versus nurture affects even time. Proper time is experienced differently by two objects starting as equals but moving apart. They observe different time lapses, including their own aging should they compare their fitness when they meet. Minkowski geometry tells us what motion makes us stay young.

In Euclidean space, a straight line between two points marks the shortest distance Δs. In space-time, a straight line between two timelike separated points, one in the future light cone of the other, marks the longest time lapse $\sqrt{-\Delta s^2}/c$. (For spacelike separated points, the straight line still marks the shortest distance, while the invariant distance between two lightlike separated events is always zero.)

Let us look at a collection of events as in Figure 2.16. Writing them in the order (ct, x), we assign coordinates $(1, 4)$ (in some length units) to event A, $(6, 4)$ to event B and $(6, 7)$ to event C. With these numbers, we compute the proper-time separations $c\Delta T_{\mathrm{AB}} = \sqrt{(c\Delta t)^2} = 5$ between A and B, and $c\Delta T_{\mathrm{AC}} = \sqrt{(c\Delta t)^2 - (\Delta x)^2} = 4$. Despite (Euclidean-misguided) appearance, the proper-time lapse between A and B is larger than the one between A and C. The physical effect is time dilation; it comes about by the minus sign in the line element making us subtract (rather than add) spatial separations between events when they contribute to proper time.

The lines from A to B and C, respectively, do not connect the same events and do not show whether a straight timelike line is of shortest or longest distance. However, we can reuse the preceding calculations if we complete the two lines to an isosceles triangle with a new event D. The dashed lines are simple reflections of the solid ones, mirrored at a horizontal line through B and C. The line BD has the same length as the line AB, and CD the same as AC. The time duration from A to D via route (1) passing B is $c\Delta T^{(1)}_{\mathrm{AD}} = c\Delta T_{\mathrm{AB}} + c\Delta T_{\mathrm{BD}} = 10$, and via route (2) passing C it is $c\Delta T^{(2)}_{\mathrm{AD}} = c\Delta T_{\mathrm{AC}} + c\Delta T_{\mathrm{CD}} = 8 < c\Delta T^{(1)}_{\mathrm{AD}}$. The straight line is longer.

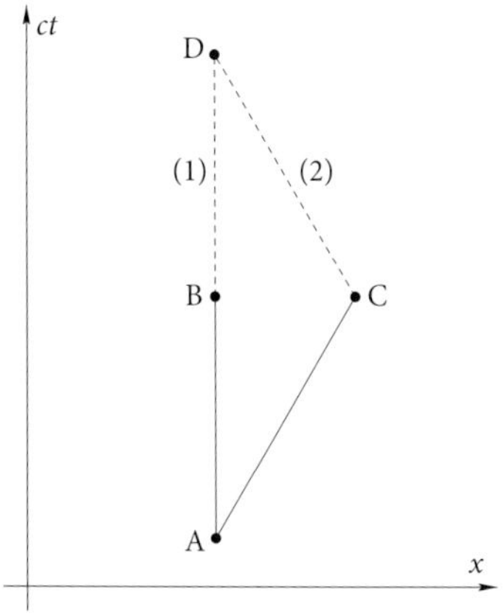

Figure 2.16 The shorter-looking line may not be the shortest. Time dilation implies that the time duration measured by the moving observer along AC is shorter than the duration along AB. The straight line from A to D via B is longer than the line via C. Two twins starting at A at the same age, but traveling along the two lines, differ in age when they meet again at D, the one staying at rest (straight vertical line via B) being older.

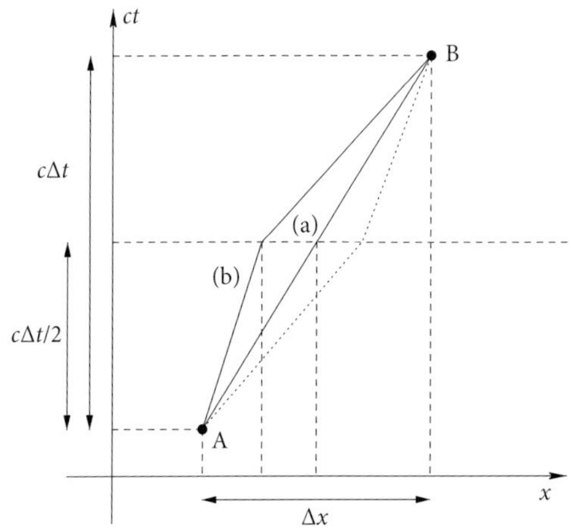

Figure 2.17 Coordinate separations along partially straight curves connecting two events A and B, with velocity changes at half-time. Proper time progresses by the largest amount if we follow the straight line.

Longest time Steady people age faster. For an unquestionable mathematical proof of this statement, we consider two events A and B and ask how a procrastinator would move between them, trying to spend the longest possible time before arriving at B. We are required to arrive at B, an event happening at a specified time. With this condition, it is not possible just to stay at A and never reach the place of event B. However, by exploiting time dilation, we can at least try to move in such a way that for our own perception, time progresses by the largest possible amount. We must find a way to move such that proper time is maximized.

Let us assume that the two events are separated by a distance Δx in space and Δt in time. As the simplest option to move, we consider a straight worldline connecting A with B, corresponding to motion at constant speed $V = \Delta x/\Delta t$. For the sake of comparison, we choose some other worldline along which the speed is not constant. Let us change the speed just once, at half-time $\Delta t/2$ after event A. If a distance X is traveled from A till half-time, the new worldline corresponds to motion at constant speed $V_1 = X/(\Delta t/2) = 2X/\Delta t$, followed by motion at another constant speed $V_2 = 2(\Delta x - X)/\Delta t$. The two velocities add up to $V_1 + V_2 = 2V$; in other words, $V = \frac{1}{2}(V_1 + V_2)$ is the average of the two velocities V_1 and V_2.

Two different amounts of proper time pass along the two worldlines, even though they both last an amount Δt as seen by the observer whose coordinate system we are using. For the fully straight line (a) in Figure 2.17, $\Delta T_1 = \sqrt{1 - V^2/c^2}\,\Delta t$, while for the partially straight line (b) $\Delta T_2 = (\sqrt{1 - V_1^2/c^2} + \sqrt{1 - V_2^2/c^2})\Delta t/2$. Since Δt and Δx, and thus V, are determined by the fixed events A and B, there is only one free parameter in the problem, V_1. By comparing ΔT_1 with ΔT_2 for different values of V_1, we see what motion maximizes proper time.

We look at $4(\Delta T_1^2 - \Delta T_2^2)/\Delta t^2 = \Delta$ and see whether it is positive or negative for different values of V_1. We have

$$\Delta = 4\left[1 - \frac{(V_1 + V_2)^2}{4c^2}\right] - \left[2 - V_1^2/c^2 - V_2^2/c^2 + 2\sqrt{(1 - V_1^2/c^2)\,(1 - V_2^2/c^2)}\right]$$

$$= 2\left[1 - V_1V_2/c^2 - \sqrt{(1 - V_1^2/c^2)\,(1 - V_2^2/c^2)}\right].$$

If we write $1 - V_1V_2/c^2$ as the square root of $(1 - V_1V_2/c^2)^2 = 1 - 2V_1V_2/c^2 + V_1^2V_2^2/c^4$, the first term in the last expression is almost identical to the square root. They differ by $V_1^2 + V_2^2 - 2V_1V_2 = (V_1 - V_2)^2 \geq 0$, a nonnegative number. The difference $\Delta T_1^2 - \Delta T_2^2$ therefore cannot be negative, and it is zero only if $V_1 = V_2$, in which case the velocity does not change at half-time. (The identity $V_1 + V_2 = 2V$ then implies that V_1 and V_2 take the value V, as realized along (a).) Proper time along the straight line (a) is longer than proper time along any piecewise straight curve with a change of pace at half-time.

A steady pace at constant velocity implies the longest lapse of proper time between two events – or, as seen by a nonmoving observer, the fastest aging. If we make a fast first move before half-time and then relax a bit to arrive at the specified time for event B, we spend less proper time and age more slowly. And so we do if we first slack off and then make up for it in the second half.

For more general ways of motion, we iterate the half-time pep talks or slack-offs and change velocity at quarter times, or even more often, by dividing the new pieces of straight lines at their half-times. As illustrated in Figure 2.18, an arbitrary worldline from A to B can be approximated in this way, along which the velocity need not be constant. Each subdivision of a straight segment decreases the amount of proper time, and so the straight line has the maximum amount among all worldlines from A to B.

Twin orthodox Critical minds are often led to distrust special relativity, not the least because twins with different ages at their reunion D (Figure 2.16) seem to contradict the notion that one can take any moving observer's view. Seen from the twin en route (2), the other twin is moving away at constant speed, then returning after reaching event B. The age difference should be reversed, but as an objective notion, it cannot depend on whether one takes the traveling twin's view or the view of the twin at rest. The older twin's hair looks more gray, no matter how we are moving.

In terms of motion, route (1) is the worldline of an observer staying at rest, while route (2) starts at the same point as route (1), but has an observer moving to the location of event C with some speed and returning to D after reversing the speed at C. Two twins starting at A at the same age will be of different age at event D and

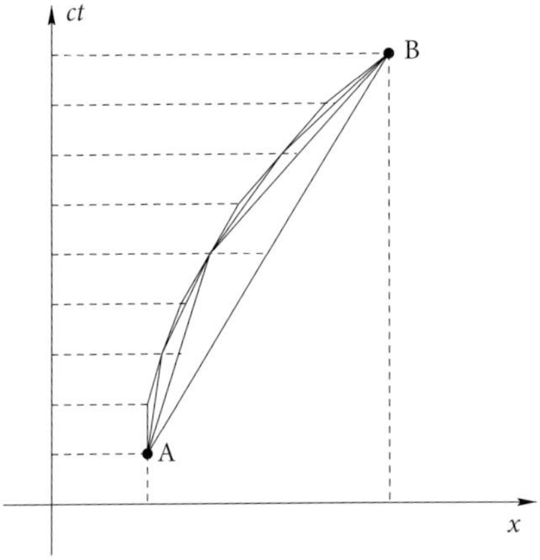

Figure 2.18 Coordinate separations along partially straight curves connecting two events A and B, with several velocity changes.

later, the traveling twin with the shorter proper-time lapse being younger. Geometrically, the traveling twin is distinguished by a line of motion not entirely straight, bending at event C where the speed is reversed. The two routes are not mirror images of each other, and there is no reason to assume that both twins age at the same rate.

Route (2) is distinguished from route (1) because a reversal of the velocity, or acceleration, is required at event C. Route (1) as seen from an observer en route (2) would also require a reversal of velocity, now at event B. However, unlike the real reversal at C, the reversal at B would only be apparent because it is obtained from the viewpoint of an observer who does not move at constant velocity. It is the observer who reverses velocity, not the twin, and an observer's antics cannot affect the age of the twin. (The twin's antics, on the other hand, do influence the aging process via proper time.) Only by viewing this observer like one at constant velocity are we introducing the apparent reversal of velocity at B. Again, we see that the two motions (1) and (2) are not symmetrically related.

To describe the situation from the viewpoint of the traveling twin, we must use two different coordinate systems in relative motion with (ct, x), each with constant velocity. While the twin is moving away, a suitable coordinate system (ct', x') is obtained by a Lorentz transformation with the speed v of the departing twin. In this system, Figure 2.19, a new time coordinate brings into play relative simultaneity. No longer are B and C simultaneous; when the moving twin reaches event C in t', the twin at rest has only come as far as event B$'$, not B. When the traveling twin reverses velocity, yet another coordinate system (ct'', x'') should be used, obtained from (ct, x) by a Lorentz transformation with speed $-v$. While the traveling twin does not move much during the reversal process, which we assume to take a negligible amount of time, the twin at rest ages considerably because event C in t'' is simultaneous with event B$''$, not B$'$ or B. The age difference of the traveling twin en route (2) is equal to the age difference of the twin at rest between events A and

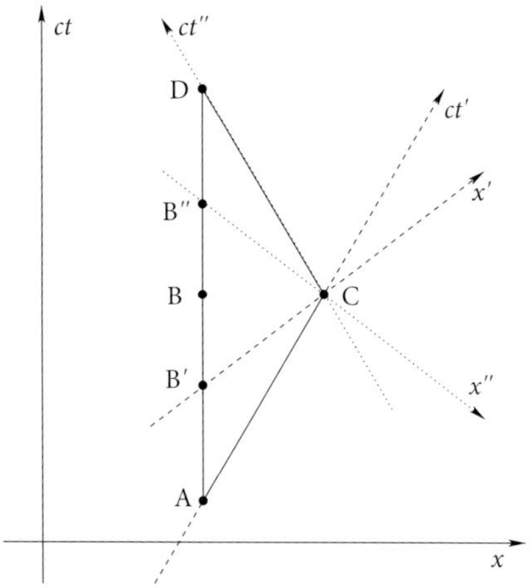

Figure 2.19 While events B and C are simultaneous for the twin staying at rest, they are not simultaneous for the traveling twin. Two new coordinate systems are introduced, one for motion along with the traveling twin departing, and one for the traveling twin arriving. These coordinate systems are characterized by time axes parallel to the motion of the traveling twin. The directions of the space axes then follow from Lorentz transformations, which require the time and space axes to have equal angles with the vertical and horizontal.

B′ plus the age difference between events B″ and D as seen by the moving twin. Since the twin at rest ages also between events B′ and B″, the older total age at the reunion event D is accounted for from the view of the traveling twin.

> With the coordinate values used in Figure 2.16, we have $v/c = 3/5$. Since the x'-axis has slope v/c, we compute $c\Delta t_{\mathrm{AB'}} = c\Delta t_{\mathrm{AB}} - (v/c)\Delta x_{\mathrm{AC}} = 16/5$. A Lorentz transformation of the displacement $(c\Delta t_{\mathrm{AB'}}, 0)$ between A and B′ gives $c\Delta t'_{\mathrm{AB'}} = 4$, in agreement with the proper-time difference $c\Delta T_{\mathrm{AC}}$ which has lapsed in the primed frame between A and C. Much of the aging $c\Delta t_{\mathrm{AD}} = 10$ of the twin at rest happens while the traveling twin is changing velocity in what seems an instant for the traveling twin: $c\Delta t_{\mathrm{B'B''}} = 10 - 32/5 = 18/5$, a significant fraction of the total aging $c\Delta t_{\mathrm{AD}} = 10$.

The reversal of velocity distinguishes the two routes, making them noninterchangeable and resolving any twin paradox. The age difference depends mainly on the velocity and the distance traveled. If the traveling twin moves far away, the travel time is much larger than the time required to reverse velocity. How can a brief acceleration event, making the routes nonsymmetric and "causing" the age difference, have such a large effect on aging? Upon closer examination, it is not just the acceleration, or the kink of worldline (2) at C, that causes the age difference, but indeed also the distance traveled: The longer the distance, the larger the portion $\Delta t_{\mathrm{B'B}}$ within Δt_{AB} in Figure 2.19. The aging effect is nonlocal: it depends on the whole extended worldline, not just one one event.

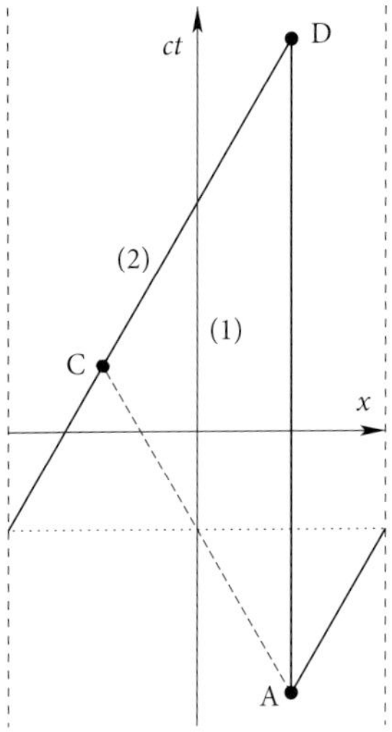

Figure 2.20 Twin travels with circular space and cylindrical space-time (the vertical dashed borders to be glued together). Both twins travel from A to D along straight lines, without acceleration. Route (2) has the shorter proper time. Time durations between A to D along route (1) and from A to C (around the cylinder) to D along route (2) are the same as before in Figure 2.16, as shown by the auxiliary line from A to C. The routes are of different proper durations, and they are not interchangeable: Route (2) always winds around the cylinder but route (1) does not, no matter how an observer is moving.

What if we change the situation so that the traveling twin never reverses velocity, but is still reunited with the twin at rest after some time? This construction may sound absurd, but it might happen if space is closed in itself so that the traveling twin, always moving in one direction, returns to the twin at rest from the other side. In our two-dimensional space-time diagrams, we would no longer have a plane but rather some version of a cylinder, with closed circular space and time along the axis. There is no reversal of velocity and no lapses of time as the one from B′ to B″, and yet the traveling twin must be older at reunion, owing to time dilation. In this case, it is the closure of space that renders the two motions inequivalent. The twin at rest just moves along the cylinder axis in space-time while the traveling twin moves once around the cylinder, two worldlines which cannot be transmuted into each other by changing our viewpoint. The line element still tells us that the traveling twin traverses the longer distance in space-time; see Figure 2.20.

Lorentz contraction Relative simultaneity is the basic phenomenon of special relativity. It explains many additional properties, as seen for the age differences of twins having traveled at different speeds. It also has consequences for length differences, as should perhaps be expected from the common home of space and time. However, as often in the space-time household, there are subtle differences between its members.

Lorentz transformations show relative simultaneity because time separation $\Delta t' = 0$ and space separation $\Delta x' \neq 0$ for two events in one coordinate system implies time separation $\Delta t = (v/c^2)\Delta x'/\sqrt{1-v^2/c^2} \neq 0$ in a system at relative velocity v. If we apply the spatial part of a Lorentz transformation, we obtain $\Delta x = \Delta x'/\sqrt{1-v^2/c^2} > \Delta x'$. Space separations between events seem larger from a moving view: they are dilated just as time separations.

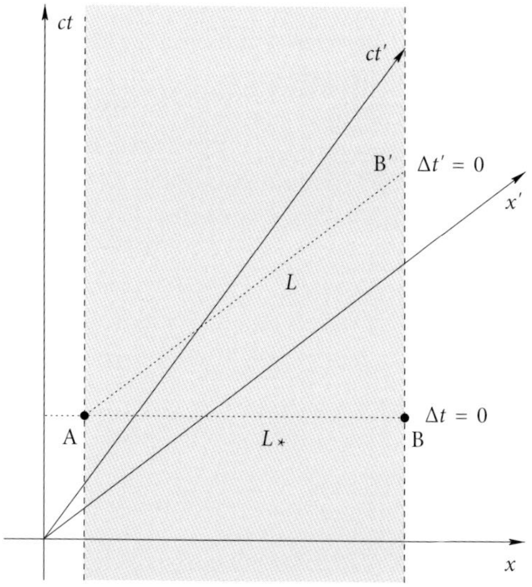

Figure 2.21 A straight object extending between events A and B in system (ct, x) and staying at rest, thus sweeping over the shaded region in space-time, has length L_*. In a system at relative velocity, event B is not simultaneous with event A, but rather B′ is. The observers attribute invariant line elements along different curves to the lengths L and L_*, respectively. From the point of view of the system at rest, the moving observer measures a shorter distance between the endpoints of the object, subtracting the distance traveled while the endpoints align with measurement marks.

However, if we measure the length of an object while moving, or if we measure the length of a flying object, it seems shorter or contracted. How can these two statements of dilated spatial separation between events and contracted length measurements not be in conflict with each other? The resolution is again relative simultaneity.

If we measure the length of an object at rest, we determine the distance between its endpoints at a given time. In a space-time diagram, Figure 2.21, the endpoints for a system (ct, x) at rest with the object are two events such as A and B. Being at different points in space, A and B are not seen as simultaneous by a moving observer with system (ct', x'). Such an observer would not take events A and B to determine the length because they seem to correspond to the endpoints of a moving object occupied at different times. A moving observer would conclude that the distance traveled by the object must be subtracted from the spatial separation between A and B so as to obtain the proper length of the object at one time t'. Or, the observer would refer to altogether different events which do happen at the same time for the observer, with $\Delta t' = 0$, and still lie on the worldlines of the two endpoints. In the figure, these would be events A and B′, not B. The invariant space-time distance according to Minkowski geometry is different between A and B′, smaller than between A and B. (Again, the distance only looks larger to our Euclidean eye.)

We calculate the invariant space-time distance between A and B′ in Figure 2.21. The expression is simple in the (ct', x') system because the events have $c\Delta t' = 0$ by our choice of those two events. The spatial separation, on the other hand, is the measured length L: $\Delta x' = L$. The invariant distance is then $\Delta s^2 = -(c\Delta t')^2 + (\Delta x')^2 = L^2$. It is invariant, and must take the same value in the (ct, x)-system where we have the spatial separation $\Delta x = L_*$ (the same as the coordinate separation between A and B, for which we measure the length L_* at rest) and nonvanishing time separation $c\Delta t$. To summarize, $L^2 = \Delta s^2 = -(c\Delta t)^2 + L_*^2$. The value of Δt between events A and B′ follows from a Lorentz transformation. In the primed system, we know that $\Delta t' = 0$, and we know that it equals $0 = \Delta t' = [\Delta t - (v/c^2)\Delta x]/\sqrt{1 - v^2/c^2}$. With $\Delta x = L_*$ the length at rest, $\Delta t = (v/c^2)\Delta x = vL_*/c^2$. Inserting this in Δs^2,

$$L = \sqrt{1 - v^2/c^2}\, L_* < L_* . \tag{2.11}$$

The effect of different length measurements in moving systems is called Lorentz contraction.[9] It refers to measurement processes, not to actual shrinking processes. The moving object occupies the same space-time region in all systems, shaded in Figure 2.21. If it shrank, it would occupy a more narrow strip when seen from a moving system, but this is not the case. Instead, the difference in measured lengths occurs because measurements in different systems refer to different events for the endpoints, or the endpoints taken at different times. The reasons for Lorentz contraction and time dilation are rather different, even though their common origin is relative simultaneity.

Motion pictures Measuring the size of an object is not the same as seeing it. Lorentz contraction is not a physical shrinking process, but rather a measurement effect, based on the method of aligning endpoints with measuring marks on a ruler. When we see an object, we do not perform such an elaborate procedure but rather interpret light signals reaching our eyes. A picture taken of an object records events different from a measurement of the object's size.

As a first guess, we might expect to see objects deformed by Lorentz contraction, in a significant way for objects moving at speeds close to that of light. Then, the distance traveled by light in any given time is comparable to the distance traveled by the object. Light reaching our eyes or a photographic film from the far end of the moving object will show considerable travel delays compared to light from the near end. The body may have moved a good distance during the extra time taken by light from the far end, making some parts appear at different places on the film. We should expect further distortions in the image we see of the body, owing to travel delays of light.

9) Looking back at the light clock, Lorentz contraction provides another reason why there is no effect if the mirrors move parallel to the light rays: it cancels out time dilation.

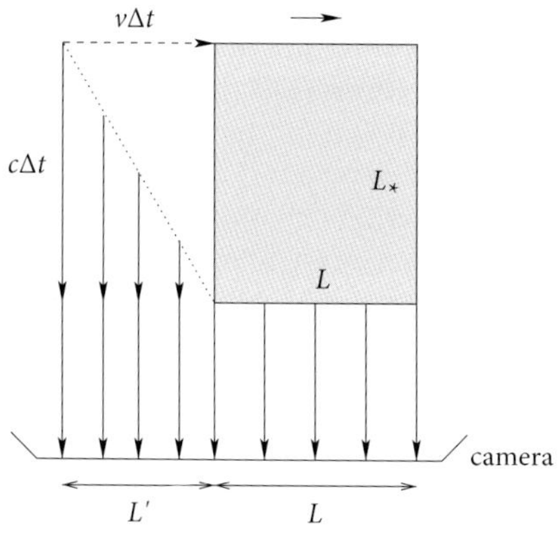

Figure 2.22 Light scattered off the back side of a cube moving at speed v can reach a camera if the body moves aside far enough while light travels. The camera then sees part of the back side of the object, in an area that depends on the velocity v. At the same time, the proportions of the object are reduced from L_* to L by Lorentz contraction along the direction of motion, but not in orthogonal directions.

In Figure 2.22, we illustrate a cube moving in the direction of one of its sides. A camera positioned with a line of sight perpendicular to the direction of motion records all light rays that arrive at its front lens at the same time, after they were scattered off the moving cube. All points on the near side of the cube are at equal distance from the lens and therefore light rays imaging this side are scattered off all at the same time. The length of the near side as seen by the camera is the Lorentz-contracted rest length $L = \sqrt{1 - v^2/c^2}\,L_*$, projected on the film.

Light that reaches the lens from the back side of the moving cube, however, has been scattered at different times, earlier than light from the near side so as to make up for the extra travel time. The farther up on the back side light is scattered, the earlier it must have been scattered to reach the lens at shot time. In order to compute the length of the region on which light from the back side will splatter, we first compute the time it takes for light from the far end of the back side to make up for the extra distance compared with the near side: light must travel a distance L_*, for which it needs time $\Delta t = L_*/c$. During this time, the cube moves a distance $v\Delta t = L_* v/c$, providing the length $L' = L_* v/c$ of the region imaged by the camera as the back side of the cube.

In order to interpret the image seen by the camera, we compare the two lengths $L' = L_* v/c$ of the imaged back side and $L = L_* \sqrt{1 - v^2/c^2}$ for the Lorentz contracted near side. Both factors, $a = v/c$ and $b = \sqrt{1 - v^2/c^2}$, are less than one, and satisfy the equation $a^2 + b^2 = 1$, the same identity obeyed by the trigonometric functions sine and cosine. We can find an angle $\sphericalangle$ such that $a = \sin(\sphericalangle)$ and $b = \cos(\sphericalangle)$. Trigonometric functions are defined and often appear in the context of rotations. Also here, the same parameters a and b can be seen as obtained by a rotation of the cube; Figure 2.23 demonstrates that the same side lengths are indeed obtained for the image projected on the camera: $L' = L_* \sin(\sphericalangle)$ and $L = L_* \cos(\sphericalangle)$ in both cases.

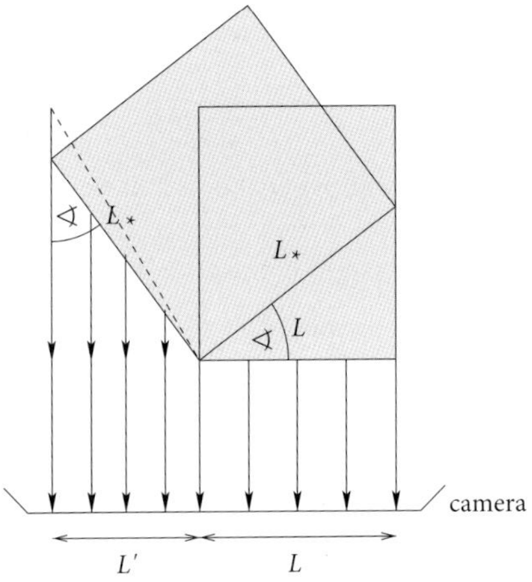

Figure 2.23 A moving cube compared with a rotated one. If $\sin(\sphericalangle) = v/c$, the side lengths of the rotated cube projected on the camera are the same as the side lengths seen of the moving cube. Based only on the projected proportions, the images of a moving cube and a rotated one are indistinguishable. However, light rays from the left side in both cases travel different distances.

When our eyes see a moving cube (or a photographic image of one), our brain, which is used to rotated objects, but not at all to ones moving at high speed, would interpret the image as rotated, without any distortions from Lorentz contraction. Only the slight differences in travel times of light rays from the back side of the moving cube compared with the same side from the rotated one could lead to differences, if there is absorption in the medium surrounding the cube. Seen through a thick fog, the back side of the rotated cube would look brighter than the back side of the moving cube.

Superluminal motion No object can travel faster than light in vacuum. If $v > c$, we would be led to taking square roots of negative numbers in many of our formulas, not resulting in real numbers. No motion can surpass the speed of light.

However, this law does not mean that we cannot *see* an object move faster than light. We see an object by light emitted from or scattered off it, into our eyes. If it is moving toward us, light sent to us at later times need not travel as far as light sent earlier because the object's own motion has shortened the distance to our eyes. Apparent velocities larger than the speed of light can be seen, even if the object does not actually travel faster than light. Such an apparent effect is similar to the apparent rotation of a moving cube, as a combination of Lorentz contraction and

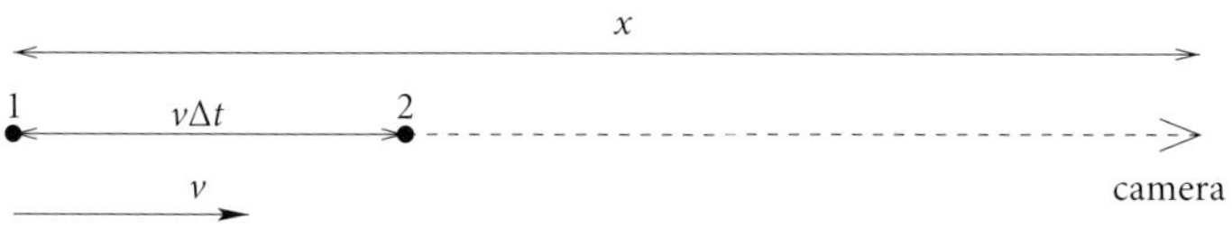

Figure 2.24 An object moves straight at a camera with speed v, emitting light along the way. Two positions 1 and 2 of the object at different times separated by Δt are shown. At 2, the initial distance x has been shortened by $v\Delta t$, a distance which light emitted from 2 will no longer have to travel through.

travel delays of light reaching a camera from different parts of the cube. The cube does not rotate, but the proportions of the image we receive make us believe so.

Let us assume that a small, almost pointlike object is moving straight toward our camera, at some speed $v < c$. If v is rather close to c, travel delays of light during the object's motion are significant. This situation is not as uncommon and contrived as it may seem. There are many examples for active stellar objects that emit matter in the form of relativistic jets, for example, matter ejected at speeds close to c in focused form along some axis. (The axis could be the rotation axis of a star.) Ejected matter is hot and emits light on its own. If the jet happens to point almost straight at Earth, our telescopes will see light emitted by particles moving toward them, at high speed.

As in Figure 2.24, we consider two different times at which the object emits light toward a camera. If the first emission is at time t_1, it is recorded at time $T_1 = t_1 + x/c$, taking into account the travel time of light. The second emission, at time $t_2 = t_1 + \Delta t$, is recorded at time $T_2 = t_1 + \Delta t + (x - v\Delta t)/c$, with a travel time reduced by the distance traveled by the object itself. From the point of view of the camera, two emissions have been recorded at a time lapse $T_2 - T_1$ apart, separated by a spatial distance of $v\Delta t$. The camera records a velocity $V = v\Delta t/(T_2 - T_1) = v/(1 - v/c)$. If v is close to the speed of light, V can be larger than c, and is unlimited as v approaches c.

In these arguments, we have not used much of special relativity, except for the crucial and basic assumption that the speed of light is the same in all reference systems, no matter whether we use it in the camera system at rest or in the moving object's system. No actual motion faster than light has occurred, but apparent velocities larger than the speed of light are recorded. Relativistic jets indeed provide confirmations of the effect.

Speed of light How can it be that light emitted by an object at speed v has the same speed c, the universally respected celebrity celerity, as light emitted by an object at rest? We expect that the object's speed must be added to the speed of light, and even though we are deriving all our concepts of space-time based on the primary assumption that the speed of light is the same in all systems, it is counterintuitive that this can come about in actual emission processes.

It cannot be correct that velocities are added if a moving object is ejecting another object, at least when velocities close to c are involved. We can answer the question of how we have to combine velocities instead by using Lorentz transformations, a well-specified quantitative basis for space-time calculations. If an object is moving with velocity $\vec{V} = \Delta\vec{x}/\Delta t$ (with all spatial components collected in vector notation), we see how to combine velocities by considering the same motion in a moving system with velocity v in the x-direction. The system's velocity is combined with the object's velocity by computing $\vec{V}' = \Delta\vec{x}'/\Delta t'$ in the coordinates of the moving system.

Velocity addition We compute the new velocity $\vec{V}'$ by components, using Lorentz transformations to relate (ct, x) to (ct', x'):

$$V^{x'} = \frac{\Delta x'}{\Delta t'} = \frac{\Delta x - v\Delta t}{\Delta t - v\Delta x/c^2} = \frac{\Delta x/\Delta t - v}{1-(v/c^2)\Delta x/\Delta t} = \frac{V^x - v}{1 - vV^x/c^2} \tag{2.12}$$

for the x-component along the motion of the system,

$$\begin{aligned} V^{y'} &= \frac{\Delta y'}{\Delta t'} = \frac{\Delta y}{[\Delta t - (v/c^2)\Delta x]/\sqrt{1-v^2/c^2}} \\ &= \frac{\sqrt{1-v^2/c^2}\,\Delta y/\Delta t}{1-(v/c^2)\Delta x/\Delta t} = \sqrt{1-v^2/c^2}\frac{V^y}{1-vV^x/c^2} \end{aligned} \tag{2.13}$$

for the y-component transversal to the motion, and

$$V^{z'} = \sqrt{1-v^2/c^2}\frac{V^z}{1-vV^x/c^2}\,. \tag{2.14}$$

Although the combination formulas are more complicated than simple addition, they satisfy the same mathematical laws (of a group) which, as we saw, are required on physical grounds. They also ensure that the combined speed cannot exceed the speed of light. For light moving in the x-direction, such that $V^x = c$, a moving observer will measure the velocity $V^{x'} = (c-v)/(1-v/c) = c$, that is, still the speed of light.

Using the same addition formulas, we see that the speed of an object which is less than c for one observer cannot be larger than or equal to c as measured by any other observer. No matter how much speed we try to add to an object, throwing it while we are standing on a moving platform, which itself might be standing on a moving platform, and so on, the speeds, with the relativistic formula, always add up to something smaller than c. No massive object can move at the speed of light, and no faster. Light sets the absolute speed limit.

To be precise, the top speed is reached by light in vacuum. In transparent matter, light still moves but usually at a speed less than c. It is then possible for massive particles to be faster than light in that medium, but the velocity is always less than the speed of light (in vacuum, which is understood if not mentioned). An impressive confirmation of particles faster than light in matter is Cherenkov radiation: Just as an airplane moving faster than sound (or waves of compression in air) produces a loud shock wave when the sound barrier is broken, particles entering a medium at a speed faster than light in the same medium emit an electromagnetic shock wave. One can see this wave, for instance, in cooling basins surrounding radioactive materials.

Four-velocity Objects move through space and time, not in space alone. So far, we have defined the velocity of an object as the ratio of the spatial distance traveled by the required time. Both quantities, the spatial distance and time, refer to coordinates in the system of one observer, and they change if one transforms them to a different system. The resulting transformation formulas for velocity components are quite complicated and nonlinear, unlike the nonrelativistic laws. In nonrelativistic physics, one only transforms the spatial coordinates in $\vec{V} = \Delta\vec{x}/\Delta t$ by translations, rotations or Galilei transformations. Not only the position components, but also the velocity components then change in a linear fashion.

It is possible to introduce linear transformation laws for velocity components in relativity, too, and at the same time take seriously the notion that objects travel through space and time when they move, not just in space as time proceeds. The key step is to refer not to a time coordinate t, but rather to the invariant proper time T, the time perceived in the moving object's system. Being invariant, T does not change if a different observer is considered. We can then use the time coordinate t, freed from its role as providing the reference for progress of time, as what it really is: a coordinate of space-time together with the spatial coordinates $\vec{x}$. We write the collection of all four coordinates as a four-component vector $\boldsymbol{x} = (ct, x, y, z)$, or a four-vector for short. To distinguish them from three-dimensional vectors such as $\vec{x}$, four-vectors will be denoted by boldface letters. In a four-vector, it is sometimes useful to consider the spatial part as a three-dimensional vector and write $\boldsymbol{x} = (ct, \vec{x})$. By convention, $\vec{x}$ is always the spatial part of $\boldsymbol{x}$ if the same letter is used in both objects.

In space-time, it is meaningful to describe worldlines of moving objects by recording their space-time position $\boldsymbol{x}(T)$ depending on proper time, as in Figure 2.25. The rate of change of this quantity is a measure for how fast the object moves in space-time, called the four-velocity $\boldsymbol{u} = \Delta\boldsymbol{x}/\Delta T$. In this equation, T does not depend on what observer is measuring the motion. While $\boldsymbol{x}$ changes, it does so in the linear way determined by Lorentz transformations. The components of the four-velocity, therefore, also change by Lorentz transformations, a much simpler behavior than the one of the spatial velocity components (2.12)–(2.14). Like $\boldsymbol{x}$, $\boldsymbol{u}$ is a four-vector and deserves its boldness.

Norm We move in all four directions of space-time, not just in the three directions of space. The four-velocity has four components, $\boldsymbol{u} = (u^t, u^x, u^y, u^z)$, while the velocities we are used to have only three components along the spatial directions. Is there extra information contained in the surplus component? Although we do move in time because time is always changing, we have no means of influencing

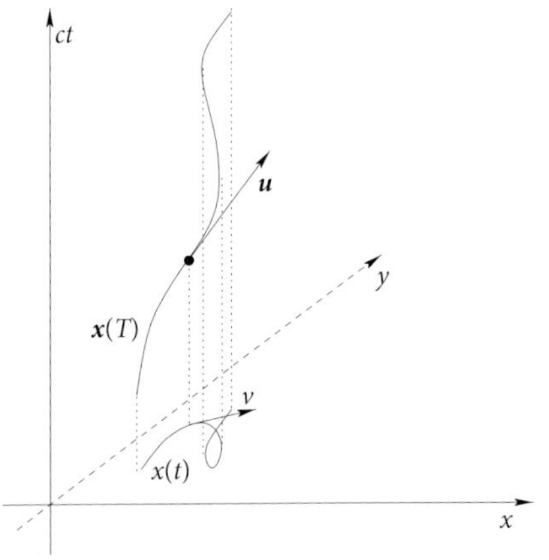

Figure 2.25 An object moves through space and time along its worldline $x(T)$, along which its proper time T changes. Taking only the spatial part $\vec{x}$ of $x = (ct, \vec{x})$ and using coordinate time t instead of T, one obtains the spatial trajectory of the object, projected from space-time onto space. The rate of change of the worldline is $\boldsymbol{u}$; the rate of change of the spatial trajectory is $\vec{v}$.

the rate $u^t = c\Delta t/\Delta T$ independently of the spatial velocity $\Delta\vec{x}/\Delta T$ we experience. There is no additional information or freedom to move because the components of the four-velocity cannot be chosen arbitrarily; they must under all circumstances fulfill one equation between them.

A direction-independent measure of motion is the magnitude of the velocity. In space, the Euclidean line element $\Delta s^2 = (\Delta x)^2 + (\Delta y)^2 + (\Delta z)^2$ tells us that the length of a spatial displacement is obtained by summing the squared coordinate displacements. We write this as $\Delta s^2 = |\Delta\vec{x}|^2$, the norm squared of the displacement vector $\Delta\vec{x} = (\Delta x, \Delta y, \Delta z)$. Since the displacement vector is the prime example for a direction or a vector, we use the same rule to compute the magnitude of any vector, for instance of the velocity: $|\vec{V}|^2 = (V^x)^2 + (V^y)^2 + (V^z)^2$.

In space-time, the Euclidean line element must be replaced by the Minkowski line element, and four-vectors now have four components. The displacement vector in space-time, $\Delta\boldsymbol{x} = (c\Delta t, \Delta x, \Delta y, \Delta z)$, again provides the prime example, whose norm squared (now written with double bars for the extra dimension) is taken from the line element: $||\Delta\boldsymbol{x}||^2 = -(c\Delta t)^2 + (\Delta x)^2 + (\Delta y)^2 + (\Delta z)^2 = \Delta s^2$.

The magnitude of the four-velocity $\boldsymbol{u} = \Delta\boldsymbol{x}/\Delta T$ is $||\boldsymbol{u}||^2 = ||\Delta\boldsymbol{x}||^2/(\Delta T)^2 = \Delta s^2/(\Delta T)^2$. Recalling the definition of proper time, $\Delta T = \sqrt{-\Delta s^2}/c$, we see that the norm squared of the four-velocity is negative the speed of light squared, a predetermined number for any object. The four-velocity $\boldsymbol{u} = (u^t, u^x, u^y, u^z)$ has just three independent components: the equation

$$||\boldsymbol{u}||^2 = -(u^t)^2 + (u^x)^2 + (u^y)^2 + (u^z)^2 = -c^2 \tag{2.15}$$

allows us to compute any one of them from the other three (at least up to a sign).

Space-time speedometer Motion in space-time is determined by motion through space. The components of the four-velocity always fulfill (2.15), resulting in a (negative) magnitude equal to the speed of light. This result may be perplexing. How can the velocity of anything that moves be the speed of light? We are not even considering light (or electromagnetic waves), but rather objects of mechanics, having some mass m and moving in space by $\vec{x}(t)$.

Using the four-velocity does not speed up motion; it quantifies it in a way different from what we perceive. The four-velocity $\boldsymbol{u}$ has one component more than the three-dimensional velocity $\vec{V}$. When we compute the magnitude, the extra component u^t is subtracted (squared), indicating that the result is even smaller. However, the spatial part $\vec{u}$ of $\boldsymbol{u}$ need not be identical to the velocity $\vec{V}$, although in some way they are related.

Four-velocity To make the relation between four-velocity and three-velocity explicit, we compute $u^t = \Delta x^t / \Delta T = c\Delta t / \Delta T = c/\sqrt{1 - V^2/c^2}$ by (2.10) for proper time. Here, $V = |\Delta \vec{x} / \Delta t|$ is the magnitude of the three-dimensional velocity, the velocity V of an object distinguished from v of an observer. Then, we have $u^x = \Delta x / \Delta T = \Delta x / (\Delta t \sqrt{1 - V^2/c^2}) = V^x / \sqrt{1 - V^2/c^2}$, $u^y = V^y / \sqrt{1 - V^2/c^2}$ and $u^z = V^z / \sqrt{1 - V^2/c^2}$. We write

$$\boldsymbol{u} = \left(c/\sqrt{1 - V^2/c^2} \, , \quad \vec{V}/\sqrt{1 - V^2/c^2} \right) \tag{2.16}$$

and see that the three components of $\vec{V}$ determine the four components of $\boldsymbol{u}$. (The case of light is problematic because the denominators in (2.16) become zero; we will discuss it at a later stage.)

The norm of the four-velocity does not tell us how fast an object moves *in space*. It tells us how fast an object moves *through space-time* and this, by (2.15) always happens to be with the speed of light. With hindsight, the results are not surprising. While we can make objects stop in space, reducing their velocity to zero, they always keep moving through space-time. Even if their spatial position does not change, time still goes on and $\Delta t / \Delta T \neq 0$. If the object stands still in space, the time component of the four-velocity in (2.16) is $u^t = c$, the speed of light. As the spatial velocity increases, other components of $\boldsymbol{u}$ become nonzero and $\boldsymbol{u}$ points in a different space-time direction. However, it always points in a timelike direction (toward the future part of the light cone) because $||\boldsymbol{u}||^2 = -c^2$ is negative.

It is the direction of $\boldsymbol{u}$ in space-time, not the magnitude of $\boldsymbol{u}$ which has information about the velocity. The four-velocity is nothing but an elegant way of encoding a real number, the magnitude of the speed, by a direction or an angle. We are used to this phenomenon from actual speedometers: the pointer changes its direction as the car accelerates, but it always has the same length.

Four-vector With the relation (2.16) between the components of the three-velocity and the four-velocity, we can rederive the composition formulas for velocities in an independent and useful way. Earlier, we obtained them by applying a Lorentz transformation to the displacement components in the velocity $\vec{V}' = \Delta\vec{x}'/\Delta t'$. In the meantime, we have seen that the components of the four-velocity $\boldsymbol{u}$ also form a four-vector, and can be transformed by a Lorentz transformation. Instead of using displacement components in (2.8), we may insert velocity components. For a boost along the x-axis by velocity v,

$$u^{t'} = \frac{u^t - vu^x/c}{\sqrt{1-v^2/c^2}}\,, \quad u^{x'} = \frac{u^x - vu^t/c}{\sqrt{1-v^2/c^2}}\,, \quad u^{y'} = u^y\,, \quad u^{z'} = u^z\,.$$

Matrices Such a transformation with new components, linear in the old ones, can be written in matrix form. Just as we arrange the vector components to be transformed in vector form, we arrange the coefficients in a square matrix:

$$\boldsymbol{u}' = \begin{pmatrix} 1/\sqrt{1-v^2/c^2} & -v/(c\sqrt{1-v^2/c^2}) & 0 & 0 \\ -v/(c\sqrt{1-v^2/c^2}) & 1/\sqrt{1-v^2/c^2} & 0 & 0 \\ 0 & 0 & 1 & 0 \\ 0 & 0 & 0 & 1 \end{pmatrix} \boldsymbol{u}\,.$$

A Lorentz transformation along the x-axis does not change the orthogonal displacements Δy and Δz, nor do the y- and z-components of $\boldsymbol{u}$. As in (2.13) and (2.14), the y- and z-components of $\vec{V}$, however, are transformed because the magnitude V of the velocity, entering via the factor $1/\sqrt{1-V^2/c^2}$, depends on the transformed x-component as well. As in this case, it is not always easy to see how three-dimensional vector components should be combined to a four-vector so as to obtain a linear transformation law with the usual Lorentz formulas. Knowing just three-vector equations such as (2.12)–(2.14), it would be hard to guess that multiplying the $\vec{V}$-components with $1/\sqrt{1-V^2/c^2}$ and including the same factor as the time component provides us with a four-vector (2.15). It is better to go back to the definitions and see how its ingredients, in this case $\Delta\vec{x}$, Δt and their ratios, can be made compatible with Lorentz transformations. This way led us to consider the four-velocity $\boldsymbol{u}$ instead of $\vec{V}$.

Four-velocity Continuing with the four-velocity, we make the transformation equations more specific by using the relation of $\boldsymbol{u}$ with $\vec{V}$:

$$\frac{c}{\sqrt{1-V'^2/c^2}} = u^{t'} = \frac{c - vV^x/c}{\sqrt{(1-v^2/c^2)(1-V^2/c^2)}} \tag{2.17}$$

$$\frac{V^{x'}}{\sqrt{1-V'^2/c^2}} = u^{x'} = \frac{V^x - v}{\sqrt{(1-v^2/c^2)(1-V^{x2}/c^2)}} \tag{2.18}$$

$$\frac{V^{y'}}{\sqrt{1-V'^2/c^2}} = u^{y'} = \frac{V^y}{\sqrt{1-V^2/c^2}} \tag{2.19}$$

$$\frac{V^{z'}}{\sqrt{1-V'^2/c^2}} = u^{z'} = \frac{V^z}{\sqrt{1-V^2/c^2}} \,. \tag{2.20}$$

Using the first equation, we find that the others are equivalent to (2.12)–(2.14).

Four-force Motion through space-time is driven by a space-time force. If we want to extend Newton's second law to space-time, relating an objects acceleration to an acting force, we must find out how the extra time direction enters and what components of the new quantities transform in the Lorentzian way.

For the acceleration, we just repeat the step practiced with the four-velocity: We take derivatives by proper time instead of coordinate time and define the four-acceleration as $\boldsymbol{a} = \Delta \boldsymbol{u}/\Delta T$. Also, the object's mass m enters Newton's second law. It is a single parameter characterizing the object, which we can treat as taking the same value for all observers, at least until we are forced to change that assumption if inconsistencies arise. It is easier to view the mass as observer-independent (somewhat like the speed of light) instead of guessing a new transformation formula or inventing three new mass components to complete m to a four-vector.

Combining mass and acceleration, we have four-vectors in the product $m\boldsymbol{a} = \boldsymbol{f}$. If Newton's second law of mechanics can be extended to special relativity, $\boldsymbol{f}$ cannot just be a short form for the not much longer $m\boldsymbol{a}$, but should receive a meaningful physical life of its own as the four-force, or the four-vector version of the three-dimensional force $\vec{F}$.

Guided by the form (2.16) of $\boldsymbol{u}$ in terms of $\vec{V}$, we may expect that the spatial components $\vec{f}$ of the four-force $\boldsymbol{f}$ are related to what we call the force $\vec{F}$ in nonrelativistic physics by $\vec{f} = \vec{F}/\sqrt{1-V^2/c^2}$. The new force components must then depend on the objects' velocity $\vec{V}$, not just on its position. Such force formulas are more complicated than ones depending only on the position, but they are not uncommon even without relativity: Whenever there is friction, so almost always in realistic cases, the force on an object depends on the object's velocity, as everyone knows who has tried to run against strong wind.

At this stage, we are making an assumption about the four-force, but if we can find a suitable time component f^t to complement $\vec{f}$ to a four-vector, we have a valid relativistic form of Newton's second law. For small velocities V, it will agree with the nonrelativistic law because $\Delta T \approx \Delta t$ and $\vec{f} \approx \vec{F}$. For velocities close to the speed of light, Newtonian expectations may be violated, but the laws of relativity will be respected because we will have a four-vector equation, valid for all observers.

Four-momentum Force is the change in time of momentum. In order to find the correct time component of the four-force, it is useful to look at a relativistic form of momentum. In classical mechanics, we multiply the velocity with the mass, a process that can also be done with relativistic notions: We interpret the product $\boldsymbol{p} = m\boldsymbol{u}$ as the four-momentum of an object with mass m. Like the four-velocity, the four-momentum has only three independent components because the speedometer equation $||\boldsymbol{u}||^2 = -c^2$ implies $||\boldsymbol{p}||^2 = -m^2c^2$.

In components, we have $p^t = mc/\sqrt{1 - V^2/c^2}$ and $\vec{p} = m\vec{V}/\sqrt{1 - V^2/c^2}$. For a nonzero mass, that is, for any object other than light and some other waves, some of the momentum components become infinite as the object's speed approaches the speed of light. The speed of light, although it is a finite value of the velocity, makes the momentum reach its maximal amount infinity. The notion of the momentum shows that a massive object cannot be accelerated to or beyond the speed of light. The speed of light can be assumed only if the mass vanishes, so as to suppress the divergence of the square-root denominator. Light, traveling at the speed of light, must be a massless phenomenon; light is lightweight to the extreme. For now, however, we continue with considerations about nonzero mass.

Nonrelativistic limit For V small, we can write $1/\sqrt{1 - V^2/c^2} \approx 1 + \frac{1}{2}V^2/c^2$ to a good approximation; see Figure 2.26. For confirmation, we square both sides of the equation: First, $(1 + \frac{1}{2}V^2/c^2)^2 = 1 + V^2/c^2 + \frac{1}{4}V^4/c^4 \approx 1 + V^2/c^2$. In the last step, we have dropped the term $\frac{1}{4}V^4/c^4$ because it is much smaller than V^2/c^2 if V is much smaller than c. (Also, V^2/c^2 is much smaller than one and could be dropped in the equation, but then we would not obtain anything of interest. Often, the trick in mathematics is to retain just enough to remain interesting.) We now must show that $1/(1 - V^2/c^2) \approx 1 + V^2/c^2$. The equation looks reasonable because dividing one by something reduced by an amount of V^2/c^2 should be a number increased above one. We confirm the equation by multiplying both sides with $1 - V^2/c^2$. On the left-hand side, we obtain just one, and on the right-hand side, $(1 - V^2/c^2)(1 + V^2/c^2) = 1 - V^4/c^4 \approx 1$ amounts to the same value in our approximation that drops terms such as V^4/c^4. The approximation is confirmed.

Energy-momentum Momentum is change of position in time; energy is change of time in time. The first part of this statement follows from the relation between momentum and velocity. The second part can sound meaningful only when one remembers that there are different notions of time in special relativity, coordinate time and proper time. If we make the statement more precise by saying that energy is change of coordinate time in proper time, we see that the notion of energy in special relativity is related with time dilation.

The faster an object moves (or the more energetic it is in kinetic terms), the more dilated its time is compared to an observer at rest, and the more significant the change of the observer's coordinate time relative to the moving object's proper time. (Of course, we could take momentum instead of energy as a measure for the

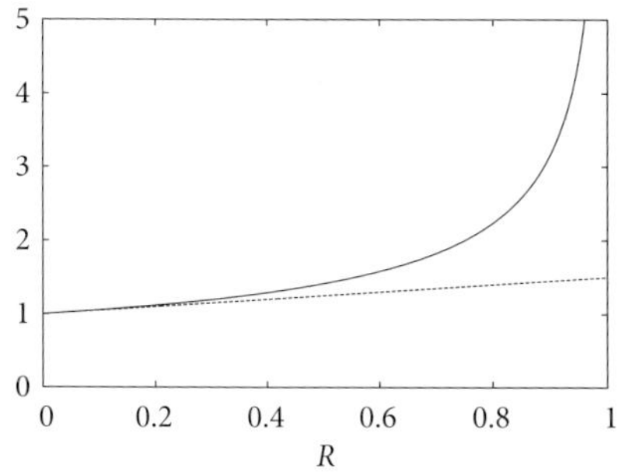

Figure 2.26 The function $1/\sqrt{1-V^2/c^2}$ (solid) is almost identical to $1+\frac{1}{2}V^2/c^2$ (dashed) for the ratio $R = V^2/c^2$ much smaller than one, an approximation called first-order Taylor expansion.

object's motion. However, the momentum is change of space in time, and so the change of time in time, another meaningful quantity in relativity theory, must be something else.)

> We keep these thoughts in mind when we go about interpreting the time component of $\boldsymbol{p}$, defined as $p^t = m\Delta(ct)/\Delta T$ or mass times the change of time in time. For small velocities, in a nonrelativistic limit, we write the t-component of the four-momentum as $p^t = mc/\sqrt{1-V^2/c^2} \approx mc + \frac{1}{2}mV^2/c$, or $cp^t \approx mc^2 + \frac{1}{2}mV^2$. In the second form of writing the equation, we recognize the kinetic energy of an object of mass m moving at velocity V. It is independent of c and agrees with the nonrelativistic formula. If we had kept additional terms in our approximation, we would obtain corrections to the nonrelativistic energy, but they are small for small velocities. Up to the constant, velocity-independent contribution mc^2, the t-component cp^t of four-momentum can therefore be interpreted as the kinetic energy of the object.

By extending momentum to space-time, we include energy cp^t, the change of time in time, in its components. For this reason, the four-momentum $\boldsymbol{p}$ is often called the energy-momentum vector. There is no potential energy in the equation for cp^t because we have not yet included the influence of a force. A potential function is difficult to formulate in relativistic terms because the position dependence of a standard potential would imply time dependence after a Lorentz boost. Formulating potential forces in a Lorentz-invariant way is complicated and depends on the type of the force. We drop the notion of a potential at this general stage and continue using the force.

Momentum and force are related by $\boldsymbol{f} = \Delta\boldsymbol{p}/\Delta T$. If the force vanishes, $\boldsymbol{p}$ does not change in time and is a conserved quantity. We recognize the familiar law of momentum conservation from the spatial components of $\boldsymbol{p}$. However, with our new result about the time component p^t, we see that energy conservation is included in the same equation. In special relativity, momentum conservation and energy conservation are one and the same law; combining space and time has not only resulted in a new physical entity, space-time, it also gives us greater economy.

$E_0 = mc^2$ An object at rest still moves through time; it still has energy. Even if there is no force and potential energy, motion in time suggests that there should be a kinetic form of energy that, unlike the Newtonian value, does not vanish even if an object remains fixed in space. If we can transform space into time by changing our viewpoint, we must also be able to transform kinetic energy of spatial motion to kinetic energy of motion only in time.

We have already identified $E = cp^t$, with the time component p^t of the energy-momentum vector, as the correct term that carries information of kinetic energy of a moving object. If the full energy-momentum vector is conserved, so is E. We should then look at the nonzero constant $E_0 = mc^2$ that remains even when $V = 0$. For an object assumed to be at rest, there are not many processes that could challenge the conservation of its energy. Yet, there are nontrivial reactions, for instance the decay of an unstable particle into different, lighter components. Energy conservation is an important law even for the rest contribution E_0.

While the sum of all contributions to cp^t from all objects involved is conserved, individual terms in the balance equation may change. The rest energy E_0 of a single unstable particle can be transmuted into the sum of contributions from its decay products including kinetic energy, adding up to the same value E_0. In particular, it must be possible for some part of rest energy E_0 to be transformed into kinetic energy (or vice versa in the reverse reaction). The contribution mc^2 to cp^t is far from being just a simple and inconsequential constant; it is an energy form of its own unknown to Newtonian physics, one related with the mass of an object rather than its motion or an acting force. We identify all of cp^t as the energy $E = mc^2/\sqrt{1 - V^2/c^2}$ of an object of mass m moving at velocity V, with $E_0 = mc^2$ for a particle at rest.

With this identification of the energy E as the time component cp^t of the four-momentum, we revisit the norm $||\boldsymbol{p}||^2 = -m^2c^2$. Written in components, $-(p^t)^2 + (p^x)^2 + (p^y)^2 + (p^z)^2 = -m^2c^2$, we obtain

$$E^2 - c^2\vec{p}^2 = m^2c^4 \tag{2.21}$$

or $E = \sqrt{m^2c^4 + \vec{p}^2c^2}$. While energy and momentum, as components of a four-vector, are not observer-independent, the combination (2.21) is determined by the invariant mass.

One often makes use of these relationships in experimental particle physics. At particle accelerators, one can measure the momentum of charged particles produced in collisions or by decay because the deflection of a charged particle in a magnetic field, according to the Lorentz force law, depends on the momentum. A particle's energy can be determined by summing up the energy deposited by all decay products in the detector. For these reasons, modern particle detectors are large (to catch and count all decay products) and infused with a magnetic field strong enough to make the deflection angle noticeable even for particles almost as fast as light). With (2.21), there is a simple way of obtaining the invariant mass from the measured energy and momentum, resulting in a characteristic quantity independent of how the particle is moving and what reference system one is using. These are the mass values quoted for the elementary particles produced and detected at colliders. With an invariant notion of mass, a value not encountered before shows the discovery of a new particle, rather than the detection of a known one just measured in a different frame.

Four-force All motion is controlled by forces. In relativity, we use a four-vector f for neater transformation properties, but its components must be determined by the three-dimensional forces $\vec{F}$ we use in Newtonian mechanics, just as motion is described by a four-vector $\boldsymbol{u}$ related to the three-dimensional velocity $\vec{V}$.

We have identified the spatial components as $\vec{f} = \vec{F}/\sqrt{1-V^2/c^2}$. The energy-momentum vector tells us that $f^t = \Delta p^t/\Delta T = \Delta E/(c\sqrt{1-V^2/c^2}\Delta t)$ for the time component. The rate of change $\Delta E/\Delta t$ of the total energy E equals the work performed on the object by the acting force. In order to move an object a distance $\Delta\vec{x}$ by applying a force $\vec{F}$, an amount $\vec{F}\cdot\Delta\vec{x} = \Delta E$ is required. In a given time interval Δt, the energy changes by the rate $\Delta E/\Delta t = \vec{F}\cdot\Delta\vec{x}/\Delta t = \vec{F}\cdot\vec{V}$. All these equations are nonrelativistic, but we can use them to write the time component of the four-force in terms of the three-force: $f^t = \vec{F}\cdot(\vec{V}/c)/\sqrt{1-V^2/c^2}$, and

$$f = \left(\frac{\vec{F}\cdot(\vec{V}/c)}{\sqrt{1-V^2/c^2}}, \frac{\vec{F}}{\sqrt{1-V^2/c^2}}\right) . \tag{2.22}$$

The four-force computed for any three-force $\vec{F}$ of interest appears in the equation of motion $m\Delta^2\boldsymbol{x}/\Delta T^2 = \Delta\boldsymbol{p}/\Delta T = f$, with solutions $\boldsymbol{x}(T)$ that tell us how an object under the influence of $\vec{F}$ is driven through space-time.

Constant force As an example, let us look at a constant force $\vec{F} = \vec{k}$. With (2.22), the spatial part of the four-force equation tells us that $\Delta\vec{p}/\Delta t = \sqrt{1-V^2/c^2}\vec{f} = \vec{k}$ is not only constant but also satisfies the same equation as in nonrelativistic mechanics. We have the same solution, a linear function $\vec{p} = \vec{k}t + \vec{p}_0$ for the spatial momentum as a function of coordinate time. In order to find the position as a function of time, we use the relationship between momentum and velocity and integrate once more, solving a first-order differential equation for $\vec{x}(t)$. This equation, however, differs from the nonrelativistic one because the spatial momentum is $\vec{p} = m\vec{V}/\sqrt{1-V^2/c^2}$ rather than just $m\vec{V}$. It ensures that the velocity $|\vec{V}| = c/\sqrt{1+m^2c^2/(|\vec{k}|t+\vec{p}_0)^2} < c$ stays below the speed of light. The integration to $\vec{x}(t)$ is more complicated, but can be performed, for instance, using a vector field as in Figure 2.27.

It may be surprising that a constant and never waning force cannot accelerate the object to arbitrarily large velocities. The explanation is that the force is constant in the initial rest frame of the object, to which we refer by using its coordinate time t. As the object speeds up, the force implies smaller and smaller accelerations in the object's rest frame, causing decreasing accelerations. One can also solve the equations directly in the object's rest frame, using proper time instead of coordinate time.

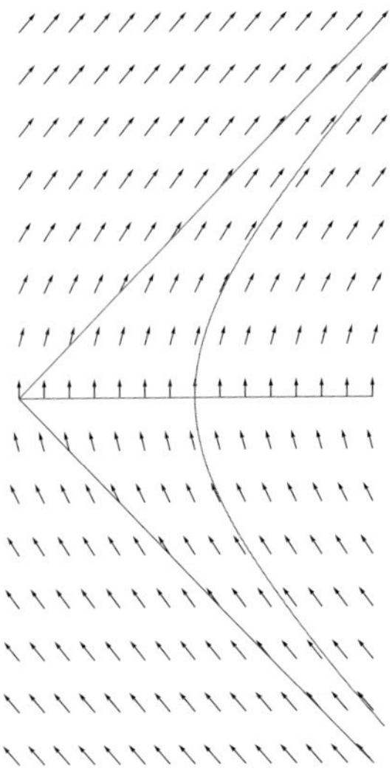

Figure 2.27 Vector field $(c\Delta t/\Delta T, \Delta x/\Delta T)$ for the differential equation of constant acceleration in space-time. The solutions are hyperbolas, approaching the light cone at early and late times. As long as the velocity is small, near the turning point, the hyperbola is close to the parabola $x(t) = kt^2/(2m)$ obtained in non-relativistic mechanics.

Light boosts Light is a force. The electromagnetic field not only has a life of its own in the form of electromagnetic waves or light, it also produces the Lorentz force on charged particles. One way to understand the peculiarities of light in special relativity is to look at the Lorentz force of a general electromagnetic field from the point of view of different observers. We write $\vec{F} = q(\vec{E} + \vec{V} \times \vec{B})$ as a four-force (2.22), apply a Lorentz boost (2.8) to transform the four-force to a new system, and read off the new three-dimensional Lorentz force $\vec{F}' = q(\vec{E}' + \vec{V}' \times \vec{B}')$ and the corresponding electromagnetic field.

We will end up with transformation equations for the electric field $\vec{E}$ and the magnetic field $\vec{B}$. The proposed derivation may seem somewhat roundabout because we do not transform a four-vector for the electromagnetic field itself. In fact, there is no four-vector for the electromagnetic field, as shown by counting components: We have six components of the electromagnetic field, that is, three in the electric field and three in the magnetic field, but a four-vector could hold only four of them. We may look for a matrix instead of a vector. However, a matrix in four dimensions has $4 \cdot 4 = 16$ components, too many for the electromagnetic field unless the components are restricted. First transforming the Lorentz force shows us how the field components must change, information that can be used to determine how to arrange them in a matrix.

Lorentz force We first consider the case in which the magnetic field vanishes in the initial system. This condition cannot be realized for light, a phenomenon of changing electric and magnetic fields, but we will still see what to expect for field transformations. The four-force to transform is then $f = (q\vec{E}\cdot(\vec{V}/c), q\vec{E})/\sqrt{1-V^2/c^2}$. After a Lorentz boost along the x-direction with velocity v, the four-force becomes

$$f' = q\begin{pmatrix} [E^x(V^x - v) + E^y V^y + E^z V^z]/\sqrt{(1-v^2/c^2)(c^2-V^2)} \\ (E^x - v\vec{E}\cdot\vec{V}/c^2)/\sqrt{(1-v^2/c^2)(1-V^2/c^2)} \\ E^y/\sqrt{1-V^2/c^2} \\ E^z/\sqrt{1-V^2/c^2} \end{pmatrix}.$$

We would like to read off the transformed force $\vec{F}'$ from $f' = \vec{F}'/\sqrt{1-V'^2/c^2}$, for which we should first trade in $\vec{V}'$ for $\vec{V}$. We do so by using the formulas (2.12)–(2.14) and (2.17) for combining velocities:

$$f' = q\begin{pmatrix} \frac{E^x V^{x'} + (E^y V^{y'} + E^z V^{z'})/\sqrt{1-v^2/c^2}}{c\sqrt{1-V'^2/c^2}} \\ \frac{E^x - c^{-2}v(E^y V^{y'} + E^z V^{z'})/\sqrt{1-v^2/c^2}}{\sqrt{1-V'^2/c^2}} \\ \frac{E^y(1+vV^{x'}/c^2)}{\sqrt{(1-v^2/c^2)(1-V'^2/c^2)}} \\ \frac{E^z(1+vV^{x'}/c^2)}{\sqrt{(1-v^2/c^2)(1-V'^2/c^2)}} \end{pmatrix} = \begin{pmatrix} \frac{\vec{F}'\cdot\vec{V}'}{c\sqrt{1-V'^2/c^2}} \\ \frac{F^{x'}}{\sqrt{1-V'^2/c^2}} \\ \frac{F^{y'}}{\sqrt{1-V'^2/c^2}} \\ \frac{F^{z'}}{\sqrt{1-V'^2/c^2}} \end{pmatrix}.$$

For $\vec{F}'$ to be of Lorentz form, the time component of f must be $f^{t'} = q\vec{E}'(\vec{V}'/c)/\sqrt{1-V'^2/c^2}$, which is the case if

$$\vec{E}' = \left(E^x, \frac{E^y}{\sqrt{1-v^2/c^2}}, \frac{E^z}{\sqrt{1-v^2/c^2}}\right). \tag{2.23}$$

The difference $\vec{F}' - q\vec{E}'$, read off from the spatial components of the force equation, is then nonzero, implying that there is a magnetic field in the transformed system. By comparing the spatial components with a Lorentz force, we see that

$$\vec{B}' = \frac{(\vec{E}/c)\times(\vec{v}/c)}{\sqrt{1-v^2/c^2}}. \tag{2.24}$$

An electric field measured by a moving observer has a magnetic component. Electric and magnetic fields cannot be seen in separation, just as space and time are not separate. There is only a combined object, the electromagnetic field in space-time.

Electromagnetic field The electromagnetic field is observer-dependent. An electric field is not transformed like components of a four-vector; (2.23) rather shows that the component E^x parallel to the direction of motion of an observer is not changed, while the two transversal components E^y and E^z do change. A four-vector such as the displacement vector $\Delta\boldsymbol{x}$ or the four-velocity $\boldsymbol{u}$ transforms in the opposite way, changing the component along the direction of motion, but not the transversal ones.

One can understand the unexpected behavior under Lorentz transformations by referring to electric fluxes. Electric flux through a surface is defined as the product of the surface area and the electric-field component perpendicular to the surface. Gauss' law, as a consequence of one of Maxwell's equations, identifies the electric flux through a closed surface with the enclosed electric charge.

Electric flux The electric field determines the force exerted by a charge. According to Maxwell's equation (2.1), the electric field caused by some amount of charge contained in a region divided by the volume, or the charge density d_{charge}, is computed by solving the differential equation $\Delta_x E^x/\Delta x + \Delta_y E^y/\Delta y + \Delta_z E^z/\Delta z = d_{\text{charge}}$. Instead of calculating all the components of $\vec{E}$, it is easier to find the combination $F_S = \mathbf{Sum}_{\vec{x}}\, E_{\text{perp}}(\vec{x})\Delta A$, the electric flux through a surface S defined by decomposing the surface in small pieces of area ΔA and summing the products of areas with the components E_{perp} of $\vec{E}$ perpendicular to the surface, taken at a point $\vec{x}$ within each small area. For S being the surface of a cube with sides of lengths L_x, L_y and L_z aligned with the coordinate axes, assuming that $\vec{E}$ does not vary much over the rectangular surface pieces, we have $F_S = [E^x(x_0 + L_x) - E^x(x_0)]L_y L_z + [E^y(y_0 + L_y) - E^y(y_0)]L_x L_z + [E^z(z_0 + L_z) - E^z(z_0)]L_x L_y$, putting one corner of the cube at coordinates (x_0, y_0, z_0). For small side lengths, we can use $L_x = \Delta x$ and so on as small intervals to compute derivatives of $\vec{E}$, and rewrite the flux as $F_S = (\Delta_x E^x/\Delta x + \Delta_y E^y/\Delta y + \Delta_z E^z/\Delta z)L_x L_y L_z$. The parenthesis is given by Gauss' law: $F_S = d_{\text{charge}} L_x L_y L_z = q$. The flux through a closed surface equals the enclosed charge.

Field-strength tensor Charge does not change. Like the mass, it is invariant under Lorentz transformations, and therefore the electric flux must be invariant. For a square area normal to the x-axis with side lengths L_y and L_z, the electric flux is $E^x L_y L_z$. A Lorentz transformation along the x-direction indeed does not change this value because each factor in the product is left unchanged. The fluxes $E^y L_x L_z$ and $E^z L_x L_y$ are left unchanged, in a less obvious way: The transformations of E^y and E^z in (2.23) are compensated for by Lorentz contraction of L_x.

Equations (2.23) and (2.24) can be identified as a matrix or tensor transformation. As already noted, the six components of the electric and magnetic fields are too many to arrange them in a four-vector, but not enough to arrange them in a general 4×4-matrix. However, an antisymmetric matrix, one that equals its negative when reflected at the diagonal, has six components. We can arrange the electromagnetic field as the field-strength tensor

$$F = \begin{pmatrix} 0 & E^x/c & E^y/c & E^z/c \\ -E^x/c & 0 & B^z & -B^y \\ -E^y/c & -B^z & 0 & B^x \\ -E^z/c & B^y & -B^x & 0 \end{pmatrix} \tag{2.25}$$

and indeed obtain the correct Lorentz transformations by matrix multiplication (from both sides of the matrix), in agreement with the transformation of the Lorentz four-force.

Electromagnetic waves An electromagnetic wave is the interplay of changing electric and magnetic fields. One manifestation, in the right range of frequencies, is light which, in vacuum, always travels at the speed of light just like any other electromagnetic wave. Not only the speed of light, but also its wavelike properties of changing electric and magnetic fields are independent of the motion of an observer doing measurements. How does this agree with the transformation behavior of the electric and magnetic fields, which do depend on the observer?

Most space-time objects depend on the state of motion of an observer, with interesting effects such as time dilation or Lorentz contraction. However, some quantities are invariant, for instance, the speed of light as (minus) the norm squared of any four-velocity, or the mass of a particle obtained from the norm of the four-momentum. The norm is the primary invariant associated with a four-vector, but the electromagnetic field is not a four-vector; it is a 4×4-matrix, the field-strength tensor. Such a tensor is associated with invariants as well, for the electromagnetic field amounting to the two quantities $\vec{E} \cdot \vec{B}$ and $\vec{E}^2/c^2 - \vec{B}^2$. For an electromagnetic wave, both invariants vanish: the electric and magnetic fields are orthogonal to each other and have the same magnitude, in all reference frames. Not only the speed of light but also the "shape" of light, in the sense of directions and magnitudes of the fields constituting it, is an observer-independent invariant. Other, nonwavelike electromagnetic fields, for instance, a static electric field, have different and nonzero values of the invariants.

Light Wavelike motion shares some similarities with particle dynamics, if we properly implement the value of the universal celerity. In several equations used so far, some factors diverge or become zero if we just take the limit of the velocity approaching c. Not all these quantities can be meaningful, or they must take specific values such as $m = 0$ for the mass.

Worldline The mathematical description of massive objects makes use of their worldlines $x(T)$, the lifeline of the object seen as a series of events x in space-time at different values of proper time. A worldline is a curve in space-time with a timelike tangent at each point, or a timelike rate of change $u(T)$ with

$||\boldsymbol{u}||^2 = -c^2 < 0$. In components, $\boldsymbol{u} = (c, \vec{V})/\sqrt{1 - V^2/c^2}$ depends on the spatial velocity $\vec{V} = \Delta\vec{x}/\Delta t$ of the object, with magnitude $V = |\vec{V}|$. Proper time T progresses at a rate $\Delta T = \sqrt{1 - V^2/c^2}\Delta t$ compared to coordinate time t.
For light, or anything moving at the speed of light, this description breaks down. Proper time for $V = c$ progresses by $\Delta T = 0$, or not at all, as light does not age, and cannot be used as a parameter along the worldline. The components of $\boldsymbol{u}$ become infinite if we let V come closer and closer to c. The norm of $\boldsymbol{u}$ depends only on c and does not change as we increase V. However, the temporal and spatial parts of $\boldsymbol{u}$, $c/\sqrt{1 - V^2/c^2}$ and $|\vec{V}|/\sqrt{1 - V^2/c^2}$, while divergent, come closer and closer to each other. In the norm, they are subtracted, indicating that light requires $||\boldsymbol{u}||^2 = 0$. The worldline of light is lightlike with a four-velocity $\boldsymbol{u}$ of zero norm.

While the norm squared of the four-velocity of a massive object is always (minus) the speed of light squared, $||\boldsymbol{u}||^2 = -c^2$, the four-velocity of light has zero norm, $||\boldsymbol{u}||^2 = 0$, and is lightlike. In space-time, massive objects have a four-velocity of magnitude given by the speed of light, while light itself has four-velocity of zero magnitude. At first thought, one might have expected the speed of light to appear in the norm of the four-velocity of light. However, seen in its own proper time, light does not travel at all through space-time because its T does not progress. Only massive observers see light move with respect to their time, in such a way that in the Minkowski distance $\Delta s^2 = -(c\Delta t)^2 + (\Delta x)^2$, any progress Δt in time is canceled by the progress $\Delta x = c\Delta t$ in space. The invariant norm of light's four-velocity is zero for all observers.

In order to describe the worldline of light, we take a lightlike curve in space-time characterized by $\Delta s^2 = 0$ along it. With this condition, the curve is lightlike, tangent to a light cone. The direction along which light moves through space-time is a lightlike vector, akin to the four-velocity $\boldsymbol{u}$. However, without any proper-time parameter, there is no unambiguous definition of an analog of $\boldsymbol{u} = \Delta\boldsymbol{x}/\Delta T$ for massive objects. For a clear distinction, we call the direction vector $\boldsymbol{w}$, with components $\boldsymbol{w} = (f/c, \vec{w})$ such that $||\boldsymbol{w}||^2 = -f^2/c^2 + |\vec{w}|^2 = 0$. Light indeed happens to be characterized by such an equation, provided only that we interpret f as its frequency and $|\vec{w}| = 1/l$ as the inverse wave length l. The lightlike condition $||\boldsymbol{w}||^2 = 0$ is then equivalent to the so-called dispersion relation $fl = c$ of light. In space-time, light is described by its wave properties contained in the wave four-vector $\boldsymbol{w}$.

It is not easy to define the four-momentum of light because we do not have an unambiguous four-velocity. Moreover, we have already seen that we should assign zero mass to light, so that $m\boldsymbol{u}$ can be an interesting object only if $m = 0$ somehow cancels the diverging factor $1/\sqrt{1 - V^2/c^2}$ in $\boldsymbol{u}$. Instead of trying to make sense of the limit of V approaching c, we start with the wave four-vector and introduce the four-momentum as a new four-vector proportional to it: $\boldsymbol{p} = h\boldsymbol{w}$ with components the energy $E = cp^t = hf$ and the spatial momentum $\vec{p} = h\vec{w}$. The constant h is chosen just to write the proportionality of $\boldsymbol{p}$ and $\boldsymbol{w}$ and must be determined, for instance, by comparing the energy of light with its frequency. Rather surprisingly, it turns out that understanding h requires quantum mechanics, another radical rethinking of our assumptions about the world. We must postpone this discussion, and continue to work with the wave four-vector $\boldsymbol{w}$ rather than the four-momentum $\boldsymbol{p}$ of light.

Doppler shift The speed of light is observer-independent; its other properties depend on an observer's state of motion. The frequency and wave length are components of the wave four-vector, which changes by a Lorentz transformation if an observer's motion changes. Different observers assign different frequencies to the same light ray or, with a reversal of viewpoints, a moving light source compared to a static one is seen to emit light of a different color. In acoustics, a similar effect is well-known from the siren of an approaching ambulance, heard at higher frequency than the one of an ambulance speeding away. In both cases, the change of frequency is called the Doppler shift.

Consider a light source moving along the x-axis with velocity V, and assume that we know the source to emit light of frequency f when it is at rest. A single light ray emitted in the (x, y)-plane at some angle $\sphericalangle$ with the x-axis has a wave four-vector $\boldsymbol{w} = (f/c, w^x, w^y, 0)$ with $cw^x = f\cos(\sphericalangle)$ and $cw^y = f\sin(\sphericalangle)$; see Figure 2.28. An observer at rest measures a different frequency f' and perhaps a different angle $\sphericalangle'$ with the x-axis. The latter assumption may not be obvious, but should be made as a precaution. After all, Lorentz contraction happens in the direction of motion, but not transversal to it, changing distance ratios that define an angle. The observer then measures some wave four-vector $\boldsymbol{w}' = (f'/c, w^{x'}, w^{y'}, 0)$ with $cw^{x'} = f'\cos(\sphericalangle')$ and $cw^{y'} = f'\sin(\sphericalangle')$.

A Lorentz boost relates frequencies and angles: for the time components, $f = (f' - Vw^{x'})/\sqrt{1 - V^2/c^2} = f'[1 - (V/c)\cos(\sphericalangle')]/\sqrt{1 - V^2/c^2}$ implies

$$f' = f\frac{\sqrt{1 - V^2/c^2}}{1 - (V/c)\cos(\sphericalangle')} \tag{2.26}$$

measured by the observer at rest for a light ray at angle $\sphericalangle'$ with the x-axis.

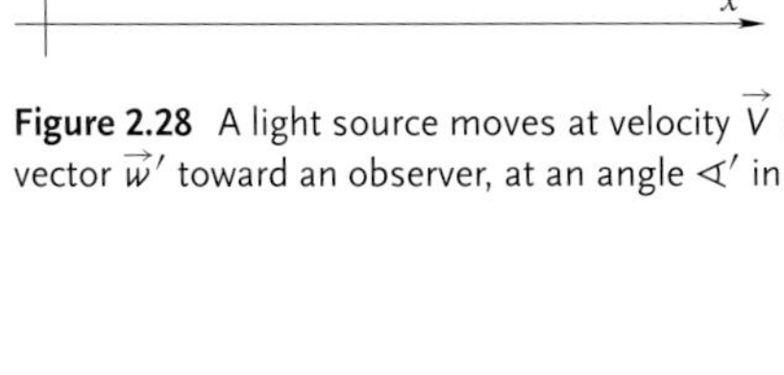

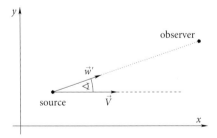

Figure 2.28 A light source moves at velocity $\vec{V}$ along the x-direction and emits light with wave vector $\vec{w}'$ toward an observer, at an angle $\sphericalangle'$ in the (x, y)-plane.

For small source velocity V and angle $\sphericalangle'$, we can approximate the factors and write $f' \sim f[1 + (V/c)\cos(\sphericalangle')]$, a formula in agreement with the acoustic or nonrelativistic Doppler shift. If the source moves toward the observer, $\sphericalangle' = 0$ and $f' > f$ with $\Delta f = f' - f = V f/c$. Larger frequency means that the light looks shifted toward the blue part of the spectrum, or blue-shifted for short. If the source moves away from the observer, $\sphericalangle' = \frac{1}{2}\bigcirc$ and $f' < f$ with shift $\Delta f = -V f/c$; light is redshifted. In the intermediate case of a light ray emitted transversal to the motion of the source, $\sphericalangle' = \frac{1}{4}\bigcirc$, we have $f' = f\sqrt{1 - V^2/c^2}$. In acoustics, no frequency shift would occur in this case. The effect is due to time dilation for the moving source, with the two time standards defining frequencies in the source and observer systems related by $1/\Delta t = \sqrt{1 - V^2/c^2}/\Delta T$.

Beaming Approaching lights beam bright and blue. We have already seen the blue shift of the frequency, and also the direction of emitted light changes: $\sphericalangle' \neq \sphericalangle$.

The spatial part of a boosted wave four-vector shows

$$w^x = \frac{w^{x'} - (V/c) f'}{\sqrt{1 - V^2/c^2}} = \frac{w^{x'}[1 - (V/c)/\cos(\sphericalangle')]}{\sqrt{1 - V^2/c^2}} ,$$

and $c w^x$ equals

$$f \cos(\sphericalangle) = \cos(\sphericalangle) f' \frac{1 - (V/c)\cos(\sphericalangle')}{\sqrt{1 - V^2/c^2}} = \frac{\cos(\sphericalangle)}{\cos(\sphericalangle')} \frac{1 - (V/c)\cos(\sphericalangle')}{\sqrt{1 - V^2/c^2}} c w^{x'}$$

with the formula for Doppler shift and the lightlike nature of $\mathbf{w}'$. Combining the rightmost sides of these equations, we cancel $w^{x'}$ and obtain

$$\cos(\sphericalangle') = \frac{\cos(\sphericalangle) + V/c}{1 + (V/c)\cos(\sphericalangle)} . \tag{2.27}$$

The emission angle changes when a light source moves. A few special choices illustrate the implications. For $\sphericalangle = 0$, we have $\sphericalangle' = 0$, and for $\sphericalangle = \frac{1}{2}\bigcirc$, $\sphericalangle' = \frac{1}{2}\bigcirc$: emission straight forward and backward is not redirected. However, an angle in between, such as $\sphericalangle = \frac{1}{4}\bigcirc$ for transversal emission in the frame of the source, is transformed to $\sphericalangle'$, satisfying $\cos(\sphericalangle') = V/c$. For motion at half the speed of light, half the energy emitted in all directions in the rest frame is beamed forward within an angle of 60° for a moving object. If V is close to c, the angle $\sphericalangle'$ is close to zero. What is emitted in a transversal direction in the source system appears to be emitted almost straight ahead for an observer at rest. If the light emitted is isotropic as seen from the source, with no preferred direction, an observer at rest sees almost half the light as a narrow beam in the forward direction as illustrated in Figure 2.29. A light source moving toward an observer at high speed appears brighter than a source at rest.

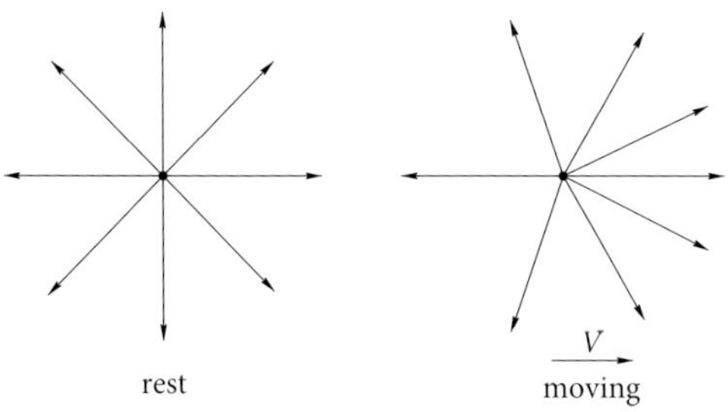

Figure 2.29 A radiant object at high speed emits more light in the forward direction, even if its emission at rest is isotropic: Relativistic beaming. The angles shown correspond to motion at half the speed of light, $V = \frac{1}{2}c$.

Significant relativistic beaming occurs for radiant objects moving at velocities close to c. These conditions, an object emitting light or other electromagnetic waves and moving at large velocity, seem difficult to realize at the same time if we think of our commercial light sources. However, there are several important situations in which relativistic beaming is significant for physics, or even in technological applications.

The first is synchrotron radiation, the emission of light by accelerated elementary particles. Particle accelerators, including the version called synchrotron, bring charged particles such as electrons or protons to high energies, with speeds near c. Moreover, a charged particle emits electromagnetic waves when it is accelerated, even if the accelerating force just keeps the particle on a circle rather than further increasing its speed. Often, such radiation is a nuisance, as the particle loses energy and cannot be stored in the accelerator for a long time. However, the electromagnetic waves emitted by a bunch of accelerated particles can be intense, and they have other useful properties such as a sharp frequency range. In combination with relativistic beaming, which makes light being emitted predominantly in the forward direction, a powerful source of synchrotron radiation results, used for instance to probe matter properties. By now, some accelerators have been built with the specific purpose of producing synchrotron radiation, rather than colliding the accelerated particles. Without relativistic beaming, radiation would not only be much less focused; it would be emitted in all directions, a potential hazard in the accelerator buildings.

In astrophysics, typical light sources are stars, but they rarely attain speeds close to c. Sometimes, however, active stars eject some of their hot matter as jets whose constituents can have large velocities. They are still hot and emit light, fulfilling both conditions for relativistic beaming. If the jet points almost straight at Earth, it seems much brighter than one would expect without beaming. Some bright sources, such as gamma-ray bursts, can be explained only with relativistic beaming, for otherwise they could not contain enough energy to produce the brightness we see.

Stars Stars emote; they radiate their inner lives. Hot matter emits light with frequencies characteristic of its constituents, to the most part hydrogen in a rather young star. On Earth, we can analyze hydrogen in the laboratory and see what frequencies of light it prefers to emit. As found by measurements in the nineteenth century, there is a specific set of frequencies that can be emitted, with lines like a fingerprint rather than a vague smeared-out range. The emitted frequencies are sharp and form a discrete subset of the real numbers. Quantum physics provides the reason; for now we make use of this property to conclude that stars radiate in specific frequencies. With a prism or a grating, one can determine which colors are represented in a sample of light from a star, and thereby find what frequencies the star emits.

When we observe stars, we are in a situation in which Doppler shift and relativistic beaming apply. Relativistic beaming is not so relevant for most stars because their velocities are not high, they are already bright to begin with, and any extra brightness due to beaming would be difficult to measure. The Doppler shift is more interesting even at small velocities because we can measure frequencies very well, using spectroscopy. Moreover, since light is emitted by hydrogen in a star at sharp and well-known frequencies, we can identify the set of emission colors and see whether they are shifted compared with the frequencies measured for hydrogen on Earth. As a test, one can analyze sunlight and find just the frequencies we see on Earth; the velocity of the Earth around the Sun is not large.

When we look at other stars, especially distant ones, there are significant frequency shifts, toward the red. For each emission color of hydrogen, we have one test frequency f which we insert in the Doppler formula to compare with the f' we measure. The shifts then indicate that the stars are moving with respect to us, at velocities larger than the orbiting velocity of the Earth around the Sun. We cannot see motion because the stars are so far away, and so the angle of sight is not known, nor is the magnitude of the velocity. However, if we take a large sample of stars, we find that their frequency shifts match the Doppler formula if we assume that they are moving away from us with a velocity proportional to the distance.

Accelerated observers What we derived so far does not apply to observations we make. We have assumed that observers move at constant velocities; otherwise, we could not use Lorentz transformations with constant v to transform between them, and extra forces would arise. We cannot make v in a Lorentz transformation time dependent to account for accelerated observers, for this would mess up the Minkowski line element. On the other hand, in astronomy, for instance, when we determine the frequency shifts of star light, we are not observers moving at constant speed with respect to the star. As observers on Earth, we are forced to follow the orbiting motion of the planet around the Sun, and the motion of the whole solar system around the center of the Milky Way. Our velocity is not constant, but changes direction and, according to Kepler's laws, speeds up or slows down depending on how far we are from the Sun.

It is not necessary to drop everything derived so far, or generalize it to complicated states of acceleration. An elegant consideration of invariants in special relativity allows us to apply this theory even to accelerated observers. As such an observer, measuring the energy of an object moving in some way or the frequency of light emitted, we determine the time component of the four-momentum or the wave four-vector. Even if our velocity is changing during our accelerated motion, at every instant of time we obtain measurement results according to the velocity we have at that time; they agree with those obtained by an observer always moving at that speed. We could perform a steady series of Lorentz boosts to find the measurement results we expect at each time during our motion. With more efficiency, we can relate those results to invariant quantities that are valid in all systems provided only that we know the object's properties and our own state of motion as an observer.

Scalar product An observer at rest identifies the energy of a moving object with cp^t, which happens to be the component of the four-momentum $\boldsymbol{p}$ along the direction of the observer's four-velocity, $\boldsymbol{u}_{\text{obs}} = (c, 0, 0, 0)$ for an observer with $\vec{V} = 0$. If we define a number $\boldsymbol{a} \cdot \boldsymbol{b} = -a^t b^t + a^x b^x + a^y b^y + a^z b^z$ for two four-vectors $\boldsymbol{a} = (a^t, a^x, a^y, a^z)$ and $\boldsymbol{b} = (b^t, b^x, b^y, b^z)$, called the scalar product of the two vectors, we have $E = -\boldsymbol{p} \cdot \boldsymbol{u}_{\text{obs}}$. The coefficients in the scalar product are chosen such that its value remains unchanged when both four-vectors are transformed by a Lorentz boost, as indicated by the resemblance with coefficients in the invariant Minkowski line element. We write the energy as an invariant scalar product of the object's four-momentum and an observer's four-velocity, and the frequency of light is $f = -\boldsymbol{w} \cdot \boldsymbol{u}_{\text{obs}}$.

Using invariants for energy and frequency seems wrong, with all we have learned about relativity. How can they be invariant if they are time components of four-vectors, which transform according to Lorentz if the state of motion changes? In the new equations, the same dependence is realized, but by direct reference to the observer's four-velocity rather than four-vector components. We capture the correct observed properties by Lorentz invariants. Compared with the previous formulation, we have the great advantage of allowing the four-velocity of an accelerated observer, while Lorentz transformations cannot be generalized to nonconstant v.

Local inertial frame If the observer moves along the x-axis with velocity V, $\boldsymbol{u}_{\text{obs}} = (c, V, 0, 0)/\sqrt{1 - V^2/c^2}$. The energy of an object at rest, with four-momentum $\boldsymbol{p} = (mc, 0, 0, 0)$, is measured as $E = -\boldsymbol{p} \cdot \boldsymbol{u}_{\text{obs}} = mc^2/\sqrt{1 - V^2/c^2}$. The frequency of a light ray in the x-direction, with frequency f_* and wave four-vector $\boldsymbol{w} = (f_*/c, f_*/c, 0, 0)$ at rest, is measured as $f = -\boldsymbol{w} \cdot \boldsymbol{u}_{\text{obs}} = f_*(1 - V/c)/\sqrt{1 - V^2/c^2}$.

The energy formula agrees with the energy measured at nonvanishing velocity, taking into account kinetic contributions in addition to the mass energy. The frequency formula agrees with the equation for Doppler shift and provides a concise rederivation. In addition to confirming the previous results, we have generalized them because V, now, need not be constant in time but may belong to an accelerated observer. An observer could be one of us measuring the frequencies of distant stars on Earth, following the planet's orbit, moving not just in one direction but rather on (nearly) a circle in a plane.

> If an observer is in circular orbit with frequency F, the four-velocity is $\boldsymbol{u}_{\text{obs}} = (c, V\cos(FT\bigcirc), V\sin(FT\bigcirc), 0)/\sqrt{1 - V^2/c^2}$. The measured frequency of a light ray moving in the x-direction, with a wave four-vector as before, equals $f = f_*[1-(V/c)\cos(FT\bigcirc)]/\sqrt{1 - V^2/c^2}$, amounting to periodic Doppler shift.

2.3 General Relativity

The ancients lived in a dangerous world; they were afraid for many reasons that we now know to be harmless. Unaware of the spherical form of Earth, they made the next-best assumption about it, considering the range of lands and seas on which they moved as disk-shaped. A disk has an edge, and if one only travels far enough one might fall off into oblivion, or whatever else awaits. Adventurers nonetheless traveled far, and found that the Earth is a globe, with a surface curved and closed in itself. A closed surface has no edge to fall off from, and there is nothing beyond except outer space.

An edgeless surface in Euclidean space must either be curved so that it can close in itself, or stretch out all the way to infinity. The second option was unimaginable to the ancients; the first one, although encountered in everyday objects such as cups, just didn't occur to them as an option for something as grand as all of Earth. The only valid assumption, although unrecognized as such, was for Earth's surface to be flat and finite, a bounded disk. As harmless as this mathematical statement may seem, it harbored abundant dangers lurking in the world of superstitious minds.

Treating Earth's surface as flat is a good starting point as long as one does not travel far. All the curving and bending does not matter much then, and one might well view landscapes as drawn like maps on a sheet of paper. However, when longer distances can be reached, travel on a curved surface differs from travel on a flat plane. Looking at a globe reveals that we change our direction in space as we travel from America to Europe (and still keep standing upright at both places). We do not rotate actively, nor do we notice it because our travels are rather slow, even with modern jets, and the rotation is gradual. Also, geometry on a globe is different from what we learn for flat Euclidean space; for one, we cannot shift a line or a triangle arbitrarily far. And, speaking of lines and triangles, what is a straight line on a globe supposed to be? If we start at some point and try to draw a straight line, it will soon leave the globe's surface and enter outer space. Such lines cannot describe triangles *on* the surface.

Curvature When we don't look far, the world seems flat. However, when long distances are involved, the curvature of space becomes noticeable by careful observations. For this reason, everyone, even the most stubborn and conservative adherent of yestersay, is now convinced that Earth is of spherical shape and has no edge to fall off from. The assumption of a flat, finite world was just that: an assumption made for the lack of better knowledge. Nowadays, we are more comfortable with the notion of the infinite, at least some forms of it, and although curvature still has some subtleties, we can make the mental transition from globe models to the whole surface of Earth. Our modern, educated, open-minded world has no room for primitive guesses about our surroundings, and no room for imagined dangers that may come with them.

With some pathos, one might say that science has defeated the demons of space, by pure thought. However, with all our progress, when it comes to space-time, we often are no better than the primitive ancients about whose fears we may laugh. Earth's surface, for sure, can be closed in itself because it curves in space. We see small objects in our day-to-day life, and notice that their curved surfaces allow them to be finite and edgeless at the same time. On the scale of the whole planet, we can even determine the radius of the sphere, to estimate on what distances curvature becomes important. Already, some of the ancients, during the philosophical awakening in Greece, undertook such measurements, steered by pioneering thoughts that allowed humankind to break through the old worldviews steeped in superstition and myths. With the rather enormous result of 6367 km for the radius, it becomes clear why it is not obvious, yet possible, to notice the curvature of Earth's surface.

We think that we understand our planet, at least in geometrical terms. It is, to a good degree, a ball in space, and spherical models can be used to do geometry on it by comparing angles and sizes of triangles. However, with statements like these, we often do what the ancients did: we are making assumptions about our surroundings. The Earth's surface is a sphere in space and curved, but what about space? We imagine Earth as an object in Euclidean space, surroundings that extend all the way to infinity and are flat. These are again assumptions that one has to justify and test. Space may very well be finite, with or without an edge. It may be curved and closed in itself, just like a three-dimensional analog of the sphere we have come to associate with Earth's surface. As we need to traverse long distances on Earth to see its curvature, we would have to do an enormous, perhaps impossible amount of space-travel to test whether space is flat. Special relativity has taught us that space is not separate from time, but rather combined to one entity, space-time. What is this four-dimensional structure? Is it finite? Is it flat or curved? How far do we have to travel in space-time to test these properties?

Space-time curvature Cosmology explores the form of space-time. While it is not possible for us, or for our probes such as satellites, to travel far across space and see whether it is curved, there are messengers, mainly electromagnetic waves such as light, that reach us from afar. If we use astronomy to map out as much of the universe around us as we can, the distribution of all cosmic signals may be used to understand the space-time region they travel through. It takes more time to travel around the bulge of a curved surface such as Earth's than it would to tunnel right through it. Light, even with its enormous and constant celerity, would take longer or shorter depending on the curvature of regions it must move through to reach us. We are dealing with the curvature of space and space-time, notions that are much more difficult to visualize than a curved surface, but whose influence on light could nevertheless be detected.

The revolution of thinking undertaken in the transition from a disk-shaped surface of Earth to a spherical one must be repeated in the next dimensions, three-dimensional space and four-dimensional space-time. In our thinking about Earth, dangerous, bottomless cliffs at the disk's boundary were eliminated just because a spherical surface does not have any edge. When humankind started thinking about space, the notion of the infinite had already become less revulsive. No edge of space had to be assumed to combine the flat with the finite; no imaginative dangers were lurking far away in outer space. We moderns think of Euclidean space or Minkowskian space-time as reaching all the way to infinity, keeping the well-known form of geometry intact no matter how far we travel, but these are assumptions about the universe that turn out to be incorrect.

Using distant light sources to probe the universe on cosmic scales, space-time does not appear Minkowskian, just as Earth's surface is not flat when seen over continental distances. Minkowski space-time, and all the consequences seen by special relativity, is an excellent description of the geometry of space-time near us. However, geometry changes as we move far into outer space or look back into the distant past. Space-time is curved, not flat. For space-time, in contrast to Earth's surface, curvature introduces new dangers and demons instead of defeating them. Curved space-time has edges; it cannot extend all the way to infinity in all directions, yet not be closed in itself. If one travels far enough, one reaches an edge beyond which space-time and its geometry not only remain unknown, they don't even exist. When we reach these edges, we do not fall off, but everything falls when no space and time exist. Such space-time edges are called singularities. The last chapters of this book will deal with the question of how these new demons can be defeated, but first we will return to more placid regions of space-time and see what consequences their curvature may have for our experience.

Gravity Curvature is a passive form of motion. If space or space-time is curved, something about us must change as we move, just as we change the direction of our bodies as we travel on Earth. Our orientation in space is specified by angles, or ratios of two distances such as the length of our body relative to its projections on the axes of some coordinate system. Directions in space-time are specified by ratios of distances to time durations. This geometrical statement has a more familiar form in terms of motion: Ratios of distances to time durations are nothing but velocities, the velocities of objects moving the distances involved in the time durations considered.

As we move on the curved surface of Earth, our bodies are forced to change their directions, or the angles they have with some coordinate axes. As we move in curved space-time, by analogy, our space-time angles change, for instance, with respect to some Minkowski coordinate system around us. Changing space-time angles mean changing velocities, and changing velocities mean acceleration. Just by shifting our position in curved space-time, we accelerate. We attribute any acceleration to the action of a force; space-time curvature implies a force. On the grand scales of the universe, on which we might expect space-time curvature to be significant, there is only one relevant force, gravity, binding together stars to galaxies and even galaxies to larger clusters.

Space-time must be related to the gravitational force, a thought of considerable magnitude. Relating a force to geometrical properties of space-time instead of introducing new fields, such as the electromagnetic one, is not only economical; it makes the force inevitable, forced upon us by general implications of curvature applied to space-time. We eliminate the geometrical assumption of space-time being flat, and are led to expect a physical force to act. Geometry and physics are intertwined. We may speak of motion in curved space-time, or we may speak of motion under the action of a force, both viewpoints will lead us to the same conclusions.

Such a view is attractive, but difficult to formulate in precise terms. There seem to be obvious problems. How can it be true that the gravitational force, which we all too easily experience on our own bodies on Earth, is related to space-time curvature, something so elusive and small that we have to travel long distances to perceive it, far into outer space? If we fall from a high cliff (a real one, not the edge of the universe), we are accelerated even though the distance we move through is far from astronomical. Why must our space-time angles change so much if we move just a short distance? In order to explore the thought of unifying the gravitational force with space-time curvature, or physics with geometry, we return to the geometrical investigations begun for flat Minkowski space-time in special relativity.

Special relativity describes how moving observers measure distances in space, durations in time, or related quantities such as energies and frequencies. Measurements of space and time determine what geometry space-time has, described by the Minkowski line element. Measurements and geometry, basic notions in physics and mathematics, respectively, are related with each other even before one notices the forceful consequences of curvature. However, in special relativity, this relationship is realized only for observers moving at constant velocities, not for realistic observers like us on Earth who must change their velocity and are accelerated. While measurements can be described for accelerated observers as well, referring to vector components along the observer's four-velocity and local inertial frames, the geometry is of Minkowski form only for nonaccelerated observers. Special relativity does not tell us what geometry we should use when we do space-time measurements from our accelerated viewpoint on Earth. Special relativity and its Minkowskian geometry are to be generalized to a new geometrical description suitable for accelerated observers, a new theory called general relativity.

Rotation Rotating observers see a strange geometry. Viewed from a merry-go-round, stationary bystanders seem to move around us. In a rotating frame, any object, just by being observed, moves around even if there is no force. If physics can be done in a rotating frame, there must be some geometry that does not allow force-free objects to stand still.

Line element Starting with familiar static geometry, we consider a coordinate transformation to a space in rotation, leaving $t' = t$ and $z' = z$ unchanged while $x' = x\cos(Ft\bigcirc) + y\sin(Ft\bigcirc)$ and $y' = -x\sin(Ft\bigcirc) + y\sin(Ft\bigcirc)$ implement time-dependent rotation with frequency F. This coordinate change is not a Lorentz transformation. It is a rotation, but one by a time-dependent angle. The dependence on t implies that the unprimed line element, if we assume the Minkowski form for the primed coordinates, is of a new form. Small coordinate differentials in $\Delta s'^2 = -(c\Delta t')^2 + (\Delta x')^2 + (\Delta y')^2 + (\Delta z')^2$ translate to $\Delta t' = \Delta t$ and $\Delta z' = \Delta z$, while

$$\Delta x' = \cos(Ft\bigcirc)\Delta x + \sin(Ft\bigcirc)\Delta y + F\bigcirc[-x\sin(Ft\bigcirc) + y\cos(Ft\bigcirc)]\Delta t$$

and

$$\Delta y' = -\sin(Ft\bigcirc)\Delta x + \cos(Ft\bigcirc)\Delta y - F\bigcirc[x\cos(Ft\bigcirc) + y\sin(Ft\bigcirc)]\Delta t$$

depend on Δt in addition to the spatial coordinate separations. The Δt-independent parts in these transformations are the same as those for a rotation with a fixed angle $Ft\bigcirc$ and therefore leave the line element invariant. The Δt-dependent terms change the line element to

$$\Delta s^2 = -[c^2 - F^2\bigcirc^2(x^2 + y^2)](\Delta t)^2 + 2F\bigcirc(y\Delta x - x\Delta y)\Delta t + (\Delta x)^2 + (\Delta y)^2 + (\Delta z)^2 .$$

One may transform the line element for any kind of accelerated motion and find the corresponding geometry seen by a noninertial observer. However, the result cannot always be correct, even for a rotation. There are points where there is no time duration left in Δs^2 because the coefficient of $(c\Delta t)^2$, $1 - F^2\bigcirc^2(x^2 + y^2)/c^2$ may vanish. Moreover, there are contributions from mixed space-time separations $\Delta t \Delta x$ not encountered before. Some of these problems are related to the fact that we assumed a rigidly rotating coordinate system. At some distance R from the rotation center, points will reach or surpass the speed of light. This breach of our old rules happens when $F\bigcirc R = F\bigcirc\sqrt{x^2 + y^2} = c$, or when $c^2 - F^2\bigcirc^2(x^2 + y^2) = 0$ and the time coefficient in the line element vanishes. We recognize at least some of the problems as a consequence of too simple a coordinate form in which we neglect the fact that only a bounded part of space should rotate, not all of it. However, combining rotation in a bounded part of space where an observer moves with a static part outside would be much more involved. An extension of space-time geometry to accelerated observers must be more complicated than the extension of local energy measurements.

Line element Geometry may depend on where we do it. General line elements have position-dependent coefficients, as obtained, for instance, by writing the Minkowski line element in coordinates not obtained by a Lorentz transformation. The corresponding geometry looks even stranger than the now-familiar space-time geometry obtained with the extra negative sign in its Minkowski form. Position-dependent coefficients imply that geometrical distances taken in different regions of space (or space-time) have different relations with coordinate separations. It is not only the state of motion of an observer that changes perceptions; just being at a new place makes us reconsider things. Such geometry is more complicated than what we encountered so far. We always have to keep track of the position or time where and when we do geometry. Geometry depends on the surroundings, a cultural relativism which so far has come about only by transforming to complicated coordinates.

As suggested by all the new terms generated by our rotational coordinate transformation, we expect a general line element to have position-dependent coefficients, and also mixed terms such as $c\Delta t\Delta x$. With generous generality, we write

$$\Delta s^2 = \mathbf{Sum}_{a,b} g_{ab}(t,x,y,z)\Delta x^a \Delta x^b \tag{2.28}$$

with the metric components $g_{ab}(t,x,y,z)$ and a and b summed over all coordinates. Many crucial effects can be seen if one restricts considerations to the "diagonal" case in which the g_{ab} are zero unless $a = b$. A diagonal line element is

$$\Delta s^2 = g_{tt}(c\Delta t)^2 + g_{xx}(\Delta x)^2 + g_{yy}(\Delta y)^2 + g_{zz}(\Delta z)^2 \,. \tag{2.29}$$

All coefficients g_{aa} may still depend on the four coordinates, and we should require g_{tt} to be negative in order to be in agreement with the space-time effects we have seen so far for constant g_{aa}. If g_{tt} is not constant, for instance, time dilation occurs not only by changing an observer's state of motion, but also by an object's own motion in space.

Sphere A sphere is the prime example of a curved surface. Its form is determined once we decide what radius it has; geometrical relationships on it then follow. A sphere is so emblematic for a curved surface that one often expresses general curvature at some point on a surface in terms of a sphere that would produce the same amount of curvature on all its points. The radius of the matching sphere is called the curvature radius at the point of the curved surface. Looking at the geometry of a sphere indeed provides a good glimpse on curvature in more general terms.

Mathematics describes the well-known shape of the sphere, a two-dimensional surface, by two angles **La** and **Lo**, the latitude and longitude as shown in Figure 2.30. With these angles fixed, a point on the sphere of radius R has Cartesian coordinates $x = R\sin(\mathbf{La})\cos(\mathbf{Lo})$, $y = R\sin(\mathbf{La})\sin(\mathbf{Lo})$ and $z = R\cos(\mathbf{La})$. Using the laws for sines and cosines, all points indeed have distance $R = \sqrt{x^2+y^2+z^2}$ from the center, irrespective of the values of **La** and **Lo**.

Coordinate relationships can be used to compute the line element that describes the geometry on the sphere. If we allow all three variables R, **La** and **Lo** to change,

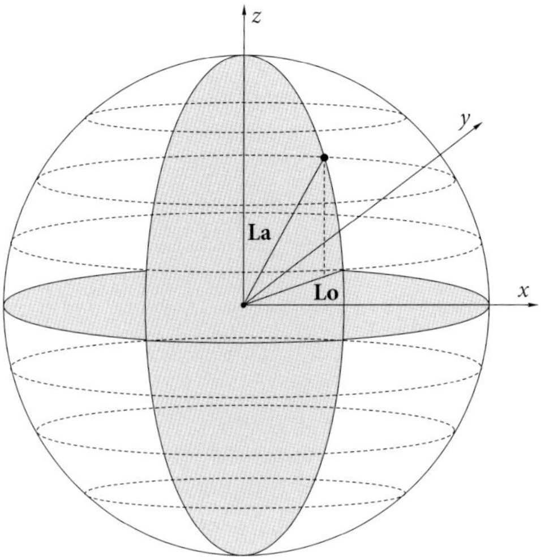

Figure 2.30 A point on a sphere is specified by two angles defined by the line connecting it with the center of the sphere: the latitude **La** between the line and another one through the center of the sphere and its North pole (with **La** constant along parallels), and the longitude **Lo** between the projection of that line to the (x, y)-plane and the x-axis (with **Lo** constant along meridians).

we have a coordinate transformation of three-dimensional space, from Cartesian coordinates x, y and z to so-called polar coordinates R, **La** and **Lo**. Computing the Cartesian coordinate separations that originally enter the Euclidean line element in terms of separations of polar coordinates, we obtain the same spatial geometry expressed in new coordinates, a process similar to transforming Minkowski space-time to a rotating frame introducing position-dependent coeffici ents. However, we are still dealing with three-dimensional space, not the surface of a sphere. To describe the two-dimensional sphere, we fix one of the coordinates, R, to have the value given by the radius of the sphere. On the spherical surface, the radius does not vary, and $\Delta R = 0$. The resulting line element only contains two coordinate separations of the angles and is two-dimensional.

With R fixed,

$$\begin{aligned}
\Delta x &= 2\pi R\cos(\mathbf{La})\cos(\mathbf{Lo})\Delta\mathbf{La}/\bigcirc - 2\pi R\sin(\mathbf{La})\sin(\mathbf{Lo})\Delta\mathbf{Lo}/\bigcirc\,,\\
\Delta y &= 2\pi R\cos(\mathbf{La})\sin(\mathbf{Lo})\Delta\mathbf{La}/\bigcirc + 2\pi R\sin(\mathbf{La})\cos(\mathbf{Lo})\Delta\mathbf{Lo}/\bigcirc\,,\\
\Delta z &= -2\pi R\sin(\mathbf{La})\Delta\mathbf{La}/\bigcirc
\end{aligned}$$

and

$$\Delta s^2 = (\Delta x)^2 + (\Delta y)^2 + (\Delta z)^2 = (2\pi/\bigcirc)^2\left[(\Delta\mathbf{La})^2 + \sin(\mathbf{La})^2(\Delta\mathbf{Lo})^2\right]\,.$$

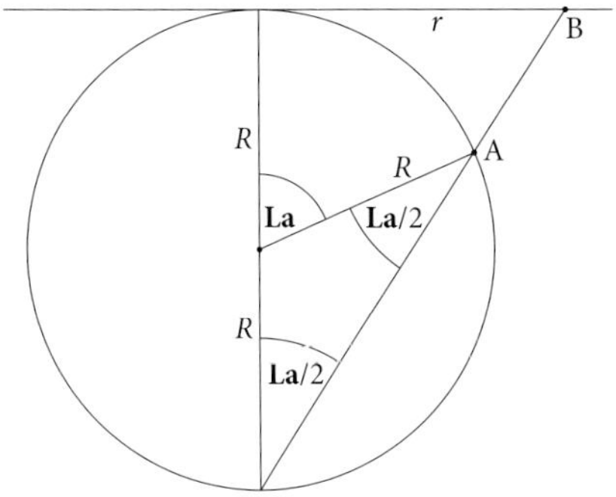

Figure 2.31 Stereographic projection identifies a point on the sphere with a point on its tangent plane through the North pole. Using Euclidean geometry, the angles of the point on the sphere are related to Cartesian coordinates of the point on the tangent plane.

Stereographic projection The plane is flat, the sphere is curved, but nonetheless they are related. It is not possible to bend a sheet of paper and obtain a perfect sphere. There are always wrinkles and fold-overs of some of its parts, or one would have to cut out some pieces and glue the rest back together. All these manipulations change the geometry of triangles drawn on the sheet of paper. If we cut out or fold over some parts of the sheet, certain distances get shortened and no longer obey the laws of planar Euclidean geometry. Spherical geometry is different from Euclidean geometry. Nevertheless, one can construct a mathematical map[10] from the plane to the sphere in somewhat more abstract terms, and see how the two different geometries are related.

The construction, called stereographic projection, amounts to placing a horizontal plane on the North pole of the sphere and shining a light from the South pole. Every point on the sphere (other than the South pole itself) casts a shadow on the plane and so do lines of geometrical objects drawn on the sphere. By casting shadows, we map the sphere to the plane.

We identify the North pole with the origin of a Cartesian coordinate system on the plane and align the x- and y-axes with the axes in the equatorial plane used in Figure 2.30. The longitude **Lo** defined for a point on a sphere and the polar angle $\sphericalangle$ of a point in the plane with Cartesian coordinates $x = r\cos(\sphericalangle)$ and $y = r\sin(\sphericalangle)$ for some r are one and the same. It remains to find the correspondence between the second spherical angle **La** and the distance $r = \sqrt{x^2 + y^2}$ between a point on the plane and the origin. As shown in Figure 2.31, Euclidean geometry tells us what this relationship is. To that end, the figure is drawn in another plane, one through the two poles of the sphere and a given point A on the sphere at some angle **La**.

10) In the nineteenth century, Samuel Rowbotham, with pseudonym "Parallax," invented and promoted "Zetetic" astronomy, a flat-earth model of a disk around the North Pole, bounded by an antarctic wall of ice. Stereographic projection shows that the spherical surface of Earth can indeed be mapped to a plane, but the geometries on both surfaces are different. Earth's surface is not planar.

We first compute the angle of a line through A and the South pole, used for the stereographic projection. The distance from a point on the sphere to its center is always the same number R, the radius. This distance is found for the point A we are mapping to the plane and for the South pole. Together with the center, these two points are the corners of an isosceles triangle with two sides of length R meeting at an angle of 180° minus **La**, the complement of the latitude. The other two angles must be equal to each other and complete the whole angle sum to 180°, as always in Euclidean geometry. If we call the unknown angle $\sphericalangle_\Delta$, we have the equation $(180^\circ - \mathbf{La}) + 2\sphericalangle_\Delta = 180^\circ$, solved by $\sphericalangle_\Delta = \mathbf{La}/2$.

We now consider the second, larger triangle seen in the figure. Its corners are the two poles and the point B cast as the shadow of A. The point B lies on the plane tangent to the North pole; the triangle is thus right-angled at the North pole. One of its side lengths, the distance between the poles, is $2R$, twice the radius of the sphere. We just determined the angle $\mathbf{La}/2$ at the South pole, so that we write the distance r of B to the North pole as $r = 2R\sin(\mathbf{La}/2)/\cos(\mathbf{La}/2)$. The mathematical mapping between the sphere and the plane is now complete: $x = r\cos(\mathbf{Lo}) = 2R\cos(\mathbf{Lo})\sin(\mathbf{La}/2)/\cos(\mathbf{La}/2)$ and $y = r\sin(\mathbf{Lo}) = 2R\sin(\mathbf{Lo})\sin(\mathbf{La}/2)/\cos(\mathbf{La}/2)$.

Sphere The sphere does not obey the laws of Euclidean geometry. We have used these laws to find equations between points on the sphere and on a plane by casting shadows, but this was geometry of triangles in the common space in which both the sphere and the plane lie. If we try to draw triangles or circles or other geometrical figures right on the surface of the sphere, we find that they do not obey what we know for those figures drawn on a plane. It is not even clear how to draw triangles, or a single straight line because those lines somehow have to follow the bending and curving of the sphere. So how can the lines remain straight if the ground they are drawn on is not? How do we plow a straight furrow on a curving field? In order to see what straight lines on a curved surface are supposed to be, we have to extract the crucial properties of what we mean by "straight" beyond our intuitive understanding of it, and then see if those properties can be realized in a more general sense that also applies to the sphere and other surfaces.

We postpone the question of straight lines and first look at properties of circles. Circles are round as the sphere, and it seems easier to draw one on the other. A circle is the set of all points having equal distance, that is, the radius, from one central point. We can measure distances on a sphere, not with a straight ruler, but with a flexible band that follows the curving. We can draw circles on a sphere and see what they look like.

There are important and well-known geometrical relationships between the various measures used to determine the size of a circle: the radius, the circumference, and the area. Euclidean geometry provides strict equations for all these numbers in terms of just one of them, for instance, the radius. The circumference of a circle of radius r is $C = 2\pi r$, its area $A = \pi r^2$. These are simple relationships, with constant ratios C/r and A/r^2 taking universal values related to the constant π.

If we draw circles on a sphere, we should expect problems with the Euclidean laws. A circle around some central point, say the North pole, still looks like the circles we know well from planar geometry. However, to draw the radius line connecting the center with the circle, we have to go along the bulge of the sphere, not being allowed to leave the surface. Stretched around the bulge, the radius curve is longer than in Euclidean space, and the relationship $C/r = 2\pi$ seems in danger. Deviations should be larger for circles of long radius, making us move along a larger piece of the bulge.

To confirm the severe danger Euclidean geometry is in, we compute the geometrical relationships of sample circles with rather simple expressions for the circumference. Let us look at the equator, a circle of constant radius around the North pole. The circumference of the equator has the usual Euclidean relationship with the radius of the whole sphere; the equator, after all, is just a circle around the center of the sphere in Euclidean space. However, the radius of the sphere is not the radius of the equator when seen as a circle on the sphere, centered at the North pole. This radius is determined by the distance from the pole to the equator. To see how long this radius is, we follow a meridian starting at the North pole. As Figure 2.32 confirms, the distance r from the North pole to the equator is the same as the distance from the equator to the South pole along the same meridian. If we continue to follow the meridian back North all the way to the starting point, we traverse four routes all of the distance r: from the North pole to the equator, from the equator to the South pole, from the South pole North to the equator, and from the equator back to the North pole. These segments of our trip complete another circle on the sphere, which is just a rotated version of the equator and has the same circumference C. (We are considering a perfect sphere. On Earth, the equator is longer than a meridian because the planet's rotation around its axis deforms it away from perfect spherical shape.)

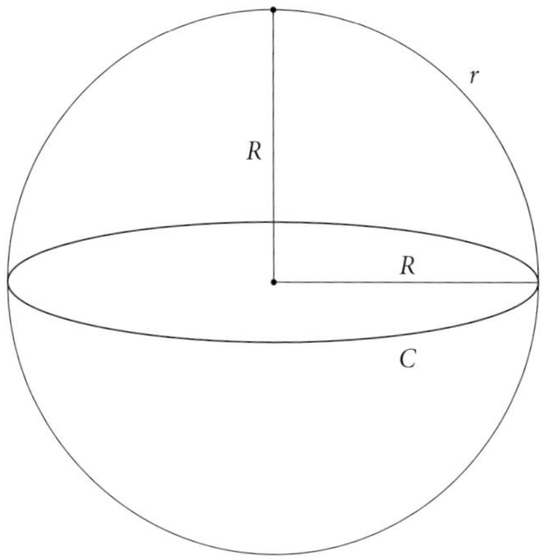

Figure 2.32 The equator as a circle around the North pole on a sphere. The radius of the circle in space is equal to the radius R of the sphere. The radius r of the circle on the sphere is the distance along a curved line from the North pole.

We now have two circles of the same circumference C, the equator and a meridian. Moreover, we have seen that the meridian is made up of four pieces of equal length, all corresponding to the radius r of the equator. The corresponding equality $C = 4r$ determines the ratio of circumference to radius, $C/r = 4 < 2\pi$. The Euclidean ratio 2π is violated for a circle drawn on the sphere. The laws of Euclidean geometry do not apply on curved surfaces; we must look for new rules that remain valid, taking into account extra distances that one might have to travel due to curvature.

Straight lines Dispensing with Euclidean geometry is a radical step. Before doing so for curved surfaces like the sphere, we should make sure that there is no other way out. We had decided to look at circles because they seem easier to draw on a sphere than straight lines. However, when we determined their proportions, we did refer to a straight line, the one along which we measure the radius. Even for circles, we are not freed from the task of understanding straight lines on the sphere.

In Euclidean space, we use straight lines to determine the radius, for we know, at least intuitively, that a straight line between two points has the shortest distance among all paths between those points. Straight lines therefore provide true measures for distances just as the crow flies, when all possible shortcuts are taken into account. Any other path would have a longer distance because it would amount to a detour.

For the equator as a circle around the North pole, we chose a meridian as a line to find the radius. The meridian looks reasonable for measuring the radius of a circle on the sphere, but what if there is a different path from the pole to the equator, one of shorter distance? If such a path existed, the radius would be smaller than $C/4$, the ratio C/r larger. Maybe one can even find a path so that the Euclidean value $C/r = 2\pi$ is redeemed.

Alas, Euclidean geometry cannot be saved. The meridian does provide the path of shortest distance from the pole to the equator, as we will demonstrate soon. For now, we provide one other, unambiguous example for a circle on the sphere violating the Euclidean laws of proportion. As the center of the circle, we again take the North pole, and the circle itself is the South pole. The South pole is just a single point, but by symmetry it is also the set of all points having equal distance from the North pole, our center. We can regard the South pole as a circle on the sphere. In order to determine its radius, we have to face the previous question of what a straight line (or the shortest distance) between the two poles is. However, whatever its value might be, the ratio C/r cannot have the Euclidean value 2π. The South pole as a circle, that is, a single point, has no circumference: $C = 0$. The distance between the poles is some nonzero number, so that the vanishing ratio $C/r = 0$ is an even worse violation of the Euclidean law than the one committed by the equator. Moreover, we see that the ratio C/r depends on the circle we look at; it no longer has a universal value as in Euclidean geometry.

Geodesics We must do what seems impossible: determine straight lines on curved surfaces. Often in mathematics, when it seems impossible to complete a task under new realities, one redefines the mission. In fact, we are not so much interested in straight lines just because they are straight. In Euclidean geometry, straight lines abound and enter traditional relationships, for instance, in statements about parallels or in defining triangles. On a curved surface, the whole perspective is not so straight anymore; one can afford to be more open-minded. Our considerations of circles and radius lines have hinted at what counts: Not being straight, but being close. In order to determine the radius of a circle on a sphere, we measure the shortest distance between the center of the circle and some point on it. In Euclidean space, we would draw a straight line without much thinking. On a sphere, the question of what a curve of shortest distance between two points should be is more complicated.

In spite of some complications, the change of perspective encountered here is a powerful one. We just forget about the impossible: trying to straighten out a curved surface so as to be able to draw straight lines on it. Stubbornness doesn't get us far, nor does the conservation of old values for their own sake. If there are situations in which the straight way is impossible, we should reconsider what matters, and that is connecting two different points by the shortest possible distance. Dealing with the traditional laws of Euclidean geometry for centuries has, for a while, made our minds too rigid. We put too much emphasis on being straight and forget that this property just happens to be what provides the shortest curves in Euclidean geometry. Faced by new circumstances, we must retreat to the more general notion of curves of shortest distance. In this general sense, without reference to any particular surface or geometry, curves of shortest distance between two points are called geodesics.

With the notion of geodesics, we can define all the usual figures we deal with in geometry. The distance between two points on a curved surface is the length of the geodesic connecting them. A circle on a curved surface is the set of all points having equal (geodesic) distance from some central point. A triangle is a set of three pieces of geodesics meeting at their endpoints. Geodesics then provide us with an unambiguous notion to formulate and probe geometrical laws on curved surfaces.

Parallels Parallels are an important and basic notion in Euclidean geometry. Two straight lines are parallel if they point in the same direction. If they are not the same line, they never intersect. Given a straight line and a point not on it, there is a unique straight line through the point and parallel to the first line.

To see how these statements may change if we use geodesics on a curved surface instead of straight lines in Euclidean space, we first look at a cylinder. Compared with the sphere, the cylinder has the advantage that it can be constructed from a flat sheet of paper, just by rolling it up and gluing together two edges. We do not deform, fold over, wrinkle or cut the sheet of paper, and so the proportions of triangles and circles or other geometrical figures do not change. A cylinder should obey most of the Euclidean theorems, except perhaps those that depend on the final gluing.

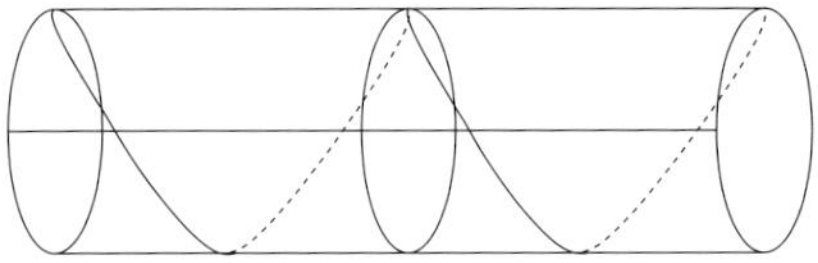

Figure 2.33 Geodesics on a cylinder are axial lines, cross-sectional circles, or helixes.

In particular, we can draw geodesics on a cylinder by drawing straight lines on the sheet of paper before rolling it up. By doing so, we obtain curves of three different types, depending on the direction of the original straight line relative to the direction of rolling up the paper, or the axis of the cylinder (Figure 2.33). A straight line along the axis is a straight line also on the cylinder. A straight line along the direction of rolling up the sheet of paper, perpendicular to the axis, becomes a circle on the cylinder, the same set as the cross-section of the cylinder with a plane perpendicular to the axis. Any other straight line with an angle with the axis not equal to 0° or 90° becomes a helical curve on the cylinder. At any given point on the cylinder, we can turn toward any direction and see how we have to move to follow a geodesic.

As seen in the figure, different geodesics, for instance, a helix and a straight axial line, can intersect more than once. One of the Euclidean properties of parallels is violated, even though we just rolled up the Euclidean plane. The reason for that is the final gluing when we construct the cylinder, making different points on the plane fall on the same point on the cylinder. Points on two parallels on the plane may fall on the same point on the cylinder, as a new intersection of the lines not realized in Euclidean geometry. Had we not glued the edges together, the helixes would end and not come back around the cylinder to intersect a straight line anew.

The second property of parallels is still satisfied for geodesics on the cylinder: Given a geodesic and a point not on it, there is a unique way of drawing a geodesic through that point not intersecting the first geodesic. We draw a geodesic through the point using the same angle with the cylinder axis as the original geodesic. These two geodesics do not intersect each other, and any other angle would produce intersection points.

Extrinsic curvature A surface can curve in different ways. We have constructed a cylinder by rolling up a sheet of paper and gluing together two edges. As a consequence, different geodesics may intersect more than once, but through any point there is still a unique geodesic not intersecting the original one anywhere. Gluing together the edges has introduced new intersection points, violating one of the Euclidean statements about parallels. However, by keeping the sheet of paper otherwise intact, we change the geometry with enough care to respect the Euclidean statement about unique parallels.

The cylinder has a mild form of curvature that does not deviate much from flat Euclidean geometry. By bending the paper without stretching or wrinkling it, we introduce what is called extrinsic curvature, that is, a form of curvature that can only be seen when we look at the surface as it lies in an external three-dimensional space. By moving on the cylinder and measuring properties of triangles or circles, on the other hand, we would not notice the curvature: These geometrical figures are not deformed by rolling up the paper, and they still obey the Euclidean laws. From within, the surface does not appear curved; it does not have what is called intrinsic curvature.

Intrinsic curvature Intrinsic curvature rests in itself; it does not need to go out to assert itself. Unlike extrinsic curvature, the intrinsic version is more difficult to spot. It does not fall in our eyes when we look at the surface lying in space. However, it modifies the internal laws of geometry and can be found by analyzing the behavior of figures on the surface.

Intrinsic curvature comes about when we have to wrinkle our sheet of paper, stretch it or perform advanced origami in order to produce a surface shape we desire. In this process, geometrical relationships for figures drawn on the paper change. Most surfaces do require such modifications and are more complicated to construct than the cylinder. A good example is the sphere. It is indeed difficult to construct a sphere in good shape from a single sheet of paper. Also, as we have already seen, it violates some of the Euclidean properties of circles. In order to study the consequences in more detail, we at last have to face the task of determining what geodesics on a sphere look like.

Great circles There is no straight line on a sphere, but a crow knows how to fly from point A to point B as the crow flies, above Earth's spherical surface. We must use the general notion of geodesics as curves providing the shortest distance between two points. Figure 2.34 shows a top view of Earth, where we assume points A and B to be of equal latitude on the Northern hemisphere, just to be specific. A crow might not want to fly over the North pole, but modern air travel does choose such routes between airports of almost opposite longitude (Figure 2.35). Such a meridian route (1) appears to be close to optimal. In the top view of Earth in Figure 2.34, any other path from A to B indeed seems longer. However, despite appearance, this view does not prove that flying along a meridian across the North pole is the shortest way; after all, the illustration does not show how Earth's surface is bending away from the sheet of paper when we move away from the North pole. Although (2) looks longer than (1), the bending-away of the surface may reduce the actual distance.

Figure 2.36 is a side view of Earth, with the North pole N on top and two pieces of the two routes from A. In contrast to the top view, the side view makes the bending down of Earth's curved surface visible, but it still is not obvious which route is shorter. We may note that the route across the North pole is special among all possible routes from A to B as the most symmetric one. It follows a planar cross-section of Earth through the center, cutting Earth in halves. If we deviate to the East as we move away from A, there is a different route of the same distance to B deviating to the West, obtained by mirror reflection on the plane through the center. The route across the North pole has either maximal or minimal distance because the distance changes in the same manner if we deviate to the East or the

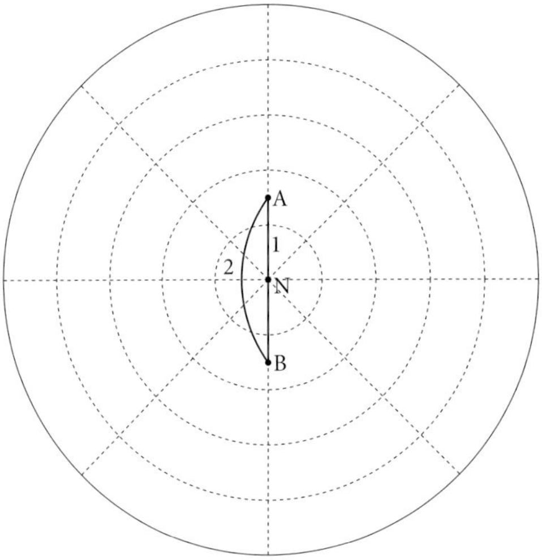

Figure 2.34 A top view of Earth on the North pole N, with two curves connecting points A and B of equal latitude. Dashed lines show some meridians and parallels for orientation.

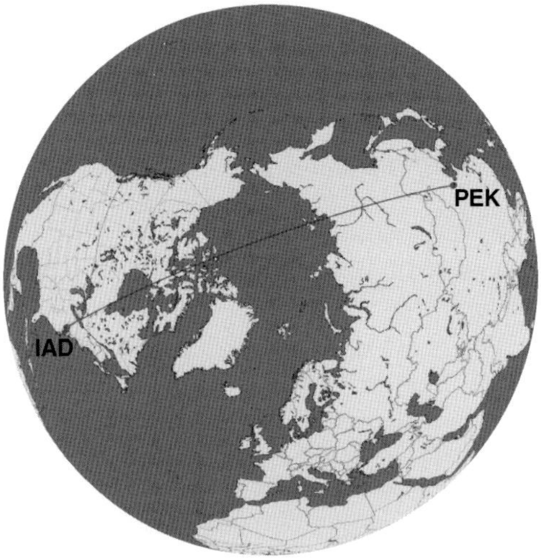

Figure 2.35 Flight route from Dulles Airport to Beijing, almost over the North Pole (www.gcmap.com, accessed 9 April 2012).

West. In order to demonstrate that the distance is minimal, we show that there are routes of longer distance.

We focus attention on paths obtained as cross-sections of the sphere with planes, but unlike the longitudinal route, with planes not through the sphere's center C. Such paths are still circle segments, with a center C′ displaced from C. Since the central cross-section of a sphere has the largest radius, routes deviating from the longitudinal one are circles of radius r smaller than the sphere radius R. If we

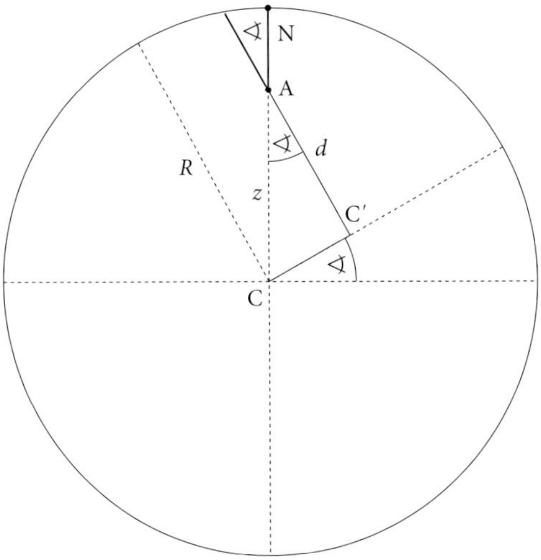

Figure 2.36 A side view of Earth with the North pole N on top. Shown in bold are parts of the two routes from A seen in Figure 2.34. The longitudinal route across the North pole is a segment of a circle around the center C of the sphere. The second route is a segment of a smaller circle with radius r, around a different center C′. The distance $d = z\cos(\sphericalangle) = R\cos(\mathbf{La})\cos(\sphericalangle)$ is the projection of r on a plane through C, C′ and N.

overlay the two planes of cross-sections, we see our routes as two circle segments from A to B of different radius, as in Figure 2.37. Now, it is evident that the route with a center C′ displaced from the sphere's center C has the longer distance: it amounts to a circle segment of smaller radius $r < R$, which bulges out more when fit between points A and B.

The distances can be compared by relating them to the tunnel distance Δ from A to B along a straight line in space. This parameter is an invariant of the problem because it does not depend on the circle route we take on the surface. If we call the actual distance along a given circular route s, it is first related to the radius r of the circle by $r = s\bigcirc/(4\pi\vee)$ if $\vee$ is half of the opening angle of the two rays from the center C′ to A and B. In the right-angled triangles made from the two radii and half the tunnel line connecting A and B, we have $r = \Delta/[2\sin(\vee)]$. These two equations imply $\Delta = s\bigcirc\sin(\vee)/(2\pi\vee)$, taking the same value for all circular routes. The function $\sin(\vee)/\vee$ decreases as $\vee$ increases from zero to $90° = \frac{1}{4}\bigcirc$. (The linear function increases faster than the sine function.) With Δ unchanged, the distance s increases as we deviate from the longitudinal curve, which has the smallest $\vee$.

To compute the lengths of circle segments, we characterize a circle by the angle $\sphericalangle$ it has with the meridian, as in Figure 2.36. The distance from A to B along such a circle is $s = 4\pi\vee r/\bigcirc$ with the angle $\vee$ and the ra-

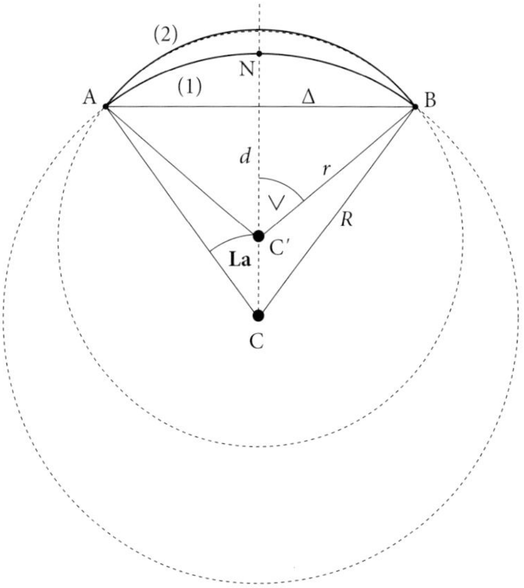

Figure 2.37 Two circle segments shown in their planes cross-secting the sphere. The circle with smaller radius $r < R$ produces a longer route from A to B. The shortest route is obtained with the largest radius, around the sphere's center C, and goes across the North pole N. (For the actual sphere, the two circles do not lie on the same plane; see Figure 2.36.)

dius r as shown in Figure 2.37. We need formulas to express $\vee$ and r in terms of $\sphericalangle$, the latitude **La** of A and B and the sphere's radius R. The length d in Figure 2.36 is $d = R\cos(\mathbf{La})\cos(\sphericalangle)$. From Figure 2.37, we have $d\sin(\vee)/\cos(\vee) = \Delta/2 = R\sin(\mathbf{La})$. Combining these two equations, $\sin(\vee)/\cos(\vee) = \sin(\mathbf{La})/[\cos(\mathbf{La})\cos(\sphericalangle)]$ can be used to determine the angle $\vee$. Using trigonometrical identities, we compute the cosine of $\vee$ from the previous equation by $\cos(\vee) = 1/\sqrt{1+\sin(\vee)^2/\cos(\vee)^2}$. With this value and $r = d/\cos(\vee)$, the distance is $s = (4\pi/\bigcirc) \vee R\sqrt{\sin(\mathbf{La})^2 + \cos(\mathbf{La})^2\cos(\sphericalangle)^2}$. (For $\sphericalangle = 0$, we indeed obtain the values $\vee = \mathbf{La}$ and $s = 4\pi R\mathbf{La}/\bigcirc$ of the meridian.)

Sphere Triangles on a sphere do not respect Euclid. Knowing what geodesics on a sphere look like, obtained as cross-sections of the sphere with planes through the center, we can construct triangles: Three segments of great circles meeting at their end points, or three corners connected by lines of shortest distance between them.

On a plane, the angles always add up to 180°, a basic statement of Euclidean geometry, but our circles have already made us wary of familiar laws when surfaces are curved. Like large circles, large triangles should best show the effects of curvature. Let us therefore consider a continent-spanning triangle, with one corner on the North pole and the other two on the equator, connected to the North pole by two meridians separated by 90°. The meridian and the equator are great circles, or geodesics, and we indeed have a triangle on the sphere. The one angle at the North pole is, by choice, 90°, and so are the two angles between the equator and the meridians. They add up to 270°, much more than the 180° allowed by Euclid.

If the angle sum is no longer required to equal 180°, it could be larger or smaller, depending on the region on the surface. When the sum is larger than 180°, we say that there is positive curvature, and negative curvature in the opposite case. The sphere is round and looks the same everywhere; it has constant positive curvature. If it is deformed, like Earth's real surface, its curvature varies as we move on the surface. It can become negative when there is an indentation that looks like a saddle: a part of the surface which bends down in one direction and up in an orthogonal one.

To determine angle sums of triangles, we need not see how the surface lies in space; we can draw a triangle and determine its angles even if we are required to stay on the surface. The difference of the angle sum to 180° is a measure of intrinsic curvature. The cylinder, which only has extrinsic curvature, always has angle sums of 180°, as its triangles are just rolled-up but undeformed planar ones.

Stereographic projection Everyone who has ever tried to tape a Halloween pattern on the curved side of a pumpkin knows that a flat plane cannot be bent wrinkle-free into spherical shape. Cartographers encounter the same problem, albeit in the opposite direction. Land and sea boundaries on Earth's curved surface cannot be mapped on a flat sheet of paper without deformations. For small regions, of sizes much smaller than the planet's radius, the deformations do not matter much, but entire-world maps make regions closer to the poles look larger than they are. Spherical and planar geometries, as signified by their different curvatures, are just too different for comparisons.

One can map a sphere on the plane, or vice versa, using stereographic projection. However, while this map faithfully assigns every point on the plane to a unique point on the sphere, geometrical sizes and proportions of extended regions are deformed, as they must be because the sphere does not respect the planar angle sum. Near the North pole, the plane is close to the sphere because we chose it to be tangent at the pole. In those regions, the geometries look rather similar. However, as we move away from the North pole, a region on the sphere casts a larger and larger shadow on the plane. At the antipode, a region around the South pole, no matter how small, is mapped to an infinitely large region on the plane: the South pole itself has its image point at infinity. By maps like this, mathematics is able to carry infinity by finite means.

A triangular region on the sphere, with edge segments of great circles, is mapped to the plane with the edge closest to the South pole magnified. The Euclidean proportions on the plane then imply non-Euclidean proportions on the sphere. If we draw a right-angled triangle and erect squares on its sides, the planar squares will obey the Pythagorean theorem while the figures on the sphere do not. The Pythagorean theorem is the underlying law of geometry captured by the Euclidean line element $\Delta s^2 = (\Delta x)^2 + (\Delta y)^2$ on the plane. For the sphere with coordinates **La** and **Lo**, deformations in the map of the plane mean that its line element is not Euclidean. We have used geometry in space to relate the spherical coordinates to the planar ones by $x = 2R\cos(\mathbf{Lo})\sin(\frac{1}{2}\mathbf{La})/\cos(\frac{1}{2}\mathbf{La})$ and $y = 2R\sin(\mathbf{Lo})\sin(\frac{1}{2}\mathbf{La})/\cos(\frac{1}{2}\mathbf{La})$. With a little more work, we can relate the geometrical distances given by line elements.

Line element Changing the latitude **La** of a point A on the sphere by $\Delta\mathbf{La}$ implies some change Δr of the distance $r = \sqrt{x^2 + y^2}$ of the shadow point B on the plane, from the origin at the North pole. The shadow of A is cast along a line from the South pole of some length l, having an angle $\frac{1}{2}\mathbf{La}$ with the vertical. The length s of one side in the small triangle at B in Figure 2.38 is $2\pi/\bigcirc$ times angle change $(\frac{1}{2}\Delta\mathbf{La})$ times distance (l), or $s = \pi l \Delta\mathbf{La}/\bigcirc$. The angle of the small triangle at B is $\frac{1}{2}\mathbf{La}$, just like the angle at the South pole, both angles spread out by pairs of mutually orthogonal lines. The distance of B from the plane's origin changes by $\Delta r = s/\cos(\frac{1}{2}\mathbf{La}) = \pi l \Delta\mathbf{La}/[\cos(\frac{1}{2}\mathbf{La})\bigcirc]$.
We have two expressions $l = \sqrt{(2R)^2 + r^2} = 2R/\cos(\frac{1}{2}\mathbf{La})$ for the side lengths of the right-angled triangle with corners at the two poles and B. Using all equations, we write $(\Delta\mathbf{La})^2 = (\bigcirc s/\pi l)^2 = (2\bigcirc R)^2(\Delta r)^2/\pi^2 l^4 = (\bigcirc/\pi)^2(\Delta r)^2/(1 + x^2 + y^2)^2$, choosing the convenient value $R = 1/2$ in the last step. With similar considerations for changes of **Lo**, the two line elements are related by

$$\left(\frac{2\pi}{\bigcirc}\right)^2 \left[(\Delta\mathbf{La})^2 + \sin(\mathbf{La})^2(\Delta\mathbf{Lo})^2\right] = \frac{4}{(1 + x^2 + y^2)^2}\left[(\Delta x)^2 + (\Delta y)^2\right] . \tag{2.30}$$

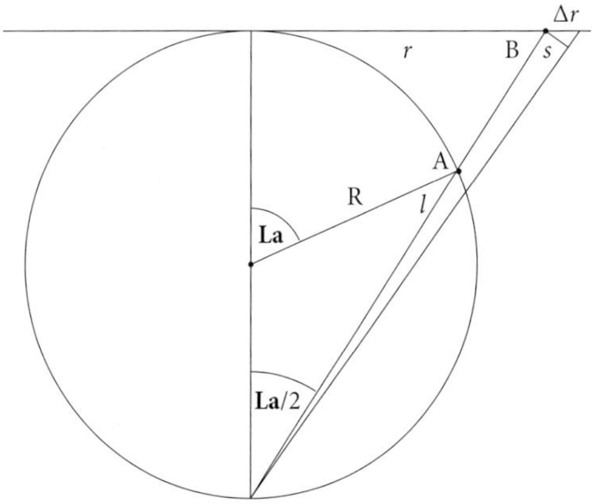

Figure 2.38 Relating the planar and spherical line elements by stereographic projection.

The factor $4/(1 + x^2 + y^2)^2$ in (2.30) shows deformations done by mapping the plane to the sphere. As we move away from the North pole, the distance $r^2 = x^2 +$

y^2 to the pole increases and the deformation factor decreases. We must squeeze regions of the plane into smaller and smaller regions on the sphere to complete the mapping. In an actual world map,[11] countries far away from the North pole, when drawn on a planar map, would appear enlarged compared to their actual size. Regions far from the North pole are squeezed so much that everything that extends to infinity on the plane is mapped to a finite region around the South pole on the sphere. Infinity can be grasped by a compact and handy sphere.

Three-sphere Three-dimensional space, in all its infinity, can be compactified by a procedure analogous to stereographic projection. We would need to draw space in a four-dimensional space in order to visualize the projection, which is difficult. However, with equations, one can follow the construction principles to obtain a map from three-dimensional space to a curved compact object called the three-dimensional sphere, or three-sphere for short. (This space is not the same as a three-dimensional ball, the surface of which is the two-dimensional sphere we considered so far. A three-dimensional sphere would be the surface of a four-dimensional ball.)

Like the three-dimensional sphere, the three-sphere is obtained by cramming all of infinity of three-dimensional space into a single point, the South pole of the three-sphere. The process of mapping space to the three-sphere deforms the geometry, so that the three-sphere has a non-Euclidean line element. One can assign angles to all points on the three-sphere, viewed as a subset of four-dimensional Euclidean space: $x = \sin(\mathbf{Li})\sin(\mathbf{La})\cos(\mathbf{Lo})$, $y = \sin(\mathbf{Li})\sin(\mathbf{La})\sin(\mathbf{Lo})$, $z = \sin(\mathbf{Li})\cos(\mathbf{La})$, $w = \cos(\mathbf{Li})$ with a new coordinate w of four-dimensional space. Now, we have three independent angles,[12] **Li**, **La** and **Lo** for our three-dimensional sphere. It follows that $x^2 + y^2 + z^2 + w^2 = 1$ on the sphere, so that all its points indeed have equal distance from the origin of four-dimensional space.

Equivalence principle Space-time is four-dimensional and may be curved. We must use a general notion of geometry, line elements, and curvature suitable for four-dimensional space-time, following the algebraic representations we have seen for the examples of the two-sphere and the three-sphere. It is difficult to visualize a three-sphere, as the surface of a four-dimensional ball, but we can compute its geometrical properties by using a suitable line element.

As a common consequence of curvature, orientation angles change as we move because space is bending. Following the analogy further, moving in curved space-time means that a space-time angle, that is, rapidity or the velocity, changes, and curvature implies acceleration or the action of a force. The complication of considering curved four-dimensional space-times has the benefit of giving us for free a mathematical description of a force about which we do not have to assume much once we know what curvature space-time should have. Curvature is determined by the line element and geometry, and its distance measures will have physical implications.

11) Common world maps choose points different from the North pole as centers of the projection.

12) Unlike **La** and **Lo** which signify latitude and longitude, **Li** is motivated by phonetics.

We do not need separate force laws once space-time geometry is known. However, how do we find out what curved geometry space-time should have, deviating from the flat and forceless Minkowski space-time used in special relativity? Just as the principle of constant, observer-independent speed of light allowed us to derive the Minkowski line element by physical considerations of the processes happening in a light clock, we now need a new principle that tells us how observers subject to different forces experience the progress of time. Noteworthy in this context is the gravitational force because it acts on the cosmic scales on which we expect curvature to be relevant.

The gravitational force depends on the mass M, but this parameter plays a dual role in physics. Mass m also appears in Newton's second law of motion where it multiplies the object's acceleration a for the product to equal the acting force F. In $ma = F$, the form and nature of the force does not matter; the equation applies equally for the gravitational force and the electromagnetic, or any other force. We call both parameters, m appearing in Newton's second law and M in Newton's formula for the gravitational force the mass of the object, because all experiments have so far resulted in the same value for both of them. There does not seem to be a difference in the values of the inertial mass m and the gravitational mass M, a fact that was established by experiments done in the nineteenth century and has been made more precise ever since.

With universal equality of inertial and gravitational mass, it is possible for acceleration and gravity to cancel each other. We are familiar with pictures of the space-station, a freely falling enclosed capsule in which the gravitational force $F = Mg$ of Earth implies an acceleration $a = F/m = gM/m$ of all objects. Any object is accelerated by the same amount as the whole station, and so their distances do not change; all objects seem at rest. If the inertial and gravitational masses were independent, there would be no reason for M/m not to depend on material properties, and objects and the station would be subject to different accelerations. However, even in the light of clear observations, the equality of inertial and gravitational mass remains enigmatic in Newtonian physics. After all, these are just two parameters in two different equations which, from a mathematical perspective, could take different values.

Instead of trying to explain the relation between both mass parameters, it is simpler, though daring, to raise their equivalence to the status of a principle. We declare that the inertial mass must equal the gravitational one under all circumstances, $m = M$, and call this statement the equivalence principle. In the following pages, we will derive far-reaching consequences. We note that already at this stage, there is a clear conceptual advantage: The principle allows us to use a notion of gravitationally active mass, independent of gravitational attraction. The inertial mass m is determined and can be measured by analyzing acceleration under an arbitrary force; interpreted as the gravitational mass $M = m$, the same value then enters the law of gravity. Our theory has become much more predictive. "The tendency of all bodies to approach one another with a strength proportion to the quantity of matter they contain, the quantity of matter they contain being ascertained by the strength of their tendency to approach one another. This is a lovely and edifying illustration of how science, having made A the proof of B, makes B the proof of A," reads the entry for "Gravitation" in Ambrose Bierce's *The Devil's dictionary*. Without the identification $m = M$, we would have no independent means to measure the gravitational mass; the role of mass in the law of gravity would be close to a tautology.

Since the inertial mass is related to accelerated motion, the equivalence principle provides us with additional information beyond special relativity, whose space-time geometry applies to observers in uniform motion. If inertial mass always equals gravitational mass, effects seen by an observer accelerated by some constant $-a$, for whom an object of mass m appears subject to a constant force $F = ma$, are indistinguishable from effects seen by an observer at rest, but placed in a region of constant gravitational acceleration $g = a$. Exploiting this identification, we can use the equivalence principle to relate distance measurements of accelerated observers.

Light clock Light lights the way not only toward Lorentz transformations in special relativity, with observers at constant speed, it also shows consequences of accelerated motion on measurements. For uniform motion, we used the constant speed of light as the main principle to derive geometrical implications, analyzing properties of a light clock. Acceleration is made accessible by the equivalence principle.

If a light clock is accelerated in the direction of its light rays, its rate does not change if its motion is uniform. Time dilation $\Delta T = \sqrt{1-v^2/c^2}\Delta t$ at speed v is compensated for by Lorentz contraction $L_* = L/\sqrt{1-v^2/c^2}$ of the mirror gap. With acceleration, the cancellation is no longer complete. In Figure 2.39, we consider a light clock in a rocket. At point A at the top of the rocket, light signals are emitted at regular intervals, the clock rate ΔT_A with respect to A's proper time. The

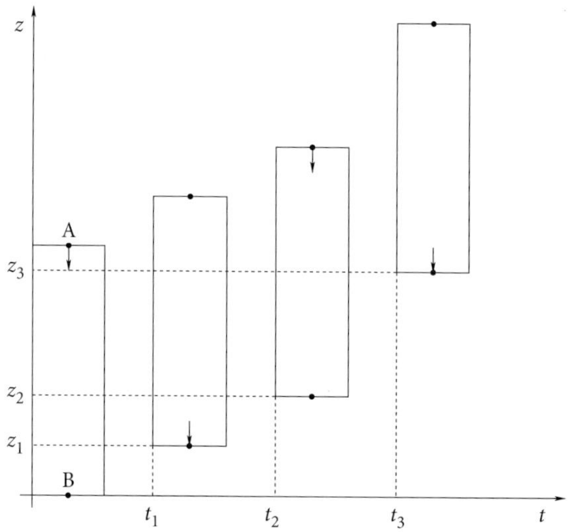

Figure 2.39 An accelerated light clock. Observers at different places along the direction of acceleration measure different time intervals between successive emissions and receptions, respectively, of light.

signals reach a detector at point B at the bottom. If the rocket stands still or moves up at constant speed, A's signals arrive at B with a constant delay, and B measures the same clock rate $\Delta T_B = \Delta T_A$. However, if the rocket accelerates upwards, its speed is different during successive emissions at A. Point B moves faster and faster toward A and the emitted light, and the travel delays of clock signals from A to B get shorter as time goes on. The rocket's motion shortens the distance for later light signals to travel, decreasing their time separation such that $\Delta T_B < \Delta T_A$. Time at different places in an accelerated rocket proceeds at different rates; by the equivalence principle, time at different altitudes proceeds at different rates.

We compute consecutive emission and reception times as shown in Figure 2.39. At time $t = 0$, the rocket is launched with constant acceleration g. At this time, its bottom still touches the ground at $z = 0$, but from then on, it moves upwards according to $z_B(t) = \frac{1}{2}gt^2$, specifying the vertical position of point B at the bottom. If h is the height of the rocket, the position of A at the top is $z_A(t) = h + \frac{1}{2}gt^2$. As the rocket is launched, the first clock signal is emitted at A, and it will be received at B some time t_1 later. The second signal is emitted at time $t_2 = \Delta T_A$, and received at time $t_3 = t_1 + \Delta T_B$. By computing the travel times of light in the accelerated rocket, we obtain expressions relating ΔT_B to ΔT_A.

In addition to the expected acceleration effect on the clock rate, the increasing velocity means that time dilation at points A and B plays an ever larger role as time goes on. Since we already know about time dilation, we will assume that the velocity of the rocket remains small during the first few emissions of light, so that we can focus on additional consequences of acceleration. The clock rate must be small for the speed $g\Delta T_A$ of the rocket after the second emission to remain small compared to the speed of light. This condition allows several approximations in our equations.

The first signal, emitted at $t = 0$, travels a distance $z_A(0) - z_B(t_1)$ before it is received at B, moved up a little bit while light was traveling down. For constant g, $z_A(0) - z_B(t_1) = h - \frac{1}{2}gt_1^2$. This distance equals ct_1, traversed by light in the same time: $h - \frac{1}{2}gt_1^2 = ct_1$. The second pulse, emitted at time ΔT_A, travels a distance $z_A(\Delta T_A) - z_B(t_1 + \Delta T_B)$, also equal to the distance $c(t_1 + \Delta T_B - \Delta T_A)$ traveled by light. Subtracting the two equations, we have $c(\Delta T_A - \Delta T_B) = gt_1\Delta T_B - \frac{1}{2}g(\Delta T_A)^2 + \frac{1}{2}g(\Delta T_B)^2$. We assume small rocket velocities, so that $t_1 \sim h/c$ (we ignore the distance traveled by the rocket for the motion of light), and small clock rates, so that $g(\Delta T_{A/B})^2$ can be ignored next to $c\Delta T_{A/B}$. Our equation simplifies to

$$\Delta T_B = \frac{\Delta T_A}{1 + gh/c^2} . \qquad (2.31)$$

We have different clock rates at different places in the rocket. The equivalence principle then implies that clocks at different altitudes have different rates: There is gravitational time dilation just as there is time dilation caused by motion.

Gravitational redshift Periodic processes at different altitudes happen with different rates. An observer measures frequencies different from the ones at an emission point of different altitude. Light is a periodic process, and its frequency determines the color. The color seen at some altitude may therefore differ from the emitted color. The conclusion applies not only to measurements at different elevations on Earth, but also to measurements of light emitted from a star surface. The gravitational force on the surface is different, much stronger than the force on Earth, and frequency shifts are expected, toward the red for stronger gravity at emission.

To compare the gravitational force on Earth with the force on the surface of a star, we introduce the gravitational potential P. Its size tells us how much the gravitational force has increased between two points. Along the vertical axis with coordinate z, a potential $P(z)$ implies a force equal to an object's mass times the negative change of P, $F = -m\Delta P/\Delta z$. (The negative sign is introduced by convention.) A constant force $F = -mg$ pointing downward, corresponding to constant upward acceleration as in the example of our rocket clock, has a linear potential: $P(z) = gz$. For the rate by which signals are emitted, or the inverse clock rate, we therefore write $1/\Delta T_\mathrm{B} = [1 + (P_\mathrm{A} - P_\mathrm{B})/c^2]/\Delta T_\mathrm{A}$. This formula can be used even if the gravitational force is not constant, or if the gravitational potential is not linear. However, we still have to respect our condition of small acceleration, or small gravitational potential.

Signals are received with larger frequency when the receiver is at lower altitude than the emitter. A clock placed higher seems to go faster when its rate is observed from below. Modern atomic clocks can detect this effect when they are elevated by just 10 m, placed at different floors of a building. Gravitational time dilation is significant when one compares and synchronizes time standards in different regions. In the US, the National Institute of Standards and Technology (NIST) is located in Boulder, Colorado, at much higher altitude than its cousins in Europe. Gravitational effects must be factored in when standards are compared. Also, the Global Positioning System (GPS) takes pains to include clock delays of its satellites at different heights over Earth's surface. If one ignored gravitational time dilation, the meter-size accuracy of position measurements would be lost in just a minute of negligent operation.

In astrophysics, the clock rate of an actual time piece is replaced by the frequency f_* of light emitted by a star, seen from Earth. Compared with the gravitational potential P_* on the star, the value of P on Earth is tiny and can be ignored in (2.31). The observed frequency is $f = (1 + P_*/c^2) f_*$. The gravitational force does not change in a constant fashion as we move away from the star. Newton's formula for the gravitational force shows that we can do calculations with the gravitational potential $P_* = -GM/R$ on the star's surface of radius R, with M being the mass of the star. We still need to assume that the potential is small, so the star cannot be too dense with small R and large M. If P is small, the observed frequency is

$$f = \left(1 - \frac{GM}{Rc^2}\right) f_* \,. \tag{2.32}$$

The frequency f seen on Earth is smaller than f_*, and thus redshifted. For many stars, including the Sun, P/c^2 is indeed small, although still larger than the potential on Earth. There are more-compact objects, called white dwarfs, with masses about as big as the Sun's, but radii of just about 1000 km. The relative frequency shift with these values is $(f - f_*)/f_* = -GM/Rc^2 \approx -0.001$, still a small number. Another class of stellar objects is even more compact: neutron stars with similar masses as white dwarfs but radii of about 10 km or less. The relative frequency shift is not that small anymore, nor is the gravitational potential. For such compact stars, we must find a more advanced derivation of gravitational time dilation and redshift that also applies to gravitational potentials or acceleration not much smaller than one.

Curved space-time The ticks of time depend on the place in a changing gravitational potential. One may think that it is the gravitational force which influences the mechanism of a clock and thereby its rate; the pendulum of a grandfather clock is an example for a time counter driven by gravity. But the gravitational potential P appears in the redshift formula (2.32), not the force obtained as a derivative of P. Two clocks placed in regions in which P is nearly constant, but taking a different value in each, experience no gravitational force, but their rates differ. Moreover, our derivation of (2.32) referred just to light as the only standard we can trust, based on the principles of constant celerity combined with the equivalence of inertial and gravitational mass. Any effect of gravity on clock rates by the motion of light must be considered as an unavoidable effect on the progress of time itself; it is independent of what clock is used.

We conclude that the gravitational force, implying different values of the gravitational potential at different places, changes the clock rate. The clock rate, or its inverse number, the clock period, is a distance measure in space-time geometry. The gravitational force changes the geometry at different regions of space-time, depending on what masses and gravitational potential we encounter. A changing, position-dependent geometry means that space-time is curved, described by a position-dependent coefficient of $(c\Delta t)^2$ in the line element. We should expect space-time to be curved, not flat, with an amount of curvature related to the gravitational force. The equivalence principle has provided us with means to compute how gravity curves space-time.

So far, we have assumed a small gravitational potential, not allowed to change quickly. Under these conditions, our formula (2.32) for gravitational redshift or time dilation shows how the invariant line element Δs^2 must look, whose timelike values are the proper-time clock rates $\Delta T_{A/B}$, with $(c\Delta T_{A/B})^2 = -\Delta s^2$.

Line element The line element expresses proper time in terms of some time coordinate t. With gravitational time dilation, we have $(c\Delta T)^2 = (1+P/c^2)^2(c\Delta t)^2$: This identification ensures that

$$\Delta T_B = \left(1 + \frac{P_B}{c^2}\right)\Delta t = \Delta T_A \frac{1 + P_B/c^2}{1 + P_A/c^2} \approx \Delta T_A \left(1 + \frac{P_B - P_A}{c^2}\right),$$

for small P/c^2, agrees with (2.32). Continuing with small P/c^2, we write $(1 + P/c^2)^2 = 1 + 2P/c^2 + P^2/c^4 \approx 1 + 2P/c^2$ because P^2/c^4 is much smaller than P/c^2 if the latter is small. The time part of the line element is therefore

$$\Delta s^2 = -(c\Delta T)^2 = -\left(1 + \frac{2P}{c^2}\right)(c\Delta t)^2 .$$

We have not determined any gravitational effects on distance measurements, and so we do not know how the coefficients in front of spatial coordinate differences in the line element react to gravity. These coefficients are not needed for our present purposes, assuming small potentials and small speeds. For any such process, Δx is well below $c\Delta t$: $\Delta x/\Delta t = v$ is and stays much smaller than c. All we can say is that there is some coefficient multiplying spatial separations in Δs^2 which for a small potential has the form $1 + KP/c^2$ with a constant K. We arrive at the line element

$$\begin{aligned}\Delta s^2 = &-\left[1 + \frac{2P(x,y,z)}{c^2}\right](c\Delta t)^2 \\ &+\left[1 + \frac{KP(x,y,z)}{c^2}\right]\left[(\Delta x)^2 + (\Delta y)^2 + (\Delta z)^2\right]\end{aligned} \qquad (2.33)$$

for curved space-time with a small gravitational potential P. In the spatial part, we have assumed that there is no direction dependence of the modification, except for the dependence of $P(x, y, z)$ on the coordinates. The constant K has not been determined by our calculations, but it turns out to equal $K = -2$ according to general relativity.

Space-time metric Newton's old force is hardly recognizable in relativity. By (2.33), we have implemented the gravitational potential and the force it describes in geometrical terms, as a component of the space-time line element. The question of how Newton's force can be described in a relativistic manner is answered, with a surprising result not at all like what we saw for the electric Coulomb force. The Coulomb force, or the Lorentz force if a magnetic field is present, is described by a four-vector, and the electric and magnetic fields form a tensor, the electromagnetic field-strength tensor. The gravitational potential, on the other hand, is a component of the line element. It does not define a new object or its own field-strength tensor; it rather changes the geometry used for distance measurements. Gravity is not independent of geometry. It is a consequence of curvature in the geometry of space-time.

When we emphasize the force aspect of curvature instead of the geometrical one, we use the space-time metric instead of the line element. The metric $g_{ab}(t, x, y, z)$ is not a new object; it just collects the coefficients of the line element $\Delta s^2 = \mathbf{Sum}_{a,b=0}^{3} g_{ab}(\Delta x^a)(\Delta x^b)$ in a matrix or tensor. With curved space-time, we can no longer use Lorentz transformations to map different coordinate systems into each other. In particular, the metric tensor is invariant under Lorentz transformations only when it corresponds to Minkowski space-time. In all other cases, the gravitational force and the metric tensor describing it depend on the observer. What is invariant and observer-independent is, as always, the line element Δs^2 and its geometry. The components g_{ab} and the coordinate separations Δx^a are observer-dependent.

Geodesics Space-time curvature implies position-dependent clock rates. Time lapses depend on where we are moving. Travel times could be shortened or extended by taking a spatial detour, just as distances measured along paths on a curved surface can be shortened by descending into curvature troughs, following valleys.

In space-time, a massive object moves in a timelike fashion, and the relevant distance measure is its invariant proper time, computed from the line element as $\Delta T = \sqrt{-\Delta s^2/c^2}$ along the worldline, with Δt and Δx in Δs^2 related to each other by the object's velocity. In Minkowski space, the straight line between two events maximizes the change of proper time among all curves. On a curved surface or in curved space-time, the notion of a straight line must be replaced by the more general concept of a geodesic, a curve maximizing the lapse of proper time between two events.

Proper time The complete proper-time lapse is the sum of all contributions $\Delta T = \sqrt{-\Delta s^2/c^2}$ along the curve. With the weakly curved line element (2.33), $\Delta T = \sqrt{(1+2P/c^2)(\Delta t)^2 - (1+KP/c^2)[(\Delta x)^2 + (\Delta y)^2 + (\Delta z)^2]/c^2}$. We factor out Δt and notice that $\Delta\vec{x}/\Delta t$ is the object's velocity $\vec{V}$. In the second term, we have the ratio $|\vec{V}|^2/c^2$ which we assume to be small, and multiplication with the term KP/c^2 makes it smaller still. We ignore this product, and as promised, the value of K is irrelevant with the current approximations. The proper-time lapse simplifies to

$$\Delta T = \left(1 - \frac{1}{2}|\vec{V}|^2/c^2 + P/c^2\right)\Delta t \tag{2.34}$$

using one of our previous approximations of powers of numbers close to one. In this equation, still valid with our assumptions of small $|\vec{V}|^2/c^2$ and P/c^2, we have combined motion-related and gravitational time dilation.

Longest time Steady people age faster, unless they are forced to move. It pays to bend to pressure instead of always insisting on one's own pace. All objects in the universe do this when they follow the gravitational force acting on them, moving along geodesics in space-time to maximize their proper time.

To show this, we follow the same maximization procedure used in Minkowski space-time, except that we have to work harder to see what curve should provide maximization. On Minkowski space-time, the straight line was a good and right guess. Now, the presence of the gravitational potential means that we can make

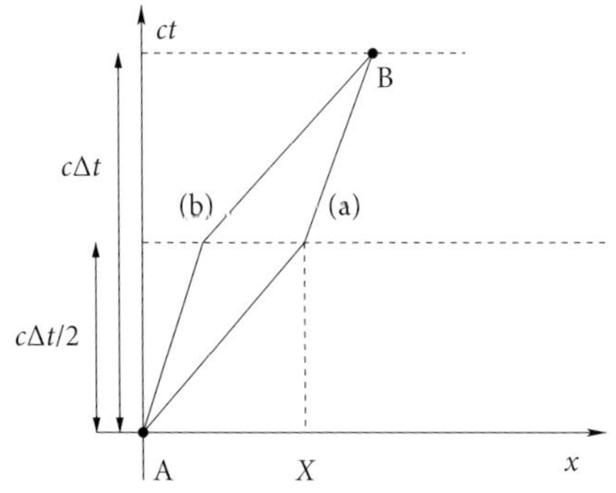

Figure 2.40 Coordinate separations along partially straight curves connecting two events A and B, the pace changing at half-time.

proper time larger by entering and dwelling in regions of good potential. The gravitational force implies acceleration, and by choosing our path wisely we are able to gain more proper time than what flat space-time would grant.

We start as in Figure 2.17, but no longer consider the straight line between events A and B as special. In fact, we do expect deviations from the straight line in curved space-time because we know that a force is acting, causing acceleration of a massive object moving from A to B along a geodesic. We therefore consider nonstraight curves. To be specific, we choose a set of curves which start at A with different velocities, then turn sharply at half-time so as to reach event B. We put A in the origin of our coordinate system, Figure 2.40, and denote the position of the turn by X, at time $\Delta t/2$.

For small velocities and gravitational potential, approximating the square root, the proper-time lapse along such a curve bending at $x = X$ is

$$\Delta T(X) = \frac{1}{2}\Delta t\left[1 - \frac{1}{2}\frac{V_1^2}{c^2} + \frac{P(0)}{c^2}\right] + \frac{1}{2}\Delta t\left[1 - \frac{1}{2}\frac{V_2^2}{c^2} + \frac{P(X)}{c^2}\right]$$

with the two velocities $V_1 = 2X/\Delta t$ and $V_2 = 2(\Delta x - X)/\Delta t$ along the two pieces of the curve.

We see why a nonstraight curve is beneficial for maximizing ΔT: An object that spends too much time in regions of small potential P is penalized by missing out on the ΔT-extending terms P/c^2, which enter ΔT with a positive sign. If P increases toward larger x, for instance, an object should fly faster through the first part between $x = 0$ and $x = X$, to have more time to stay in the region of larger P between $x = X$ and $x = \Delta x$. The velocity V_1 should be larger than V_2, and X should shift toward the position of event B at Δx. If the velocity V_1 becomes too large, on the other hand, the term $-\frac{1}{2}V_1^2/c^2$ now punishes the object by reducing its proper time. We must find the right balance between increasing the positive contribution $P(X)$ and keeping the negative contribution $-\frac{1}{2}V_1^2/c^2$ in check.

We interpret these findings as follows: an increasing P slows down the object, with V_2 smaller than V_1. There is a force that reduces the velocity, one of negative magnitude acting against the object's motion to larger x. A negative force F is indeed expected for an increasing potential, the two being related by $F/m = -\Delta P/\Delta x$. With this encouraging consistency, we turn to more detailed calculations to find a quantitative relationship between the force and the object's change of velocity, or its acceleration.

We combine several terms in the proper-time difference $\Delta T(X)$ and write

$$\Delta T(X) = \Delta t\left[1 - \frac{X^2 + (\Delta x - X)^2}{(c\Delta t)^2} + \frac{1}{2}\frac{P(0) + P(X)}{c^2}\right]$$

with the two velocities V_1 and V_2 inserted in terms of X. We have to find a value of X such that $\Delta T(X)$ is maximal, a maximization problem for which we can apply Fermat's procedure: We take two nearby values X and $X + h$ with h small, and look for values of X such that $\Delta_h T(x) = \Delta T(X + h) - \Delta T(X)$ is almost zero:

$$\Delta_h T(X) = \frac{X^2 - (X+h)^2 + (\Delta x - X)^2 - (\Delta x - X - h)^2}{c^2\Delta t} + \frac{1}{2c^2}\Delta t[P(X+h) - P(X)]$$

should vanish not only when $h = 0$, but should remain zero even after dividing by the tiny h. Taking the squares and canceling terms, we write

$$\frac{\Delta_h T(X)}{h} \approx \frac{2\Delta x - 4X}{c^2\Delta t} + \frac{1}{2c^2}\Delta t\frac{P(X+h) - P(X)}{h}\,.$$

In the first term, we rearrange as

$$\frac{2\Delta x - 4X}{(\Delta t)^2} = -\frac{2X/\Delta t - 2(\Delta x - X)/\Delta t}{\Delta t} = \frac{V_2 - V_1}{\Delta t} = \frac{1}{2}A$$

with the acceleration $A = \Delta V/(\Delta t/2)$. The second term is $\frac{1}{2}[P(X+h) - P(X)]/h = \frac{1}{2}\Delta P/\Delta x$, or the derivative of P. Proper time is maximized for $\Delta_h T(X)/h = 0$ or $A = -\Delta P/\Delta x$. The acceleration is the negative derivative of the gravitational potential, the correct result for the gravitational force divided by the mass.

Gravitational force Gravity is a gentle force. By making the time component of the metric position-dependent, multiplying $(c\Delta t)^2$ in the line element, we have forced curvature upon space-time. Geodesics of longest proper time are no longer straight or linear in terms of Cartesian coordinates; they are curved like the space-time they traverse, obeying the gravitational law of motion according to Newton when the potential is not large. There is no need anymore for a force field or other entities that would sit patiently at every point in space until a massive object comes by, to tell it where to turn. It is space-time itself which, by its own curvature, gently nudges a massive object to follow a curve as if a force were acting. We have geometrized the gravitational force.

There is no action at a distance when two massive objects attract each other. If two objects are far apart and hardly influence each other, they both curve space-time around them, but not much at the other object. As they move closer, they drag along the curvature they induce until the curved regions overlap. The objects begin to feel the gravitational force exerted by the other one. If we disregard curvature, there seems to be a force acting at a distance. With curvature following the curving object, the effect does not occur instantly over large distances.

Space-time curvature resolves the conceptual uneasiness related to action at a distance, without a medium transmitting the force. There are other advantages of using the rather unfamiliar notion of curvature instead of a force, a vector pointing with different strengths and in different directions at all points in space. In our equation for geodesics, space-time curvature, in the form of the rate of change of the small gravitational potential, determines the acceleration of an object of some mass m. The mass itself does not appear in our equations at all because gravitational acceleration, in contrast to the force, is independent of it. In Newton's law, on the other hand, we have an equation for the gravitational force equaling mA. The mass parameter which appears in the force drops out of the equation, provided that the inertial mass in mA equals the gravitational mass in the force. Curvature, requiring changing space-time angles or velocities as one moves, is more closely related to acceleration than to force. There are no mass parameters in the geodesic acceleration equation it implies, explaining why Newton's force law must lead to cancellation of inertial and gravitational mass.

We have to keep in mind our assumptions in the past derivations. We assume small gravitational potentials and small velocities. The space-time geometries and curvatures we consider are then nearly flat, and objects cannot move too fast for our equations to apply. In such situations, we recover Newton's gravitational law. However, if we try to go beyond these simplifying assumptions, the equations change and Newton's law ceases to be exact; it no longer agrees with observations when applied to strong gravitational fields. A new gravitational law arises, the one of general relativity. It cannot be described by a single function like the gravitational potential; it is described by the whole metric with several independent functions g_{ab}. The gravitational force is more complicated and multifaceted than the Newtonian case indicates.

Space-time metric Gravity curves space-time, deforming metrics. We are no longer dealing with the Minkowskian line element, but rather have invariant distances expressed, in quite general terms, as $\Delta s^2 = \mathbf{Sum}_{a,b} g_{ab} \Delta x^a \Delta x^b$. Coordinate separations Δx^a do not equal actual distances; correction factors g_{ab} are needed. With position-dependent g_{ab}, the same coordinate separation at a different point means different distances, as experienced when we move around a curved surface.

Metric functions $g_{ab}(t, x, y, z)$ are not always easy to deal with. They may look complicated just because of an unfortunate choice of coordinates. The rotating coordinate system used earlier, for instance, made Minkowski space look difficult, with position-dependent metric components. By changing coordinates, we do not modify the geometry or curvature, even though those changes and curvature can both imply position-dependence. Coordinates can mask the appearance of curvature, and we have to work harder to find unambiguous measures, namely, combinations of the metric components that are, unlike the components themselves, independent of the choice of coordinates.

Several observer-independent notions, not necessarily related to curvature, can be defined by making use of the invariant line element. Two events in curved space-time, for instance, can be related to each other in characteristic, observer-independent ways. As in Minkowski space-time, we say that two events are timelike separated if $\Delta s^2 < 0$ between them, lightlike separated for $\Delta s^2 = 0$, and spacelike separated for $\Delta s^2 > 0$. All observers agree on these classifications because Δs^2 is invariant. A picture of causal structure results in terms of light cones, separating the present from the past and future. The light cones are no longer spanned by straight lines, with $\Delta s^2 = 0$ along them implying distortions if metric components in Δs^2 change along the curve. Even light must obey the laws of curvature. It is deflected by masses, a new phenomenon to be explored in more detail later. (Newtonian mechanics can mimic this effect by the energy associated with light, but it misses the correct quantitative factors.)

Geometry The metric informs geometry. If metric components depend on the position, we must adapt our geometrical laws as we move around in space-time. The line element contains information about short distances and angles, used to compute the sizes of geometrical objects such as the areas of surfaces or the volumes of regions. For simplicity, we assume that the metric is diagonal and does not mix space and time or the different spatial dimensions: it is of the form $\Delta s^2 = g_{tt}(c\Delta t)^2 + g_{xx}(\Delta x)^2 + g_{yy}(\Delta y)^2 + g_{zz}(\Delta z)^2$ with $g_{tt} < 0$ and the other components being positive.

If we consider a small piece of a curve obtained by varying just the coordinate x by an amount Δx, its length, in terms of Δs, is $\Delta L_x = \Delta s = \sqrt{g_{xx}}\Delta x$. Similar formulas apply for varying y and z. If we vary t, we have a timelike curve along which some proper-time difference $\Delta L_t = c\Delta T = \sqrt{-\Delta s^2} = \sqrt{-g_{tt}}c\Delta t$ passes.

If we change two spatial coordinates, say x in a range of Δx and y in a range of Δy, the set of all points obtained forms an approximate square with its four edges being the x- and y-segments. At the corners, we have right angles because there are no cross-terms in the line element mixing the dimensions; while the coordinate lines are curved, they remain independent and orthogonal to one another. The square's area is the product of the side lengths, $\Delta A = \Delta L_x \Delta L_y = \sqrt{g_{xx}g_{yy}}\Delta x \Delta y$. Varying all three spatial directions gives us a cube with volume

$\Delta V = \Delta L_x \Delta L_y \Delta L_z = \sqrt{g_{xx} g_{yy} g_{zz}} \Delta x \Delta y \Delta z$. If we vary all four space-time coordinates, we obtain a four-dimensional "hypercube" with four-volume $\Delta v = \Delta L_t \Delta L_x \Delta L_y \Delta L_z = \sqrt{-g_{tt} g_{xx} g_{yy} g_{zz}} c \Delta t \Delta x \Delta y \Delta z$.

Sphere If the coordinate changes are not small, or if the regions we are interested in are not squares or cubes, we must subdivide everything into smaller pieces that do have small coordinate separations, and then sum up all contributions, or integrate. For instance, we compute the sizes of different spheres in Minkowski space by using the line element in polar coordinates: the radial distance *r* from the origin and two angles **La** and **Lo** as used on a single sphere. We obtain a two-dimensional sphere by setting $r = R$ constant (as well as t constant). The line element in these coordinates, while position-dependent, is not complicated: $\Delta s^2 = -(c\Delta t)^2 + (\Delta r)^2 + r^2 (2\pi/\bigcirc)^2 [(\Delta \mathbf{La})^2 + \sin(\mathbf{La})^2 (\Delta \mathbf{Lo})^2]$; see (2.30). The area of a small patch on the sphere obtained by varying **La** and **Lo** is $\Delta A = \sqrt{g_{\mathbf{LaLa}} g_{\mathbf{LoLo}}} \Delta \mathbf{La} \Delta \mathbf{Lo} = R^2 (2\pi/\bigcirc)^2 \sin(\mathbf{La}) \Delta \mathbf{La} \Delta \mathbf{Lo}$. If we sum up all these pieces to cover the whole sphere, with **La** ranging from zero to $\frac{1}{2}\bigcirc$ and **Lo** from zero to $\bigcirc$, we obtain $A = 4\pi R^2$. (The area under the curve $\sin(\mathbf{La})$ between $\mathbf{La} = 0$ and $\mathbf{La} = \pi$ is $-\cos(\frac{1}{2}\bigcirc) + \cos(0) = 2$.) The volume of the ball enclosed by the two-sphere is obtained by summing the sizes $\Delta V = \sqrt{g_{rr} g_{\mathbf{LaLa}} g_{\mathbf{LoLo}}} \Delta r \Delta \mathbf{La} \Delta \mathbf{Lo} = r^2 (2\pi/\bigcirc)^2 \sin(\mathbf{La}) \Delta r \Delta \mathbf{La} \Delta \mathbf{Lo}$ of small cubes, with *r* ranging from zero to *R*: $V = 4\pi R^3/3$.

By computing the sizes of different regions and comparing various measures, such as the radius of a circle with its circumference, we obtain information about the curvature of the geometry used. On a sphere like Earth's surface, for instance, we cover more general situations compared to what we found earlier for specific curves, such as the equator as a circle around the North pole. Distance measures provided by the geometry can be used to find observer-independent evidence for curvature. These notions, based on the invariant line element, are insensitive to coordinate changes.

Sphere in a sphere Like surfaces in space, space itself can be curved. We repeat the previous exercise for a sphere in curved space-time. If we modify the space-time line element in polar spatial coordinates to

$$\Delta s^2 = -(c\Delta t)^2 + \frac{(\Delta r)^2}{1 - r^2/r_{\max}^2} + \left(\frac{2\pi r}{\bigcirc}\right)^2 [(\Delta \mathbf{La})^2 + \sin(\mathbf{La})^2 (\Delta \mathbf{Lo})^2],$$

with some constant $r_{\max}$ and with *r* restricted to $0 \le r < r_{\max}$, the new *r*-dependent factor changes radial distances compared to Δr. This spatial geometry happens to be the one of a three-sphere with constant positive curvature, the analog of a sphere carried over to three-dimensions. Space in our universe, at any given time, may be close to a large version of the three-sphere, with galaxies placed on it like continents on the Earth's two-sphere surface.

A sphere in the three-sphere is obtained by setting t constant and $r = R$ constant as well. It has an area of $4\pi R^2$ because the new factor in the line element does not matter when r is not allowed to change. In addition to the whole sphere, we can consider smaller regions or curves on it, and thereby probe the geometry of the sphere itself as well as the geometry of the surrounding space of a three-sphere. The equator of the sphere has $\mathbf{La} = \frac{1}{4}\bigcirc$ constant, so that only **Lo** remains free along this curve. The equator is a circle with circumference the sum of all $\Delta L_{\mathbf{Lo}} = \sqrt{g_{\mathbf{LoLo}}}\Delta\mathbf{Lo} = 2\pi R\Delta\mathbf{Lo}/\bigcirc$. (At $\mathbf{La} = \frac{1}{4}\bigcirc$, $\sin(\mathbf{La}) = 1$.) With **Lo** ranging between zero and $\bigcirc$, the circumference is $C = 2\pi R$ with the expected relation to R. However, R is not the radius of the equator on the two-sphere, which is rather given by the distance D from the North pole to the equator on the sphere. This number is another sum, of contributions $\Delta L_{\mathbf{La}} = 2\pi R\Delta\mathbf{La}/\bigcirc$ for **La** ranging from zero (the pole) to $\frac{1}{4}\bigcirc$ (the equator). We have a distance $D = \frac{1}{2}\pi R$, indeed a quarter of the equator's circumference as found earlier. The ratio $C/D = 4$ is not equal to the Euclidean 2π: the two-sphere has curvature.

So far, we have rederived known results. Now, considering a sphere in a three-sphere, additional geometrical relations change. The equator of a sphere is a circle also around the sphere's center, with a radius R if Euclidean geometry were used. Though now, with a new r-dependent term in the metric, the distance from the center at $r = 0$ to $r = R$ is not just R, but rather the sum of all $\Delta L_r = \Delta r/\sqrt{1 - r^2/r_{\max}^2}$ along the way. The sum is not easy to compute, but it can be found and equals $S = r_{\max}\arcsin(R/r_{\max})$, the inverse function arcsin of the sine with a graph obtained by reflecting the sine along a diagonal. The ratio of circumference by radius of a circle in the three-sphere is $C/S = 2\pi(R/r_{\max})/\arcsin(R/r_{\max})$. For R much smaller than $r_{\max}$, $\arcsin(R/r_{\max})$ is close to $R/r_{\max}$ and the ratio is almost the Euclidean 2π. For larger values, when R approaches $r_{\max}$, deviations occur. One may consider $r_{\max}$ as a measure of the three-sphere's curvature, or its radius. For circles almost as wide as the whole sphere, deviations from Euclidean geometry are most pronounced.

Geodesics A sphere is symmetric, a round version of a curved surface, with great circles as geodesics. On a surface, or space, or space-time with nonuniform curvature, geodesics bend differently at different places, according to the position-dependence of the metric g_{ab} in the line element Δs^2 they maximize or minimize. A massive object always moves in curved space-time so that the sum of proper-time lapses $\Delta T = \sqrt{-\Delta s^2}/c$ along its worldline takes the maximum value among all possible worldlines between two events.

Instead of looking for curves that maximize the sum of motion-dependent and gravitational time dilation in $1 - \frac{1}{2}|\vec{V}|^2/c^2 + P$ in weak gravitational fields, we must now find curves maximizing the sum of

$$\Delta T = \sqrt{-g_{tt}(\Delta t)^2 + \frac{g_{xx}(\Delta x)^2 + g_{yy}(\Delta y)^2 + g_{zz}(\Delta z)^2}{c^2}}\,.$$

There are velocity-dependent terms such as $\Delta x/\Delta t = V^x$ once we factor out Δt, as well as position-dependent effects from the metric coefficients. It is advantageous for a massive object to spend more time, at low speed $|\vec{V}|$, in regions of larger metric components because high speeds in small-metric regions will be compensated for by the metric. However, the velocity should not increase too much and overpower the metric; again a balancing effect results with one optimal worldline that we have to find.

The general equations for a geodesic, with all metric components allowed to be nonzero, are quite involved. Compared with the equations for a small gravitational potential, where we were able to ignore products of small velocities with the small potential, we now expect force terms that also depend on the velocity of the object. The result is an equation rather different from Newtonian mechanics.

Christoffel symbol The maximization procedure followed earlier requires us to compare differences of square roots in ΔT at different places, a lengthy calculation. For a general metric, $c\Delta T = \Delta t\sqrt{-\mathbf{Sum}_{a,b}g_{ab}(\Delta x^a/\Delta t)(\Delta x^b/\Delta t)}$ depends on terms $\Delta x^a/\Delta t$ proportional to the four-velocity $u^a = \Delta x^a/\Delta T$. In a strong gravitational field, velocities can become large, and so we should not split the relativistic four-velocity into its time component and the three-velocity $\vec{V}$. Terms to be compared for different worldlines, for instance those of two straight pieces with a midpoint shifted to a variable value X of one of the coordinates, say x^c, are then

$$\Delta T(X)^2 = -(\Delta T)^2\left\{g_{cc}\frac{X^2+(\Delta x^c - X)^2}{(c\Delta T)^2}\right.$$
$$\left. + \mathbf{Sum}_{a,b}[g_{ab}(x^c) + g_{ab}(x^c+X)]\frac{\Delta x^a}{c\Delta T}\frac{\Delta x^b}{c\Delta T}\right\} .$$

We obtain an acceleration term $A_a = \mathbf{Sum}_{a,b}\Delta(g_{ab}u^a u^b)$. Instead of the gradient of the potential, we must evaluate $\mathbf{Sum}_{a,b}(\Delta g_{ab}/\Delta x^c)u^a u^b$, comparing the metric at different places x^c, and the other variations $\mathbf{Sum}_{a,b}g_{ab}u^a\Delta u^b/\Delta T$ can be combined with the first velocity variation. After rearranging a rather messy set of terms, the acceleration and metric terms cancel each other, producing a maximizing worldline, if $\Delta u^a/(c\Delta T) = -\mathbf{Sum}_{b,c}\Gamma^a_{bc}u^b u^c$, all metric terms collected in the Christoffel symbol

$$\mathbf{Sum}_d g_{ad}\Gamma^d_{bc} = \frac{1}{2}\left(\frac{\Delta g_{ab}}{\Delta x^c} + \frac{\Delta g_{ac}}{\Delta x^b} - \frac{\Delta g_{bc}}{\Delta x^a}\right) .$$

Conserved quantities Ups and downs are more complicated than the constants of life. Like classical mechanics, general relativity gives rise to conserved quantities unchanging along every geodesic. In mechanics, we had to require the potential energy not to depend on time for the total energy to be conserved, or the potential to depend only on the distance to some fixed center for angular momentum to be conserved. In curved space-time, there are conservation laws provided the metric components remain unchanged in some direction in space-time, for instance, by being independent of time or some other coordinate. If the metric is time-independent, energy is conserved along geodesics; if it is independent of the polar angle **Lo**, angular momentum is conserved.

Calculating the precise form of a conserved quantity follows the lines of our maximization equations. If the metric is independent of t, say, changing the coordinate t to $t + h$ at one event on a worldline changes the proper-time contributions $(c\Delta T)^2$ by $h\Delta\left(\mathbf{Sum}_b g_{tb}(\Delta x^b/\Delta T)\right)/\Delta T$. By Fermat's argument, this term must vanish even after we divide by the small displacement h, and therefore the quantity $\mathbf{Sum}_b g_{tb} u^b$ does not change along the geodesic. In Minkowski space-time, u^t, or the energy $E = mcu^t$, is conserved. We obtain a conserved quantity for angular momentum if the metric, as in our previous examples, does not depend on **Lo**: With $g_{\mathbf{LoLo}} = r^2 \sin(\mathbf{La})^2$, $L = 2\pi r^2 \sin(\mathbf{La})^2(\Delta\mathbf{Lo}/\Delta T)/\bigcirc$ is conserved. As in classical mechanics, knowledge of conserved quantities makes it easier to solve geodesic equations.

Space-time dynamics Curved space-time implies a force on objects moving in it. However, what is it that makes space-time curved, and how can we compute the correct curvature for a space-time of the form we see around us? With flexible metric coefficients, space-time has become a changeable and dynamical object of its own; it exerts force on the motion of masses but also reacts to their distribution. A large mass implies a strong gravitational force; it curves space-time more than a small one. Just like the motion of objects, the motion of space-time, understood in the sense of it getting more or less curved, is determined by equations of motion imposed on the metric coefficients g_{ab}.

Before looking at specific equations, it is perhaps more interesting to ask where they might come from. Newton's laws of motion have been extracted from a long series of observations of moving objects. We do not perceive the motion and curvature of space-time, but we see its effects on other objects subject to the gravitational force. Instead of trying to see the curvature in motion, we can perform a series of mathematical observations and deduce how space-time must look so as to provide the correct force in all cases. We have brought the force in contact with curvature by appealing to the equivalence principle used thus far for weak forces. The same insight can guide us to the equations for space-time in general.

The equivalence principle relates gravity to space-time by identifying motion of uniform acceleration with motion in a constant gravitational force, motivated by the equality of inertial and gravitational mass. We go beyond uniform motion at constant velocity which, as humble as such motion may be, gave us all the effects of special relativity. Space-time geometry is realized in special relativity by the Minkowski line element. Observers moving at different constant speeds experience space-time differently, but their measurements can be translated into one another by Lorentz transformations. We may turn the axes of a Cartesian coordinate system associated with one observer so as to find the reference system of another one. While the Minkowski line element and Lorentz transformations no longer apply in curved situations of general relativity, we can still turn the axes. A generalization of Lorentz transformations then leads to the dynamical laws of space-time.

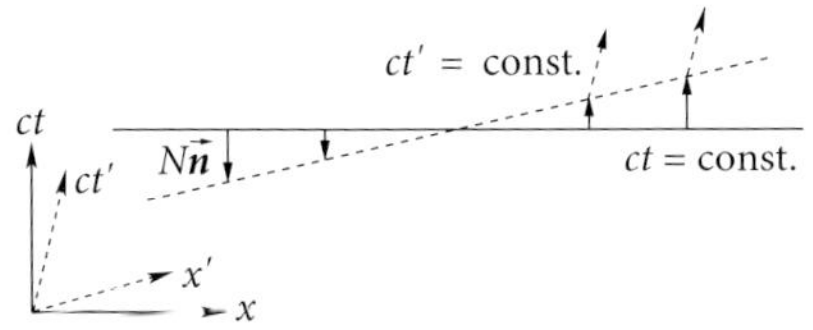

Figure 2.41 The time change of a Lorentz transformation from t to t' can be interpreted as a linear deformation of magnitude $N(x) = c\Delta t + vx/c$ along the normal $\vec{n}$ to surfaces of constant t.

A characteristic feature of special relativity, compared with Newtonian physics, is that even the time coordinate changes when an observer moves at a different speed. We focus on implications of time transformations in order to uncover the crucial properties of space-time. The time part of a Lorentz transformation, $ct' = (ct - vx/c)/\sqrt{1 - v^2/c^2}$, implies that the set of all events with the same value of t' satisfies the equation $ct = vx/c + A$ with a constant A, a linear relationship between the coordinates x and t. The new x'-axis, as the set of events with the same $t' = 0$, is a straight line with slope given by the velocity ratio v/c. We have used this feature before to draw what different observers see in Minkowski diagrams. Now, as in Figure 2.41, we reinterpret the transformation as a deformation of the old x-axis, or any surface of constant t, to a new axis, or a surface of constant t'. From every point on the old surface, we move in the vertical direction orthogonal to the surface, by a linear displacement $N(x) = c\Delta t + vx/c$. The first term is a constant shift by $A = c\Delta t$, corresponding not to a Lorentz boost but to a simple time lag.

Hypersurface deformations Motion implies deformed space, depending on the velocity and time lags. The laws of motion can be recovered if we consider the results of successive deformations. If we deform first by $N_1(x) = vx/c$, then along the new surface of constant t' by $N_2 = c\Delta t - vx/c$ with the same v but a negative sign, we undo the linear deformation in the second step and move the initial surface by $c\Delta t$. We first boosted to velocity v, then waited some time Δt and boosted back to velocity zero. In the time between the boosts, any object, as seen by our observer, moves an extra amount $\Delta x = v\Delta t$ on top of its own motion. If we draw the two deformations, as in Figure 2.42, the result is indeed a new horizontal surface but with all points shifted by some Δx.

In order to subtract off any intrinsic motion, we consider not just two deformations performed in a row, but rather the differences entailed by performing both deformations in the two different orderings. We call a single time deformation $T[N]$, with the linear deformation function $N(x)$. By comparing different time deformations, we are led to spatial deformations along a direction $\vec{w}$, which we denote as $S[\vec{w}]$. We write different deformations performed in a row as a product $T[N_1]T[N_2]$. It is not commutative: two successive time deformations $T[N_1]T[N_2]$ and $T[N_2]T[N_1]$ in the two orders yield the same surface only up to a spatial shift that depends on the constant part in $N_2(x)$ and the slope of $N_1(x)$. In the example

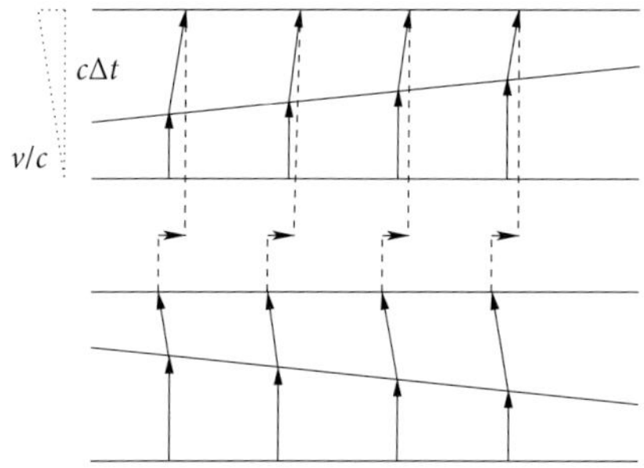

Figure 2.42 Two linear deformations that in combination amount to a simple shift in time. Performing the two deformations in different orderings (top and bottom) equals a spatial translation (middle). The dotted triangle, with angle v/c and one side $c\Delta t$, shows that the spatial translation (if small) is $v\Delta t$.

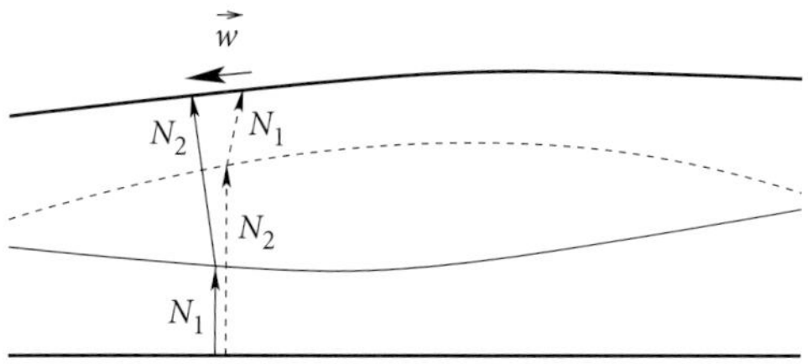

Figure 2.43 Two arbitrary surface deformations commute up to a spatial shift.

of Figure 2.42, the shift is $\Delta x = (v/c)(c\Delta t) = v\Delta t$, as follows from the triangles in the figure, and also from basic laws of motion. For more generality, we write $\Delta x = N_2(\Delta N_1/\Delta x) - N_1(\Delta N_2/\Delta x)$ such that $T[N_1]T[N_2] - T[N_2]T[N_1] = S[N_2(\Delta N_1/\Delta x) - N_1(\Delta N_2/\Delta x)]$.

We have geometrized the laws of uniform motion: they follow from linear deformations of surfaces, that is, linear deformations for constant velocities. If we allow for accelerated observers or reference systems, motion is no longer along straight lines. We generalize our deformations to arbitrary functions $N(x)$. Again, one can work out the geometry of successive normal deformations and find that they commute up to a spatial shift; Figure 2.43. With nonlinear $N(x)$, the previous algebraic relation still applies. Even if we use a general space-time geometry with metric g_{ab} instead of deformations in Minkowski space-time, the relationship does not change much:

$$T[N_1]T[N_2] - T[N_2]T[N_1] = S\left[\frac{1}{g_{xx}}\left(N_2\frac{\Delta N_1}{\Delta x} - N_1\frac{\Delta N_2}{\Delta x}\right)\right] \qquad (2.35)$$

called the hypersurface-deformation algebra.

Einstein's equation Nonlinear hypersurface deformations imply general relativity, with its unrestricted coordinate transformations and observer motions. As linear deformations geometrize uniform motion, nonlinear deformations geometrize accelerated motion. By the equivalence principle, acceleration is equivalent to the gravitational force. A deep property of space-time physics is that the form of hypersurface deformations (2.35) does not only encode motion with a gravitational force (through a region with space-time curvature); it even determines the dynamical laws of space-time itself. Having the relationship (2.35) for arbitrary deformations is such a strong requirement that only specific space-time changes between different points can occur. The geometrized laws of general motion determine the dynamical space-time equations.

Hamiltonian For hypersurface deformations to have the required form, $T[N]$ must be the sum of

$$\frac{V(\vec{x})N(\vec{x})c^2}{16\pi G}\left[\mathbf{Sum}_{i,j}(h_{ik}h_{jl} - h_{ij}h_{kl})K^{ij}K^{kl} - R_3(\vec{x}) + 2\Lambda\right] \quad (2.36)$$

summed over spatial coordinates x^i, with two constants G (Newton's constant) and Λ (a version of the rocket symbol because it can speed up the universe, pronounced as the Greek letter Lambda). The contributions (2.36) are summed over all small pieces of space, on one of the hypersurfaces being deformed, located at positions $\vec{x}$, with volume $V(\vec{x})$. The metric h_{ij} describes the geometry of space (rather than space-time), and its rate of change by an orthogonal deformation is K^{ij}, some kind of metric velocity. With K^{ij}, we have a measure of extrinsic curvature of the hypersurface in space-time, for its bending influences how the geometry changes as we move orthogonally. Finally, $R_3(\vec{x})$ depends, via h_{ij}, on the intrinsic geometry of space.

The expression (2.36), with its relation to changes $T[N]$ along a timelike direction, plays the role of an energy of the metric h_{ij}, or rather, since we do not have a physical grasp of the energy of geometry, a more general notion called a Hamiltonian. It is quadratic in the rate of change K^{ij}, a kinetic term, and has a metric-dependent potential $V(\vec{x})(-R_3(\vec{x})+2\Lambda)$. The metric must accelerate according to a force computed from (negative) derivatives of the potential by h_{ij}. These are the dynamical (and complicated) equations of general relativity. Curvature R_3 determines not only the motion of matter in space-time, but also the motion of geometry itself.

Riemann tensor Intrinsic curvature, in its most general form, has many independent components. It is constructed from the Christoffel symbols Γ^a_{bc} and their derivatives as the Riemann tensor

$$R_{abc}{}^d = \frac{\Delta\Gamma^d_{ac}}{\Delta x^b} - \frac{\Delta\Gamma^d_{bc}}{\Delta x^a} + \mathbf{Sum}_e\left(\Gamma^e_{ac}\Gamma^d_{eb} - \Gamma^e_{bc}\Gamma^d_{ea}\right). \quad (2.37)$$

One often uses a reduced version, the Ricci tensor $R_{ac} = \mathbf{Sum}_b R_{abc}{}^b$. These definitions are valid for space and space-time (or space with any other dimension). Depending on the dimension, we sum from zero to three for space-time or one to three for space, and use the space-time metric g_{ab} or the space metric h_{ij} in the Christoffel symbols. The potential in the gravitational Hamiltonian uses the Ricci scalar of space, $R_3 = \mathbf{Sum}_{i,j} h^{ij} R_{ij}$. With upper indices, h^{ij} is the matrix inverse of h_{ij}.

Hypersurface deformations and the Hamiltonian were not the starting point of general relativity; their role was realized only in the mid 1970s, especially by Karel Kuchař and Claudio Teitelboim. In 1915, when general relativity was first formulated, relations between space-time curvature and physical observables stood in the foreground. Observations do not depend on coordinate choices, and so the curvature quantities used for the dynamical equations of space-time should be insensitive. We need an analog of proper time, that is, an invariant measure of change along a worldline, now applied to the change of space-time as a whole instead of the position of one object. We do not sum up contributions of proper-time lapses along a curve, but invariant curvature contributions from small regions in all of space-time. To complete the analogy, as we maximize proper time along a physical worldline in the presence of the gravitational force, we need to maximize our curvature expression in order to find the dynamical behavior of space-time.

Einstein–Hilbert action The space-time Ricci scalar is a function that can be summed over all of space-time, with contributions $V_4(\mathbf{Sum}_{a,b} g^{ab} R_{ab} + \Lambda)$ with four-volume V_4, to obtain the Einstein–Hilbert action. If we also add energies of all matter components, the Einstein–Hilbert action is maximized when the Ricci tensor satisfies Einstein's equation

$$R_{ab} - \frac{1}{2} g_{ab} \mathbf{Sum}_{c,d} g^{cd} R_{cd} + \Lambda g_{ab} = \frac{8\pi G}{c^2} T_{ab} \,. \tag{2.38}$$

On the right-hand side, T_{ab}, the stress-energy tensor, contains information about matter. Its components correspond to the energy density T_{tt}, the energy flux T_{tx} and momentum density T_{xt}, and the spatial stress tensor. As in special relativity, energy is related to momentum and combined to one space-time tensor, but here, stress enters as well. Stress is a measure for the reaction of an elastic body to different forces. It contains pressure quantities T_{xx} and sometimes also anisotropic stress components T_{xy} if a force in one direction implies a deformation of the body in a different direction.

3
The Universe II

The universe around us is mostly space and time. On average, matter, as we know and deal with on a day-to-day basis, is so diluted that its density amounts to just about one atom per cubic meter. On Earth and in our own bodies, we are used to much higher densities. How does this uneven distribution, a form of cosmic capitalism, come about?

Gravity, acting over millions of years, concentrates matter because it makes all masses attract one another, without any neutralization as realized for the electric force. If matter is not evenly distributed, even if just slightly, regions denser than their surroundings, perhaps at some beginning stage, attract more matter and, left undisturbed long enough, gorge themselves at the expense of their minor peers. All spoils are split into dense spheres of gravitational influence, forming the progenitors of our galaxies, demarcated by vast no-matter's lands of empty space. However, even within those spheres, the fight goes on; again, some parts are denser than others and grab ever more. Some theaters of this internecine war can become so dense and hot that nuclear explosions flash in them, kindling the stars. Balkanization still goes on, and in the more dilute, not-so-hot matter clouds around the stars some denser centers form. Some may start burning as another smaller star, orbiting the central one in a binary star system. In most cases, there is not enough matter left far from the center. No new star will form, but concentrated balls, the planets, start orbiting around the central star.

While the armies of matter fight for dominance, they drag along the lands of space and time. As a region becomes denser, it exerts stronger gravitational forces, attracting even more. Gravity is exerted by matter not by direct, hard impacts, but through the deformed shape of space-time. Dense matter enslaves space-time; it bends and curves it so that it will enforce the matter master's will. Space-time around a massive object is not flat, not of Minkowski form; it is curved as per a modified geometry, with the weak-field line element (2.33) when the force is not strong. General relativity is the law of the land that holds sway over how matter can curve space-time, and how space-time reacts by making matter move. The codex Einstein, or Einstein's equation, encodes the first part of this interplay, and geodesic motion the latter. Einstein's equation tells us what space-time geometry we must use around a central mass such as a star. Space-time, curved by matter, takes revenge by making matter age. It forces matter to move forward in time, with

The Universe: A View from Classical and Quantum Gravity, First Edition. Martin Bojowald.

maximum proper time. Matter in curved space-time must move along geodesics, motion we experience as the action of the gravitational force. Only by expending other forces, such as electromagnetic ones, can matter slow down its aging process, but just a little bit.

Space-time, in the cosmic play, has become an actor in its own right. In many cases, it can be seen as obeying the will of matter, serving as a messenger. But sometimes, it can free itself and change or move or interact without the help of matter. Its regions of influence are wide, much larger than those of the more short-sighted matter. Over large distances, the roles of matter and space-time are reversed. Then, matter is the bystander without much action of its own, forced to follow the expansionary interests of space-time; matter can only react by slowing down or perhaps supporting the expansion. To watch these stories unfold, we must analyze solutions of Einstein's equation.

3.1 Planets and Stars

A star curves surrounding space-time just by its mass. If it is not dense, the weak-field line element describes the geometry. However, a new form must be used when gravity gets stronger, or when the term $-P/c^2 = GM/Rc^2$ takes a value close to one. Einstein's equation is a more complete law to derive the metric around a star.

Schwarzschild space-time Symmetry simplifies. Most stars rotate around an axis and are not spherical. Still, many important consequences of curved space-time can already be seen in the simpler case of a nonrotating, spherical central mass. The strength of the gravitational force then depends only on the distance from the center, and the star as well as space-time around it are time-independent. We may assume the metric to depend only on a radial coordinate r, not on time or the angle around an axis. Einstein's equation then implies a line element of Schwarzschild form

$$\Delta s^2 = -\left(1 - \frac{2GM}{c^2 r}\right)(c\Delta t)^2 + \frac{1}{1 - 2GM/(c^2 r)}(\Delta r)^2 + \left(\frac{2\pi r}{\bigcirc}\right)^2 [(\Delta \mathbf{La})^2 + \sin(\mathbf{La})^2(\Delta \mathbf{Lo})^2] \,. \tag{3.1}$$

Spherical symmetry For a spherical central mass, the line element of its surrounding space-time remains unchanged under rotations around the center. With one free function $L(t, r)$, the spatial part then has the form $\Delta s^2 = L(t, r)^2(\Delta r)^2 + (2\pi r/\bigcirc)^2[(\Delta \mathbf{La})^2 + \sin(\mathbf{La})^2(\Delta \mathbf{Lo})^2]$. If we further assume the geometry to be time-independent in the absence of rotation, the function L can depend on r only. Moreover, the kinetic terms K_{ij} in the (2.36) vanish if nothing depends on time, and the metric potential, taking the form $R_3[L(r)] =$

$2/r^2 - 2/(rL)^2 + [4/(rL^3)]\Delta L/\Delta r$, must vanish too. If $\Lambda = 0$, in vacuum we have a differential equation for $L(r)$ solved by $L(r) = (1 + C/r)^{-1}$ with some constant C. Using other parts of Einstein's equation, the momenta can be constant in time only if the time-component $g_{tt} = -N^2$ of the metric satisfies the equation $N^{-1}\Delta N/\Delta r = -L^{-1}\Delta L/\Delta r$, which requires $N = 1/L$. For weak forces, the line element must be of the Newtonian form, for which $C = -2GM/c^2$. With this constant determined by the central mass, the solution (3.1) follows.

With all metric coefficients independent of t and **Lo**, (3.1) has two symmetries, according to time independence and rotational invariance. Energy and angular momentum are therefore conserved along geodesics. The coordinate r, however, does appear in the coefficients, and the geometrical distance R between some point at r and the center at $r = 0$ is not simply r; rather, R must be computed by adding up invariant contributions $\Delta s = \Delta r/\sqrt{1 - 2GM/(c^2 r)}$ along a radial line. The value of r is not the radius R of a sphere at some constant r, but the spherical area is the usual $A = 4\pi r^2$, the angle coefficients of the metric being the same as in the Euclidean line element. Schwarzschild space-time is curved: the Euclidean relation $A/R = 4\pi$ is violated.

Schwarzschild radius The Schwarzschild line element carries the sign of doom. For small $GM/(c^2 r)$, it is close to the Newtonian one at weak fields, with a gravitational potential $P(r) = -GM/r$. As a complete solution of Einstein's equation, the Schwarzschild line element can be used also for strong fields when the metric coefficients deviate more from their Euclidean values. Even dense stars can be described by this geometry, with interesting space-time effects. The line element remains valid at small r as long as we do not cross the star's surface, where matter would change the metric according to the stress-energy tensor in Einstein's equation. When $2GM/(c^2 r)$ approaches the value one, space-time distortions seem large: some of the coefficients of the line element may become almost zero or near-infinite. Time-dilation effects, for instance, are then extreme, and space-time does not at all look as we know it.

At an r-value of $R_S = 2GM/c^2$, called the Schwarzschild radius, important effects happen. Some metric components vanish or diverge. If we are able to cross the line of $r = R_S$, the time and radial components of the metric switch roles according to their negative or positive nature. It is the term $(\Delta r)^2$ that has a negative coefficient for $r < 2GM/c^2$, while the coefficient of the time term $(c\Delta t)^2$ is made positive. The signs, and not our nomenclature of choosing the letter t, determine what space and what time is in space-time. With changing signs, the effects of mass on space-time are so strong that r now appears timelike; it determines causal notions, and it varies along the axis of light cones. The coordinate r, originally spacelike, must now change toward the future, while we are free to choose our direction in t, no longer playing the role of time. We have entered this space-time region by decreasing our position in r, and in this region, we must keep moving by reducing it even further. Once we approach $r = 0$, metric components diverge or vanish again; another strange place is reached.

Space-time regions like these are so alien to our intuition that we should approach them with care. They are not very relevant for most stars because a star's surface lies outside the Schwarzschild radius determined by its mass. The Sun, for instance, has a Schwarzschild radius of a mere 3 km, much smaller than the 700 000 km of its radius. Also the Earth, another almost spherical mass, has a small Schwarzschild radius of less than 1 cm. Within the surface of a star or a planet, the Schwarzschild line element is modified by matter terms from the stress-energy tensor, and they make it regular, without zeros and infinities. For dense stars, however, the surface may come close to the Schwarzschild radius. Space-time effects then have to be analyzed carefully.

Gravitational redshift Gravity dilates time and shifts the frequencies of light. With the Schwarzschild line element as a solution to Einstein's equation, we can extend our previous calculations, based on the equivalence principle, to stronger gravitational forces. In an astronomical situation, we are observers far away from a star and at a much smaller value of the gravitational potential. Light emitted at the surface of the star is subject to different time dilation than we are, and the emitted frequency f_* is redshifted to a smaller $f < f_*$ once light reaches our telescopes.

The line element determines time dilation by the position dependence of its time component. Once emitted, light moves outward along a lightlike geodesic whose form follows from the position dependence of the line element. Lightlike geodesics obey $\Delta s^2 = 0$, and a radial one with constant angles must satisfy the relationship $c\Delta t = \Delta r/[1 - 2GM/(c^2 r)]^2$. For r much larger than the Schwarzschild radius, this equation is close to the one $c\Delta t = \Delta r$ for a straight line of 45°. However, at smaller r, near the star surface, the slope is larger than 45°; the worldlines of light bend toward the worldline of the surface at constant r, as indicated in Figure 3.1.

Another implication of the position-dependent line element is time dilation even for observers at rest. Proper time at vanishing velocity is related to coordinate time by $\Delta T = \sqrt{-\Delta s^2}/c = \sqrt{-g_{tt}}\Delta t$. For the Schwarzschild line element, $\Delta T = \sqrt{1 - 2GM/(c^2 r)}\Delta t$. On the surface of a star, proper time progresses more slowly than coordinate time of an outside observer. (Time stops at the Schwarzschild radius, another ominous sign.) Light is an oscillating phenomenon unfolding in proper time; delayed proper time thus means a smaller, redshifted frequency.

To quantify redshift, we express time dilation by a modified time component of an observer's four-velocity. The time component of $\boldsymbol{u}_{\text{obs}} = \Delta \boldsymbol{x}/\Delta T$ is $u^t_{\text{obs}} = c\Delta t/\Delta T = c/\sqrt{1 - 2GM/(c^2 r)}$, depending on the displacement r from the center. We use the four-velocity of an observer in order to compute observed energies of objects or frequencies of light by means of a local inertial frame. For light, the frequency is $f = -\boldsymbol{w} \cdot \boldsymbol{u}_{\text{obs}}$ with the wave four-vector $\boldsymbol{w}$ on the lightlike geodesic. Both $\boldsymbol{w}$ and $\boldsymbol{u}_{\text{obs}}$ change as we move along the geodesic, providing the frequency of light along its way from the star. The frequency changes, but $\mathbf{Sum}_b g_{tb} w^b$ with $K = (1, 0, 0, 0)$ is a conserved quantity unchanged along the geodesic, for K points in a direction in which the metric components do not change. Since u_{obs} for an observer at rest is proportional to K, we can compute the shifted frequencies by relating $w \cdot u_{\text{obs}}$ to $w \cdot K$, writing $f_* = -w \cdot u_{\text{obs}}(R) = -w \cdot K/\sqrt{1 - 2GM/(c^2 R)}$ and use $f = -w \cdot u_{\text{obs}}(r) = -w \cdot K$ for r much larger than the Schwarzschild radius, far away from the star's surface:

$$f = \sqrt{1 - \frac{2GM}{c^2 R}} f_* < f_* . \tag{3.2}$$

For small $2GM/(c^2 R)$, we recognize our previous formula for gravitational redshift as a good approximation, but deviations occur for R closer to R_S.

Energy Gravitational redshift looks deceptively like energy loss of light moving out of the gravitational potential. Both the frequency of light and the energy of a massive particle are related to the time component of a four-vector, the frequency f of the wave 4-vector w and the energy E of the four-momentum p. We have already seen that the dispersion relation $||w||^2 = -f^2/c^2 + |\vec{w}|^2 = 0$ is consistent with the massless nature of light, $||p||^2 = 0$, if w is proportional to $p = hw$ with some constant h. The formula for gravitational redshift could be rewritten as an energy-loss equation $E = hf = hf_*(1 - GM/c^2 R) = E_* - Gm_* M/R$ if we associate the mass $m_* = E_*/c^2$ to the energy. In order to obtain the energy E at the position of an observer, we subtract the gravitational energy $-m_* P(R) = Gm_* M/R$ from it.

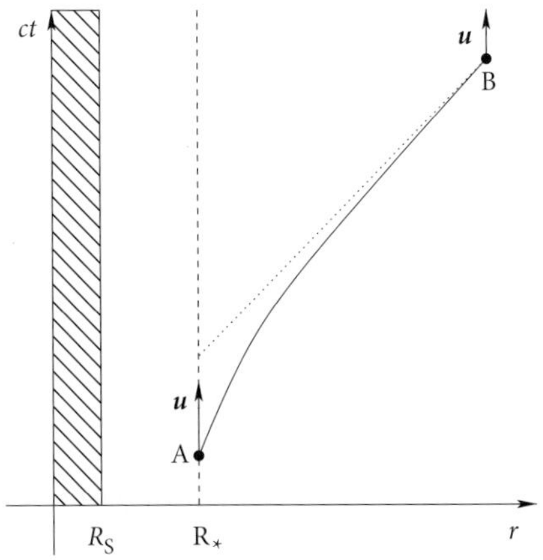

Figure 3.1 A space-time diagram with the radial displacement r from the center of a star. The star surface at $r = R_*$, outside of its Schwarzschild radius R_S, emits light outwards, seen by distant observers. Light moves along lightlike geodesics, obeying $\Delta s^2 = 0$ along them. In the coordinates r and ct, these lines are not straight, but bent according to the r-dependent metric coefficients. The r-dependence also implies that the four-velocity u has a larger time-component closer to the surface.

Newton could have told us this, for the gravitational potential provides the potential energy of an object when multiplied with its mass.

We did use one essential new ingredient in the arguments, namely, that the gravitational mass of an object is related to its energy by $E = mc^2$. Moreover, we are relating the frequency of light to an energy, and then use formulas derived for massive objects. Gravitational redshift as an energy-loss statement is therefore not supported much. And while we produce a Newtonian energy loss for light sent out radially, it turns out that nonradial motion breaks the analogy. Gravitational redshift is a statement about measurements of observers at different places. Different frequencies or energies may be measured because time standards change, not by an actual loss. Relativity is consistent with energy conservation; there are, after all, the same types of conserved quantities in its equations. What we learn from gravitational redshift is that space-time curvature requires a more careful attitude toward energy.

We have used a conserved quantity, $\boldsymbol{w} \cdot \boldsymbol{K}$, in our derivation of gravitational redshift, related to the time independence of our metric. In classical mechanics, time independence is responsible for energy conservation, but the relativistic energy is proportional to $\boldsymbol{w} \cdot \boldsymbol{u}_{\text{obs}}$, taking into account space-time effects. Time dilation implies that $\boldsymbol{u}_{\text{obs}}$, while proportional to $\boldsymbol{K}$, is not identical to this symmetry vector. The mismatch is the reason why energy conservation in the naive sense appears problematic in general relativity. There is a conserved quantity related to time independence, but the physical energy measured by observers at different places in space-time depends on the observers' systems and is affected by time dilation and other effects.

Newtonian considerations cannot be applied for general space-times, for the gravitational potential is meaningful only in the Newtonian limit of general relativity with weak gravitational forces. If the force is strong, several independent metric components must be considered, which can no longer be arranged as a simple potential for an energy-balance law. Instead, energy conservation, which remains an important law of physics also in relativity, is incorporated in a more interactive way, encoded by the interplay of matter and its motion with curvature of space-time. Geodesic motion of a small mass or of light in a given space-time, ignoring how the energy acts back on space-time by curving it, cannot show the full energy equation.

In general relativity, energy is computed not just for objects moving in space-time, but for whole space-time regions in which we have curvature as well as matter. In order to measure all energy contributions and keep track of how they balance out, an observer must oversee a large region. Matter energy is represented by the stress-energy tensor, and the gravitational analog of energy follows from curvature matched with matter terms in Einstein's equation. In this general context, energy is conserved because the energy contained in some region can change only by the flow of energy moving through the region's boundary. There is no additional change of energy, and so energy is not lost or created; it merely moves between different space-time regions.

Planetary orbits Planets have mass; they follow timelike geodesics. In space-time around a spherical mass, only two of the four polar space-time coordinates appear in the metric, while t and **Lo** do not. This symmetry implies two conserved quantities, the energy $e/c = -\boldsymbol{u} \cdot \boldsymbol{K}$ as measured by an observer far away from the planet with four-velocity $\boldsymbol{u}$, and the angular momentum $\ell = \boldsymbol{u} \cdot \boldsymbol{H}$. Here, $\boldsymbol{K} = (1, 0, 0, 0)$ is the symmetry four-vector in the time direction, and $\boldsymbol{H} = (0, 0, 0, 1)$ the one in the **Lo**-direction. Given the metric, we compute these quantities as $e = (c^2 - 2GM/r)\Delta t/\Delta T$ and $\ell = (2\pi/\bigcirc)r^2 \sin(\mathbf{La})^2 \Delta\mathbf{Lo}/\Delta T$. Both quantities disregard the object's mass, which turns out to be immaterial for an orbit, thanks to the equality of inertial and gravitational mass, or the equivalence principle.

Conservation of angular momentum, as in classical mechanics, implies that the planet moves in a plane. We can make a simple choice of our angles, setting $\mathbf{La} = \frac{1}{4}\bigcirc$, or choosing the poles of our coordinate system such that the planet's motion unfolds in the equatorial plane. With $\sin(\frac{1}{4}\bigcirc) = 1$, several factors disappear. Moreover, $\Delta\mathbf{La}/\Delta T = 0$ when **La** is constant, and we have only three nonzero four-velocity components, $c\Delta t/\Delta T$, $\Delta r/\Delta T$ and $\Delta\mathbf{Lo}/\Delta T$. These components are related to one another by the speedometer equation of the four-velocity, $||\boldsymbol{u}||^2 = -c^2$.

Effective potential With the Schwarzschild metric and our components of $\boldsymbol{u}$, the speedometer says

$$-\left(1 - \frac{2GM}{c^2 r}\right)\left(\frac{c\Delta t}{\Delta T}\right)^2 + \frac{1}{1 - \frac{2GM}{c^2 r}}\left(\frac{\Delta r}{\Delta T}\right)^2 + r^2\left(\frac{2\pi\Delta\mathbf{La}}{\bigcirc\Delta T}\right)^2 = -c^2 . \tag{3.3}$$

We do not know any of the four-velocity components yet, but we can replace two of them with the conserved quantities e and ℓ. These two numbers are constant, even if we are not sure what values they take. The speedometer equation then reads

$$-\frac{e^2}{c^2 - 2GM/r} + \frac{(\Delta r/\Delta T)^2}{1 - 2GM/(c^2 r)} + \frac{\ell^2}{r^2} = -c^2 . \tag{3.4}$$

All terms in this equation only depend on r and its rate of change with respect to T. We have arrived at a differential equation for $r(T)$ which we can try to solve. First, however, it is useful to rewrite this equation as

$$\frac{e^2 - c^4}{2c^2} = \frac{1}{2}\left(\frac{\Delta r}{\Delta T}\right)^2 - \frac{GM}{r} + \frac{\ell^2}{2r^2} - \frac{GM\ell^2}{c^2 r^3} . \tag{3.5}$$

Conservation of energy and angular momentum turns laws of motion into equalities between rates of change and positions, in the form of a kinetic-like energy term given by the square of the rate of change of r, and a potential of r-dependent terms.

In the concrete case here, two terms look just like the potential and centrifugal energy in classical mechanics with a central-mass potential, while another term is new. We may interpret the equation as an energy-balance law, at least formally because it was derived from the speedometer, not an energy equation. The constant on the left-hand side of (3.5), $\frac{1}{2}(e^2/c^2 - c^2)$, is related to the nonrelativistic energy if we approximate $e - c^2 = c^2\sqrt{1 + (e^2/c^4 - 1)} - c^2 \approx \frac{1}{2}c^2(e^2/c^4 - 1)$ for small $e^2 - c^4$. The energy for a planet of mass m is then $E = me \approx mc^2 + \frac{1}{2}m(e^2/c^2 - c^2)$, and therefore (3.5) is the total energy minus the rest energy, divided by the mass. We should expect this quantity to equal kinetic plus potential energy, as realized by our formula.

In compact form, the speedometer/energy equation for a planet of mass m is

$$E - mc^2 = \frac{1}{2}m\left(\frac{\Delta r}{\Delta T}\right)^2 + m P_{\text{eff}}(r) \tag{3.6}$$

with the effective potential

$$P_{\text{eff}}(r) = -\frac{GM}{r} + \frac{\ell^2}{2r^2} - \frac{GM\ell^2}{c^2 r^3} \,. \tag{3.7}$$

We can solve for the motion with the technique of potential landscapes, as in classical mechanics. While the method is the same as nonrelativistically, solutions will differ because we have the new term $-GM\ell^2/(c^2 r^3)$ in the effective potential. It is important for small r, for instance, for the planet Mercury, closest to the Sun.

Potential landscape Planets orbit quickly around the Sun. We parameterize the speed by taking the ratio of $R_S = 2GM/c^2$ of the Sun's Schwarzschild radius to a length parameter ℓ/c defined using the planet's angular momentum per mass, $\ell = L/m$. A planet's orbital velocity v is much less than the speed of light, and so $\ell/c \approx rv/c$ is smaller than the orbit radius r. Kepler's laws imply that $\ell/c \approx \sqrt{R_S r}$, and therefore the ratio we are interested in, $2GM/c\ell \approx \sqrt{R_S/r}$ is small. For such ratios, the effective potential looks as shown in Figure 3.2.

The potential landscape has one trough in which a planet's energy would be located. Since the potential cannot be larger than the total energy, there are barriers to smaller and larger r, forces that block the planet as it moves toward them. The orbit has a maximum and a minimum distance to the Sun, reached periodically along the orbit. The closer the energy is to the bottom of the trough, the smaller the difference between maximum and minimum radius, and the more circular the orbit. These properties are identical in relativistic and Newtonian mechanics, but the precise orbiting behavior does change because the potentials are not the same.

There is one characteristic difference between the two functions: the relativistic potential has a maximum at a small value of r and drops down to negative infinity near $r = 0$, while the Newtonian one grows unbounded and approaches positive infinity at $r = 0$. The Newtonian potential always has a barrier that keeps a planet from falling all the way to $r = 0$ (unless $\ell = 0$), even if its energy is increased. In general relativity, this barrier is still present, but is of finite height. It is therefore

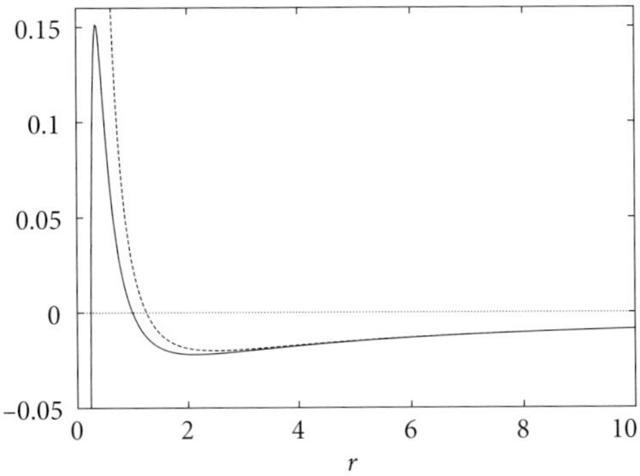

Figure 3.2 Effective potential for planetary orbits in general relativity (solid) and for Newtonian gravity (dashed). Deviations become more pronounced at small r. Planetary orbits have energies near the minimum of the potential, with r not changing much along a near-circular trajectory. (The mass and angular momentum in this example satisfy $GM/c\ell = 1/5$.)

easier to fall to small r, even though the barrier is still high. In Newtonian gravity, an object spiraling toward the Sun would spin up more and more and reach very high velocities that prevent it from falling further down. In relativity, the maximum speed is the speed of light; at some point, the spinning cannot be increased more to prevent falling further.

If the ratio $GM/c\ell$ is increased and reaches 1/4, the height of the potential barrier shrinks, the maximum reaching the value zero. An example for a ratio larger than 1/4 is shown in Figure 3.3. If the ratio is increased to $1/(2\sqrt{3}) \approx 0.29$, there is no hill at all; the potential just drops to smaller values whenever we move to smaller r; see Figure 3.4. The trough disappears, and no bound orbit is possible for

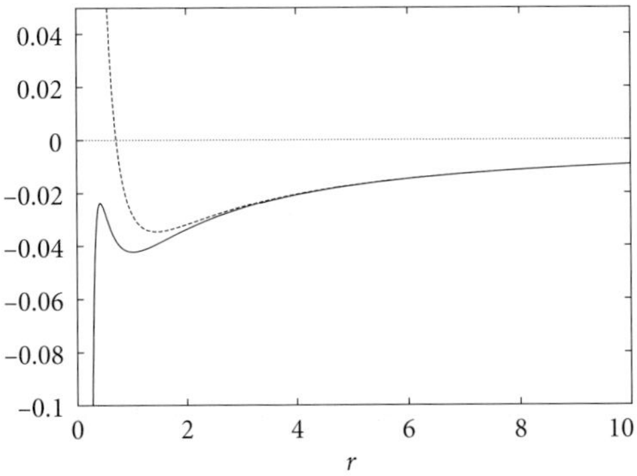

Figure 3.3 Effective potential for orbits of small angular momentum in general relativity (solid) and in Newtonian mechanics (dashed). The potential is negative for all values of r, while the form of the Newtonian potential is unchanged. (Mass and angular momentum in this example satisfy $GM/c\ell = 1/3.8$.)

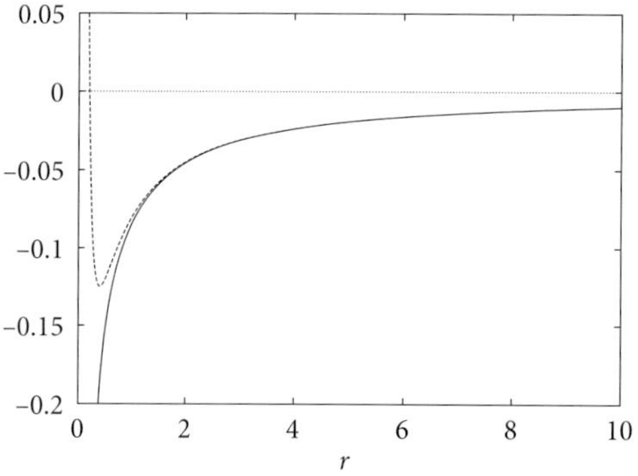

Figure 3.4 Effective potential for orbits of very small angular momentum in general relativity (solid) and in Newtonian mechanics (dashed). The potential is negative for all values of r, and does not have a trough, while the Newtonian potential is qualitatively unchanged. (Mass and angular momentum in this example satisfy $GM/c\ell = 1/2$.)

too small an angular momentum. In Newtonian physics, on the other hand, bound orbits always exist as long as there is any angular momentum, $\ell \neq 0$.

Planets, comets, flares The solar system has some variety, consisting of different objects that move along orbits controlled by the effective potential. For a small value of $GM/c\ell$, the potential landscape as in Figure 3.2 leads to orbits of four types. If the energy equals the value of P_{eff} at the bottom of its trough (or the top of its hill at small r) there is no radial force an the planet. The radius stays constant and the planet moves on a circle around the Sun. If we raise the total energy, there is some kinetic energy left at the trough position, the difference of the total and the potential energy. The rate of change $\Delta r/\Delta T$ is nonzero, and the radius does vary. It stops changing only when the potential walls of energy are reached, turning the planet's radius back toward the trough radius. The radius oscillates around the trough position, with an orbit of near-elliptic form. (For the Newtonian potential, an exact ellipse results, as in Figure 1.17.) The planet is bound to the Sun and cannot escape to large values of r.

If the energy is raised from the trough bottom so much that it becomes positive, the orbit is no longer bound. There may still be a barrier keeping the object away from small r, but no force prevents it from moving to larger and larger r. It approaches the Sun only once, and is then scattered back to a position far away. Some comets have unbound scattering orbits of this type. (In Newtonian gravity, these orbits have the form of a parabola or hyperbola.)

If the energy is increased further so that it becomes larger even than the top hill energy, there is no barrier whatsoever to stop the radial motion. If an object with such an energy falls toward the Sun, it will not be held back until it smashes into the surface. These plunge orbits, with large radial energy, do not exist for the Newtonian potential, which has no maximum; a barrier at small r always exists and a plunge is possible only if the energy is so high that the Sun's radius is reached before the energy barrier. For small angular momentum ℓ, there are plunges even in the trough of the potential if the Sun's radius is near the trough minimum. Matter ejected from the Sun, such as solar flares, starts at small r with large radial velocity, then hits the potential barrier and falls back to the surface. There is another type of plunge orbits for small r if the energy is less than the top hill. This range, however, usually falls within the Sun's bulk.

Escape velocity Escape is the opposite of plunge. Plunge orbits allow us to compute how much energy a solar flare must have for the ejected matter to reach the planets as solar wind and to escape the solar system. Applied to the gravity on Earth, the same formula tells us how much energy we have to provide for a rocket to leave the Earth.

We consider a rocket launched straight up, with radial motion and unchanging angles. The angular momentum $\ell = 0$ then vanishes. The energy must be at least so large that the rocket comes to rest far from the center, when r is almost infinite. The four-velocity at such a position is $\boldsymbol{u} = (c, 0, 0, 0)$, resulting in a conserved energy per mass of $e = (c^2 - 2GM/r)\Delta t/\Delta T = cu^t = c^2$. With these conserved quantities, the radial velocity, obeying the general speedometer equation, satisfies $\frac{1}{2}(\Delta r/\Delta T)^2 - GM/r = 0$, or $u^r = \Delta r/\Delta T = \sqrt{2GM/r}$ for the radial component of the four-velocity. We take the positive square root for outward radial motion of a fired rocket. With unchanging angles, the angular components of the four-velocity vanish, $u^{\mathrm{La}} = 0 = u^{\mathrm{Lo}}$. The time component for all r follows from our conserved energy: $c^2 = e = (c^2 - 2GM/r)u^t/c$ implies $u^t = c/[1 - 2GM/(c^2 r)]$. As a four-vector,

$$\boldsymbol{u}(r) = \left[\frac{c}{1 - 2GM/(c^2 r)}, \sqrt{\frac{2GM}{c^2 r}}, 0, 0\right] . \tag{3.8}$$

The required energy as measured by an observer staying at the launch pad, at Earth's radius $r = R$, is $E_R = -m\boldsymbol{u}(R) \cdot \boldsymbol{u}_{\mathrm{obs}}$ with the rocket mass m. The observer is at rest, and therefore has a normalized four-velocity $\boldsymbol{u}_{\mathrm{obs}} = [c/\sqrt{1 - 2GM/(c^2 R)}, 0, 0, 0]$. The energy equals $E_R = -m g_{tt} u^t(R) u^t_{\mathrm{obs}} = mc^2/\sqrt{1 - 2GM/(c^2 R)}$. Since the radius R is much larger than the Earth's Schwarzschild radius, the energy can be approximated as $E_R \approx mc^2 + GmM/R$, the rest energy plus the Newtonian potential energy.

A rocket of mass m needs a total energy $E_R = mc^2/\sqrt{1 - 2GM/(c^2 R)}$ to escape the gravitational bounds of Earth, with M and R the planet's mass and radius. An observer at rest would attribute an energy of $E_R = mc^2/\sqrt{1 - V^2/c^2}$ to the rocket moving at speed V. By direct comparison, the escape velocity of a rocket is $V = \sqrt{2GM/R} = c\sqrt{R_S/R}$ with the Schwarzschild radius R_S of Earth. The ratio of radii is small: $V \approx 0.000\,03c$, an escape velocity much less than c.

For a denser stellar object, a much larger escape velocity close to the speed of light is required. As another ominous sign of strong space-time bending, we notice that this tough escape happens when the surface radius is close to the Schwarzschild radius. At the Schwarzschild radius, the escape velocity equals the speed of light. No massive object can be that fast; no matter can escape from the Schwarzschild radius.

Kepler's third law A planet's orbit is almost circular. The orbit radius is close to the effective trough bottom for given values of M (the solar mass) and ℓ (the planet's angular momentum), located at a radius of $\ell^2\{1 + \sqrt{1 - 12[GM/(c\ell)]^2}\}/(2GM)$. The smallest possible value, reached for $c\ell/(GM) = 2\sqrt{3}$, is $r = 6GM/c^2 = 3R_S$, called the radius of the innermost stable circular orbit (or ISCO). Circular motion is impossible for $r < 3R_S$, a range which usually falls within the star's interior. However, dense stars can have a surface radius less than $3R_S$, still larger than the Schwarzschild radius. Sometimes, the ISCO radius is observable by matter orbiting the compact object. For instance, an accretion disk of debris left over from a companion star's collapse can be stable only in a range of radii larger than $3R_S$. Matter scattered to smaller radii does not orbit but rather falls all the way to the star's surface or the Schwarzschild radius.

For radii larger than $3R_S$, stable circular orbits are possible on which a planet may move around the star for a long time. The period of the orbit is obtained from the orbiting frequency F as $1/F$. Measured by an observer far away from the star, the angular velocity is $F = \Delta(\mathbf{La}/\bigcirc)/\Delta t$, the change of angle per full circle relative to the change of the Cartesian time coordinate t. We relate F to our conserved quantities by writing $F = [\Delta(\mathbf{La}/\bigcirc)/\Delta T]/(\Delta t/\Delta T)$ and using $\Delta(2\pi\mathbf{La}/\bigcirc)/\Delta T = \ell/r^2$ and $\Delta t/\Delta T = e/(c^2 - 2GM/r)$. Thus, $2\pi F = r^{-2}(c^2 - 2GM/r)\ell/e$.

The radius of a circular orbit does not change. The speedometer equation then reads $e = c^2\sqrt{[1 - 2GM/(c^2 r)][1 + \ell^2/(cr)^2]}$, and the orbit radius $r = (\ell^2/2GM)\{1 + \sqrt{1 - 12[GM/(c\ell)]^2}\}$ implies $\ell^2 = GMr/[1 - 3GM/(c^2 r)]$. Putting everything together, $\ell/e = \sqrt{GMr}/(c^2 - 2GM/r)$, or $4\pi^2 F^2 = GM/r^3$.

The angular velocity $2\pi F = \sqrt{GM/r^3}$ implies a squared orbit period of $1/F^2 = 4\pi^2 r^3/(GM)$, behaving like the cube of the orbit radius. Planets farther away from the Sun take longer to complete one orbit. This law, derived here from general relativity, is the same as Kepler's third law. There are no correction terms compared to the Newtonian result, provided we determine the period referring to the Cartesian time coordinate t of an observer far from the Sun. The orbit period measured on a planet such as Earth does not refer to t but rather to the planet's proper time T. If the planet's velocity is small, it does not give rise to much time dilation, and if it is far away from the Schwarzschild radius of the Sun, gravitational time dilation is not significant either. The formula of Kepler's third law is an excellent approximation for all solar planets.

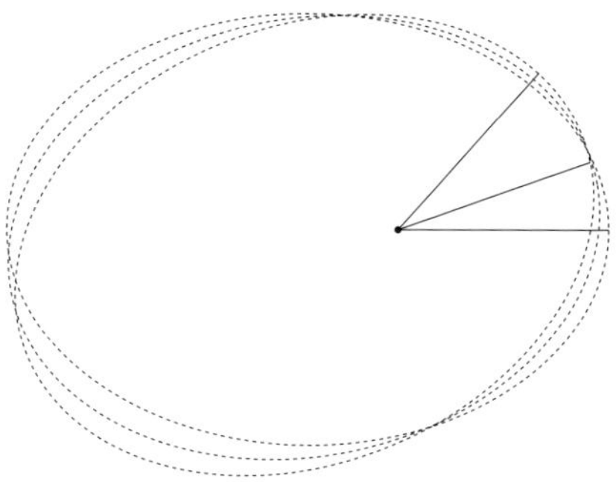

Figure 3.5 Perihelion shift: The Kepler ellipse of a planet (or of the Moon around Earth) rotates slowly, so that the point of closest encounter with the Sun, the perihelion, shifts from orbit to orbit.

Perihelion shift Mercury had a dual role: the ancient Greeks' messenger god, and the patron of rebels and thieves. Mercury the planet, closest to the Sun, rebels against the old laws of gravity and carries messages from space-time curvature, felt the strongest near the Sun. Its orbit is an interesting test case to see whether Newton's law of gravity or Einstein's theory of general relativity provide a better fit. Mercury, close to the Sun, is not easy to observe, and its orbit was not known well for a long time. By 1855, observations had accumulated to a long period covering many orbits. Looking at the data, the astronomer Urbain Le Verrier noticed that Mercury defies Kepler's laws and does not move along an exact ellipse. When observed over many of its periods, the orbit rotates by shifting its point closest to the Sun, the perihelion, by small amounts (Figure 3.5). In one century, the angular position of the perihelion rotates around the Sun by just 1.5°.

Perihelion shifts are not uncommon. The moon shifts its perigee, the analog as the closest encounter with Earth. One can explain deviations from Kepler's strict ellipse by the Sun's influence and the Earth's nonspherical form, with centrifugal bulges around the equator. The Newtonian laws of gravity need not be changed; with more precise observations, we just have to consider more details of mass distributions that cause gravity. The Sun is not exactly spherical either, accounting for some of the perihelion shift. Also, other planets, foremost Mercury's neighbor Venus, attract the planet toward them, a promiscuous source of gravity not taken into account by Kepler's ellipses for motion around the Sun. Unlike the moon, however, Mercury insists on a small amount of 0.012° per century as its perihelion shift even if all those effects are added.

To account for the remaining shift, some astronomers suggested that there might be another, melancholy planet even closer to the Sun, not much outgoing and hiding in the intense solar glare. However, it was difficult to find a culprit. General relativity, on the other hand, proffers an elegant explanation without the need of new assumptions. Its effective potential modifies Newton's force, so that noncircular orbits are no longer closed, but have their points displaced by a small amount after completing one orbit.

The effective potential suggests to view noncircular orbits as small oscillations around circular ones for a radius at the bottom of the trough. The frequency of these oscillations depends on the force, the slope of the potential around the minimum; if the potential is modified the frequency changes. For Newton's force, the oscillation frequency happens to be identical to the orbital frequency F. When a planet completes one orbit around the Sun, radial oscillations have returned to their starting point. The radius swings back to its value at the beginning of the orbit just in time to close the orbit and continue anew, in periodic fashion. The correction from general relativity brings the two frequencies out of tune. The orbit does not close; the perihelion shifts.

If we are interested in the behavior of small deviations from a circular orbit, located at the bottom of the potential trough, the general equation can be solved by approximation. We call the circular radius, for given M and ℓ, $r_\circ$, and write the radial displacement along the noncircular orbit as $r = r_\circ + d$ with $d = r - r_\circ$. Treating d as a small number allows simplifications to derive the oscillation frequency.

Harmonic oscillator We first consider the Newtonian effective potential $P_{\text{eff}}(r) = -GM/r + \frac{1}{2}\ell^2/r^2$ and write it as a function of d:

$$P_{\text{eff}}(d) = -\frac{GM}{r_\circ + d} + \frac{\ell^2}{2(r_\circ + d)^2} = \frac{-2GM(r_\circ + d) + \ell^2}{2r_\circ^2} \frac{1}{(1 + d/r_\circ)^2} .$$

When $d/r_\circ$ is small, near the circular orbit, the last fraction can be approximated by a simpler polynomial: The product of the denominator $(1 + d/r_\circ)^2$ with a quadratic polynomial $1 + Ad/r_\circ + Bd^2/r_\circ^2$, evaluated as $1 + (2 + A)d/r_\circ + (1 + 2A + B)d^2/r_\circ^2 + (A + 2B)d^3/r_\circ^3 + Bd^4/r_\circ^4$, is closest to one when $A = -2$ and $B = -1 - 2A = 3$. The linear and quadratic terms then disappear, and the cubic and quartic ones are small corrections. We therefore approximate $1/(1 + d/r_\circ)^2 \approx 1 - 2d/r_\circ + 3d^2/r_\circ^2$ and write

$$P_{\text{eff}}(d) = \frac{1}{2r_\circ^2}\left(\ell^2 - 2GMr_\circ - 2\frac{\ell^2 - GMr_\circ}{r_\circ} d + \frac{3\ell^2 - 2GMr_\circ}{r_\circ^2} d^2 + \cdots\right) .$$

The term of P_{eff} linear in d vanishes if $r_\circ$ is the radius at the trough bottom, where the potential is horizontal: $r_\circ = \ell^2/GM$ is the Newtonian circular radius. The coefficient of the quadratic term in P_{eff} is $4\pi^2 f_{\text{rad}}^2 = \ell^2/r_\circ^4$. A quadratic potential, or a linear force, implies harmonic oscillations $d(t) \sim d_0 \sin(\bigcirc f_{\text{rad}} t)$ with frequency f_{rad} around the circular radius. This frequency matches with the Kepler frequency $2\pi F = \sqrt{GM/r_\circ^3}$ for one period of the orbit. The orbit is closed, even if it is not circular.

The agreement of f_{rad} and F, computed for Newton's potential, is rather coincidental. For the effective potential of general relativity, there is an extra term to be taken into account. If we again write the potential in terms of the small deviation d

from the circular radius, a mismatch of $f_{\rm rad}$ and F arises. Oscillations around the circular orbit are out of tune with the orbit period. The orbit does not close, shifting the perihelion.

> **Perihelion precession** The circular orbit in general relativity has radius $r_\circ = [\ell^2/(2GM)]\{1+\sqrt{1-12GM/(c\ell)^2}\}$, and the radial frequency for oscillations around it amounts to $4\pi^2 f_{\rm rad} = \ell^2[1-6GM/(c^2 r_\circ)]/r_\circ^4$. There is a new term bringing the frequency out of tune with the circular period: the difference $f_{\rm prec} = F - f_{\rm rad} = [\ell/(2\pi r_\circ^2)][1-\sqrt{1-6GM/(c^2 r_\circ)}] \approx 3GM\ell/(2\pi c^2 r_\circ^3)$ is not zero and gives us the angular rate of change of the perihelion, or the precession frequency. The angular velocity implies a change of angle $2\pi\Delta\mathbf{Lo} = \bigcirc f_{\rm prec}/F = 3\bigcirc GM/(c^2 r_\circ)$ per orbit.

Impact Light follows lightlike geodesics in space-time and, just like massive objects, experiences the effects of curvature and gravity. Even though light is not massive in the usual sense, it has energy and therefore an acting gravitational mass $m = E/c^2$. Light passing by a heavy mass like a star is deflected from a straight line.

We describe lightlike geodesics by conserved quantities $e = -c\boldsymbol{w}\cdot\boldsymbol{K} = [1-2GM/(c^2 r)]c^2\Delta t/\Delta L$ and $\ell = \boldsymbol{w}\cdot\boldsymbol{H} = (2\pi/\bigcirc)r^2\sin(\mathbf{La})^2\Delta\mathbf{Lo}/\Delta L$ with the wave 4-vector $\boldsymbol{w}$, just as in the massive case with four-velocity $\boldsymbol{u}$. Instead of proper time T, which does not progress at the speed of light, we use some curve parameter L. We replace the speedometer equation $||\boldsymbol{u}||^2 = -1$ with the dispersion relation $||\boldsymbol{w}||^2 = 0$.

> **Effective potential** Inserting the conserved quantities for rates of change of t and **Lo** in the dispersion relation, setting $\mathbf{La} = \frac{1}{4}\bigcirc$ constant for an equatorial orbit, implies
>
> $$-\frac{e^2/c^2}{1-2GM/(c^2r)} + \frac{1}{1-2GM/(c^2r)}\left(\frac{\Delta r}{\Delta L}\right)^2 + \frac{\ell^2}{r^2} = 0\,.$$
>
> Light motion has no Newtonian analog, but we can write its law by an effective potential:
>
> $$\frac{1}{b^2} = \frac{1}{\ell^2}\left(\frac{\Delta r}{\Delta L}\right)^2 + W_{\rm eff}(r) \qquad (3.9)$$
>
> with $b^2 = c^2\ell^2/e^2$ and $W_{\rm eff} = [1-2GM/(c^2r)]/r^2$.

Instead of the particle energy, we use a parameter $b = c\ell/e$ with units of length. Called the impact parameter, b is the distance by which a light ray would miss the central mass if it were to move along a straight line, as illustrated in Figure 3.6. If b

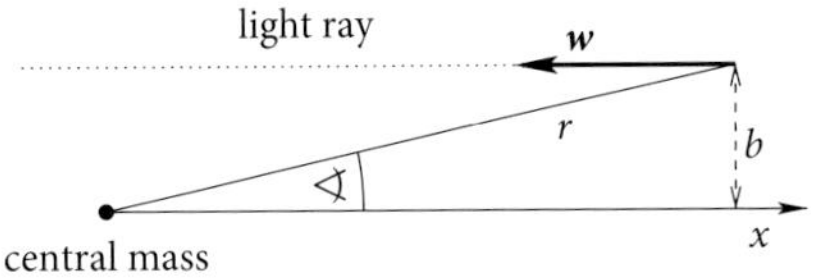

Figure 3.6 The impact parameter b by which a straight light ray would miss the central mass.

is much smaller than the displacement r, a line connecting the central mass to a point on the light ray is almost parallel to the light ray. The change of r in time is then almost equal to the speed of light, or the negative $\Delta r/\Delta t \approx -c$ when light is approaching the central mass. The connecting line is not exactly parallel, however; it has a small angle $\sphericalangle \approx \bigcirc b/(2\pi r)$ with it. The angle changes as r changes, while b remains constant. We have $\Delta\sphericalangle/\Delta t = (\Delta\sphericalangle/\Delta r)(\Delta r/\Delta t) = \bigcirc bc/(2\pi r^2)$. We obtain the same result from the conserved quantities, again for large displacement r compared to b: $b = |c\ell/e| \approx (2\pi/\bigcirc c)r^2(\Delta\sphericalangle/\Delta L)/(\Delta t/\Delta L) = (2\pi/\bigcirc c)r^2\Delta\sphericalangle/\Delta t$.

If light moved along a straight line passing by some center, the position x along the ray would change by $\Delta x/\Delta t = -c$ on approach. The displacement r of the tip of the light ray from the star, by Euclidean laws, obeys $r^2 = x^2 + b^2$ and changes according to $2r\Delta r/\Delta t = 2x\Delta x/\Delta t$, or $\Delta r/\Delta t = (x/r)\Delta x/\Delta t = -c\sqrt{r^2 - b^2}/r$. We write this equation as $(\Delta r/\Delta t)^2/(bc)^2 + 1/r^2 = 1/b^2$, just like (3.9) with an effective potential $1/r^2$ (and $L = bct/\ell = c^2t/e$). The central mass M changes the effective potential, adding $-2GM/(c^2r^3)$ and making light rays be deflected when they pass close by.

Deflection of light Even light light can be attracted to a massive star. For a light ray to pass a star like the Sun, the impact parameter b must be larger than the surface radius R of the star, which in turn is much larger than the Schwarzschild radius $R_S = 2GM/c^2$.

With this condition, the angle changes by an amount of $\sphericalangle \approx \frac{1}{2}\bigcirc + 4\bigcirc GM/(2\pi c^2 b)$, obtained by adding up $\Delta\sphericalangle/\Delta r = (\Delta\sphericalangle/\Delta t)/(\Delta r/\Delta t)$ from the minimum radius to infinity. As illustrated in Figure 3.7, the change $\frac{1}{2}\bigcirc$ corresponds to a straight line, while the rest is the deflection angle $\sphericalangle_{\text{deflect}} = \sphericalangle - \frac{1}{2}\bigcirc \approx 4\bigcirc GM/(2\pi c^2 b)$. For a light ray grazing the surface of the Sun, $b = R$ is the solar radius, and $\sphericalangle_{\text{deflect}}$ equals approximately $0.0005°$. This tiny change of the direction

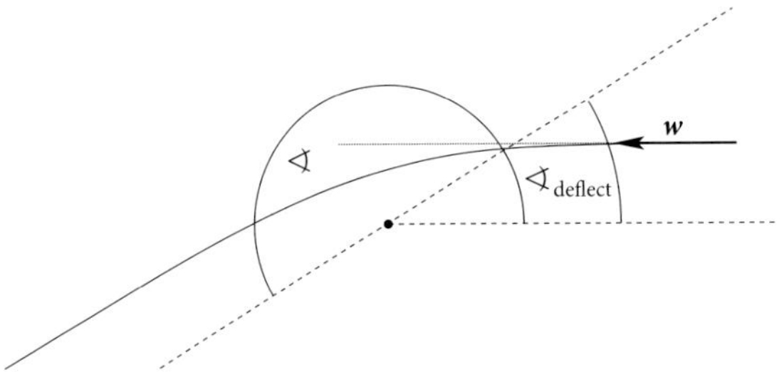

Figure 3.7 The deflection angle $\sphericalangle_{\text{deflect}}$ and the total angular change $\sphericalangle = \frac{1}{2}\bigcirc + \sphericalangle_{\text{deflect}}$.

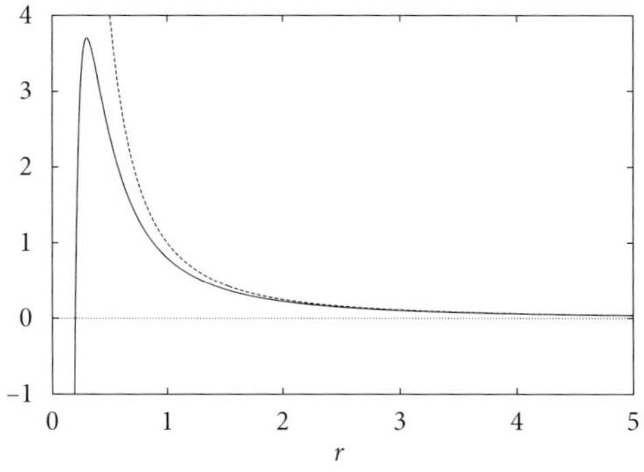

Figure 3.8 The effective potential for radial motion of light.

of sight of stars was first measured during a total solar eclipse in 1919, by a team led by Arthur Eddington.

Light orbits Light can be attracted and trapped. For large impact parameter b, light is deflected, but always keeps a safe distance from the star. In the potential landscape of Figure 3.8, any ray with small $1/b^2$ encounters a potential barrier. Coming from larger r, it is turned back to increasing r after reaching a minimum distance to the star. When b is smaller, however, $1/b^2$ may be larger than the potential at its peak, and no barrier limits r, realized when b^2 increases above $27(GM/c^2)^2$. The light ray then reaches any small r, or hits the star surface at $r = R$. Gravitational attraction can capture light passing close-by, as an analogy of plunge orbits of massive objects.

If the impact parameter equals $b = \sqrt{27}GM/c^2$, $1/b^2$ is at the maximum of the effective potential. The dispersion-relation/energy equation can then be fulfilled with vanishing $\Delta r/\Delta L$, or with a constant r. A constant displacement from the star means that such a light ray is in a circular orbit; light is captured and orbits around the star almost like a planet. The radius of the orbit is small, with $r = 3GM/c^2$ just above the Schwarzschild radius. For the Sun, this place would be well inside its bulk where light cannot move without being scattered, but it may be outside of the surface of a dense star. Even so, the orbit is unstable: a slight displacement to smaller or larger r implies forces pulling the light ray away from the circular orbit. One would have to shine one's light in just the right direction, staying exactly at the circular radius in order to see it come back after one orbit around the star.

Shapiro time delay Gravitational time dilation delays light. Although time does not progress for light itself, an observer will see different travel delays for light rays sent out, depending on what gravitational forces they encounter along their ways. The travel time can be computed by relating $\Delta t/\Delta L$, obtained from the conserved quantity e, to $\Delta r/\Delta L$ from the potential equation. Just as with light deflection, travel delays are the strongest for light passing by near the surface of the Sun.

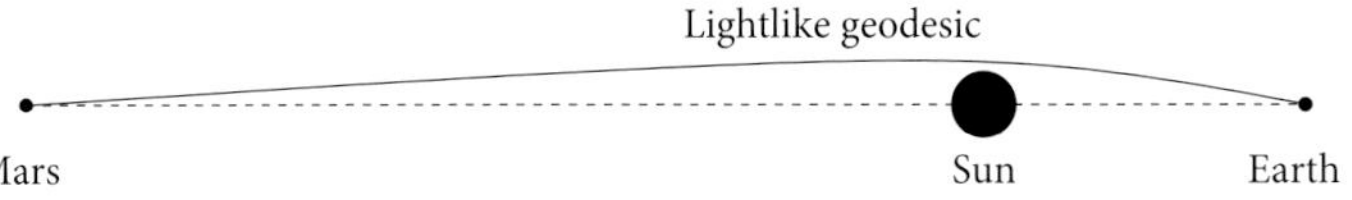

Figure 3.9 Radar signals pass near the Sun along a lightlike geodesic, and are reflected back from Mars. Gravitational time dilation implies that the trip takes longer than the distance from Mars to the Earth would imply.

One cannot use star light to measure time delays: these are steady sources of light without a clear beginning of their travel. Even bursts like supernova explosions are not helpful in this case because we do not have independent information about the time when they started. Instead, we can try to send our own signal out, at some time of our choosing, and have it reflected back to us. If we know the path traveled by light, we can compute the expected delay and compare with observations. It is not easy to have a light ray reflected back to us from far away and still be able to identify it among the sun light reflected by the same object. Radar signals can be identified with more ease, and it became possible to have them reflected back from Mars after the Viking lander placed a radar mirror there. In 1977, measurements undertaken by Stuart Shapiro confirmed the gravitational time delay as predicted by general relativity (Figure 3.9).

3.2 Black Holes

There are ominous signs for what might happen when the surface radius of a star, of mass M, reaches the Schwarzschild radius $R_S = 2GM/c^2$. Gravitational redshift and time dilation become infinite: all processes on the surface, as seen by an observer afar, stop. And, even if light could show us a picture of this frozen age, we could not see it, for any image would suffer from infinite redshift. All colors would be shifted out of the visible or even infrared and radar ranges. Light, a process of oscillating electromagnetic fields, terminates. No oscillations occur if the wave length is shifted by an infinite amount.

If this weren't enough, the world closer to the center at $r = 0$ is stranger yet. At a place with $r = R_S$, the escape velocity, required for anything, signals or rescue capsules, to reach positions far away, is the speed of light. At best, light could leave and bring us notice (if it weren't infinitely redshifted), but no one else can flee. If we go deeper still, the escape velocity increases above the speed of light, we are shut off. Nothing at all can leave the region of $r < R_S$. Without messengers, such regions cannot be observed from afar, except perhaps by the gravitational forces they exert.

We may ask how a hypothetical observer, falling to smaller r and crossing $r = R_S$, would experience what happens there. Such a daring adventurer could not tell us anything afterwards, for no information will get out, but we could still try to predict how the trip would go. Such questions may not seem relevant to us, at least at present. Dense objects with a surface position near their Schwarzschild radius do exist in the universe: neutron stars. However, they are far away, too far for us to send satellites (or astronauts) there to do observations for us. (Even if we could build rocket engines to send satellites that far, strong gravitational forces, which could deform even atomic nuclei, would make observations challenging.) Yet, we could probe regions near R_S if we see a collapsing star shrink to a surface radius below R_S, after it burned through most of its nuclear fuel. Or, if we find compact objects with surface radius already below R_S, black holes, additional hot matter might fall in, emitting light. As it approaches R_S, redshift and other effects should affect emitted light we observe.

What we would expect to see is this: As radiating matter falls in, gravitational redshift changes the color of emitted light to the red side of the spectrum, or even into the infrared. Time dilation also means that the falling-in process appears to slow down, and light is emitted less frequently. Matter seems to darken even while we expect it to shine more brightly as it heats up. Close to the Schwarzschild radius, time dilation and redshift make any radiation so dim and of low frequency that no distant detector could see it. The light of infalling or collapsing matter would fade out as it approaches the Schwarzschild radius; we would never see it cross this radius.

If we want to explore the region within $r = R_S$, we cannot avoid doing the trip ourselves, falling toward small r of a black hole or standing on the surface of a collapsing star. Let us now go on such a trip in the logical dream world of mathematics.

Inside of a black hole Geometry warns us of collapse. The line element (3.1) around a spherical mass M, of the Schwarzschild form, has several vanishing or infinite coefficients when $r = 2GM/c^2$. If there are infinities, it is no longer meaningful to work with such a line element. However, vanishing metric coefficients need not be such a bad thing. A two-sphere has a vanishing metric coefficient at the poles, where $\sin(\mathbf{La}) = 0$. Nothing special happens there; in fact, the geometry looks just the same as around any point on the sphere. What goes bad at the poles is our coordinates: meridians intersect and no longer determine a unique value of the longitude **Lo**. Changing **Lo** means that we jump between meridians, but we are not moving at all at a place where they intersect. The line element multiplies the nonzero $(\Delta\mathbf{Lo})^2$ with a vanishing number, $\sin(\mathbf{La})^2$ at $\mathbf{La} = 0$ and $\mathbf{La} = \frac{1}{2}\bigcirc$, to account for the zero distance. The zero is just a coordinate effect because we can avoid it by using new coordinates, for instance, by rotating the sphere, shifting the zero elsewhere, or by using stereographic projection.

When metric components vanish or diverge, one should always check whether this is a consequence of bad coordinates or something physical. The metric coefficients depend on what coordinates we choose, not just on physical processes. For the Schwarzschild radius, it turns out that its badness can be forgiven; it is not owed to the geometry, but rather to coordinates. Forgiveness may be difficult to test, but for the Schwarzschild metric, one can find honest coordinates whose confession removes all troubles.

We suspect that matter can fall through the Schwarzschild radius, and that it is only redshift and time dilation which make the interior region unobservable for telescopes positioned afar. At r much larger than R_S, observers refer to t used in the Schwarzschild line element as the time coordinate of their reference system. When they see interior physical processes stop, a nonvanishing Δt (an observer's time lapse) must imply $\Delta s = 0$ at $r = R_S$. In Δs, $(c\Delta t)^2$ is indeed multiplied with zero, the metric coefficient $-[1 - 2GM/(c^2 r)]$ evaluated at $r = R_S$, just as $(\Delta \mathbf{Lo})^2$ on the two-sphere is multiplied with the vanishing $\sin(\mathbf{La})^2$ at the poles. New, more suitable coordinates can be found by considering the motion of an observer falling in along a timelike geodesic. Proper time T along such a curve can be computed as a function of t and r. If we replace t by T, this new coordinate should provide a well-defined geometry even around $r = R_S$. Different choices exist for new coordinates, depending on what infalling observers one considers. A convenient one is Eddington–Finkelstein coordinates.

Eddington–Finkelstein metric We use a new coordinate $v(t, r) = ct + r + (2GM/c^2)\log|c^2 r/(2GM) - 1|$ instead of t (keeping r unchanged). As v remains finite, $r = R_S = 2GM/c^2$ can be reached only for infinite t. We can therefore hope that the new coordinate will allow us to cover a larger space-time region, including $r < R_S$, by using v to go beyond the infinity of t.
Coordinate separations satisfy $c\Delta t = \Delta v - \Delta r - \Delta r/(r/R_S - 1) = \Delta v - \Delta r/(1 - R_S/r)$. Inserted in the Schwarzschild line element, this produces

$$\Delta s^2 = -\left(1 - \frac{R_S}{r}\right)(\Delta v)^2 + 2\Delta v \Delta r + \left(\frac{2\pi r}{\bigcirc}\right)^2 \left[(\Delta \mathbf{La})^2 + \sin(\mathbf{La})^2 (\Delta \mathbf{Lo})^2\right],$$

a new line element which is not diagonal and therefore more complicated to deal with. However, even though the coefficient of $(\Delta v)^2$ still vanishes at the Schwarzschild radius, distances corresponding to changes Δv are not reduced to zero: The mixed term $2\Delta v \Delta r$ can still provide nonzero Δs^2 if r changes too. The off-diagonal nature saves our geometry near the Schwarzschild radius.

Light cone Light is trapped in a black hole, for even its speed is not enough as an escape velocity. Light cones in a space-time of Eddington–Finkelstein form show the same effect. By drawing light cones at different events in space-time, we explore the causal structure and see how observers in different regions can communicate.

We will focus our attention on radial light rays moving straight toward or away from the center without angular change. They come in two different types, those toward the center at small r and those pointing away, to larger r. They solve the lightlike equation $\Delta s^2 = 0$ with the conditions $\Delta \mathbf{La} = 0 = \Delta \mathbf{Lo}$: an equation $-(1 - R_S/r)(\Delta v)^2 + 2\Delta v \Delta r = 0$ that relates changes Δv to changes Δr.

One solution is easy to see: $\Delta v = 0$. If v is constant while r changes, we move along a light ray. The new coordinate v therefore is not a time variable, labeling space as different instants of time, but rather a variable that labels different light rays. If Δv is not zero, the lightlike equation can be solved only if $-(1 - R_S/r)\Delta v + 2\Delta r = 0$, which happens to be fulfilled for $v - 2(r + R_S \log|r/R_S - 1|)$ a constant. There is one more solution, belonging to constant Δr provided that constant is $r = 2GM/c^2 = R_S$ (staying at the Schwarzschild radius means that we have to move like a light ray). Only light can stay put at the Schwarzschild radius.

Space-time diagram A space-time diagram can show us what space and time are. In special relativity, space and time are not separate from each other but combined to one entity, space-time. Still, there is a distinction between these two markings of events. Our own experience tells us that we can move back and forth in space, the direction left to our own choosing, most of the time at least. In time, however, we are bound to keep moving forward. While special relativity does not explain the strict directedness of time, it models our understanding of causality by the different parts of light cones. Motion, starting from one event, can only happen toward the future part of the light cone of the event. Nothing can leave the light cone or move back to the past light cone.

The form of light cones, a property of space-time geometry, is encoded by the line element, Minkowskian in special relativity. Here, time intervals are characterized by the negative coefficient they carry in Δs^2, while all spatial intervals have positive coefficients. The coordinate with a negative coefficient points along the axis of light cones, and therefore plays the role of time.

In general relativity, the same understanding of causality is used. Light cones are determined by the line element of curved space-time, and show in which space-time direction one must keep moving forward. Metric coefficients now depend on space or time coordinates; if we move in space-time, some of the signs may change. What looks like a time coordinate in one region of space-time may no longer play the role of time in another region. The Schwarzschild line element is one example: When r crosses $r = R_S$, the metric coefficient $-(1 - R_S/r)$ multiplying $(c\Delta t)^2$ turns positive, and t is not time. Instead, the coefficient of $(\Delta r)^2$ then carries the minus sign. Once we cross the Schwarzschild radius, we must consider r, not t, as the time coordinate.

This strong sense of malleability of space and time, allowing them to switch their roles, can make the drawing of space-time diagrams tricky. If we want to illustrate interior and exterior regions around the Schwarzschild radius in a single diagram, we must use a notion of time valid in both. We cannot use either t or r because each one looks like time in only one of the regions.

Eddington–Finkelstein time We search for a better time, by organizing space. A good time coordinate must be such that the rest of the line element, once our time is fixed, includes only space. A spatial line element has its coefficients all positive, a condition that can be checked for different time candidates. For t itself, for instance, we confirm that it is no good time within the Schwarzschild radius: $(\Delta s^2)_{t\,\text{constant}} = (\Delta r)^2/(1 - R_S/r) + (2\pi r/\bigcirc)^2 \left[(\Delta \mathbf{La})^2 + \sin(\mathbf{La})^2 (\Delta \mathbf{Lo})^2\right]$ has a negative metric coefficient for $r < R_S$; and r certainly is no time for $r > R_S$.

Eddington–Finkelstein coordinates give us more options. We have already seen that they remove some pathological features of our original coordinates at the Schwarzschild radius. However, we cannot use the new coordinate v, constant along light rays not along space, as time: the expression

$$(\Delta s^2)_{v\ \mathrm{constant}} = \left(\frac{2\pi r}{\bigcirc}\right)^2 \left[(\Delta \mathbf{La})^2 + \sin(\mathbf{La})^2(\Delta \mathbf{Lo})^2\right] ,$$

while it does not have negative coefficients, has a vanishing coefficient of the third spatial separation $(\Delta r)^2$. There is a simple modification of v so as to make it a good time: a constant $v - r$ means that $\Delta v = \Delta r$, and gives $(\Delta s^2)_{v-r\ \mathrm{constant}} = (1 + R_S/r)(\Delta r)^2 + (2\pi r/\bigcirc)^2 \left[(\Delta \mathbf{La})^2 + \sin(\mathbf{La})^2(\Delta \mathbf{Lo})^2\right]$ with all three spatial metric coefficients positive. We can draw space-time diagrams covering the interior as well as exterior region of the Schwarzschild radius by using $v - r$ as time and keeping r as the spatial coordinate for radial separations.

Light cone tipping Having a good time with light cones makes us see space-time. Using coordinates $v - r$ and r for space-time diagrams, our equations for light rays show one set of solutions for constant v, so that time $v - r$ increases as much as r decreases when a light ray moves inwards. These light rays are lines of 45° in the space-time diagram, Figure 3.10. Outgoing light rays obey a more complicated equation, with constant $(v - r) - r - 2R_S \log|r/R_S - 1|$. For r much larger than R_S, the logarithm is much smaller than the linear r. These light rays are at almost 45° for large r, where $(v - r) - r$ is almost constant: time $v - r$ grows like a constant plus r.

When we get closer to the Schwarzschild radius the logarithm becomes the dominant term. It becomes negative infinity for $r = R_S$ because the log is negative

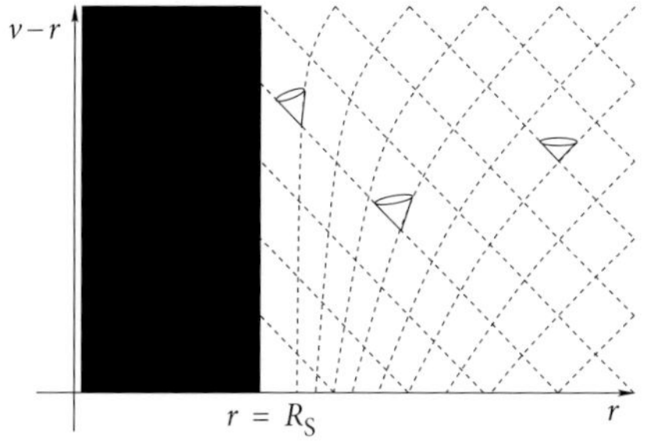

Figure 3.10 A space-time diagram of the Schwarzschild geometry outside of the Schwarzschild radius. Space-time coordinates are adapted from Eddington–Finkelstein form, with the interior region of radii smaller than the Schwarzschild radius blacked out. One set of radial light rays (moving toward smaller r) has constant v and is diagonal. Outgoing light rays move along curved lines which are nearly diagonal for large r, but turn almost vertical near R_S. They never intersect the line $r = R_S$, which itself is a light ray. Three examples for light cones are shown, demonstrating how they tip toward the Schwarzschild radius at small r. Near R_S, it becomes more and more difficult for light to move outwards: most of the future light cone points toward smaller r.

infinity for vanishing argument. For $(v - r) - r - 2R_S \log|r/R_S - 1|$ to remain constant, $v - r$ must approach negative infinity as well. The whole past range in time for these light rays is already covered when r approaches R_S from larger values. The rays never intersect the Schwarzschild radius, even though they get closer and closer at early times.

With the drawing in Figure 3.10, the bending light rays show that light cones tip over compared to their form in flat space. Near the Schwarzschild radius, most of a future light cone points to the left, toward R_S. Most light rays starting near R_S move to smaller r, and only a minority can move out to large r. An observer far outside cannot see much of the light emitted near R_S; even if hot matter falling in keeps shining brightly, it appears to dim, as seen from outside. An observer falling in with the hot matter, on the other hand, would still receive the full blast of light.

Black hole Inside the Schwarzschild radius, light cones are introverted. With $v - r$ remaining a time coordinate even in this region, we continue our space-time diagram as in Figure 3.11. All light rays point toward smaller r; none can escape the region and cross the Schwarzschild radius. Light cones are tipped to the left so much that their entire future comes to lie within the Schwarzschild radius. Within this region, the future means even smaller radii, a directedness that is consistent with the fact that r is a time coordinate when it is less than R_S. Just as our time always increases toward the future, at $r < R_S$, there is a radius always decreasing toward the future. At the border line $r = R_S$, the light cone is tipped such that all but one light ray point to smaller r, the sole exception a light ray staying fixed at $r = R_S$. Indeed, as we have seen earlier, $\Delta s^2 = 0$ along $r = R_S$, identifying the line as a light ray.

No light ray can escape from the region of $r < R_S$. If we stay outside, we will never see this region; it appears black: a black hole. The line $r = R_S$ is the border between the region of no escape at smaller r and the outside from where one could still avoid falling in. At $r = R_S$, all massive objects are doomed to fall in, and light can at best stay at this surface and avoid falling in, but will never be able to move outside.

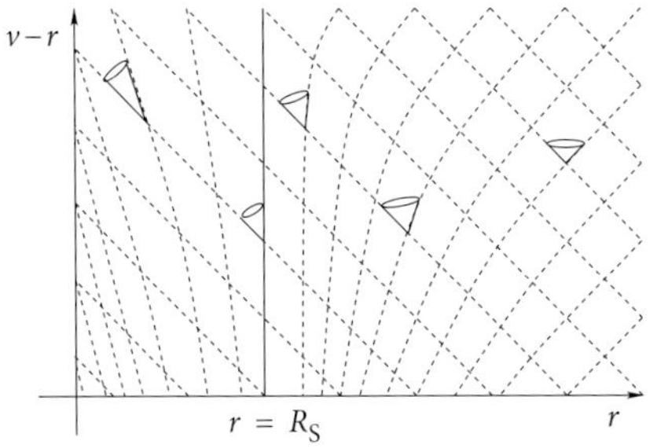

Figure 3.11 A space-time diagram with light cones inside and outside of the Schwarzschild radius. For $r < R_S$, the tipping of light cones is so strong that no light ray points outside. No light can escape the interior region, which appears black, that is, a black hole.

An observer falling in with collapsing matter, crossing the Schwarzschild radius, perceives nothing particularly dangerous. We notice the tipping of light cones only with respect to our nearly Minkowskian reference system at large r. An infalling observer would not base observations on such coordinates. In a small surrounding region, one could rotate the space-time diagram, thereby defining new coordinates good for local measurements. Such coordinates would not be useful for an observer staying outside because light rays far away from R_S seem unnecessarily tilted. However, our infalling observer does not care about coordinates far outside; for such an observer, we can always use coordinates that do not show the tipping of light cones.

For an observer staying at large values of r, the Schwarzschild radius seems like an insurmountable barrier, blocking any message from the other side. An observer falling to smaller r and crossing R_S does not notice any barrier, neither a material one like the surface of a star nor strong forces that could not be withstood. Such an observer moves right across the Schwarzschild radius, and from then on, must keep moving to ever smaller r. The Schwarzschild radius is not a material surface; it rather appears like the horizon we see on the Earth's spherical surface. It seems the end of the world if we stand still, but it reveals what is behind it when we walk toward and cross it.

The Schwarzschild radius, by somewhat of an analogy, is called the horizon of the black hole. It marks the region from which escape is impossible for anything, including light. It is not a material surface; in fact, it is not even a spatial surface. It is the place of the last light ray that is just barely able to prevent falling in to smaller r. The horizon is a lightlike region, unlike any surface bordering the interior of a star. We cannot land on the horizon, stay there for a while and take off again, for any material object, once it reaches the horizon, falls right through it to smaller r.

Black hole singularity A black hole has no surface and no center. Once objects or observers cross the horizon, they must keep moving to smaller r. They will reach $r = 0$, at which place the line element again becomes problematic. For a star, $r = 0$ is where its center is located, with the surface nearly spherical around it. A star contains matter within its surface, so that the geometry inside is no longer of Schwarzschild form, a strict vacuum region. For a black hole, the Schwarzschild metric applies for all r.

If we use the Schwarzschild metric throughout the interior, we must make sense of its components at $r = 0$, some of which become zero or infinite. At the horizon, Eddington–Finkelstein coordinates have provided a regular description of the geometry, but they fail to remove infinities at $r = 0$. We should look for yet another set of coordinates that may remain valid even there. However, any coordinate system fails at the center, for $r = 0$ is a singularity of space-time, that is, a place of infinite metric components not removable by a better coordinate choice. It would take long to go through all possible coordinate systems and test whether any one can provide finite, nonzero metric coefficients. Instead, we once again make use of invariants that do not depend on what coordinates we choose. What we need now is a curvature invariant, a measure of space-time geometry insensitive to coordinate changes.

Curvature is a geometrical property that can be noticed by the gravitational force it implies. All observers would feel the force independent of the coordinates they choose to use, and so there must be an invariant measure of the corresponding curvature. Geometry gives us examples for such quantities, and one of them has the value $R_2(r) = 2\sqrt{3}R_S/r^3$ for Schwarzschild geometry. It has units of inverse length squared, like all curvature quantities, and it depends on the radial distance in such a way that it is infinite for $r = 0$. This is an infinity that cannot be removed by changing coordinates, for R_2 is coordinate invariant. The place $r = 0$ is a singularity, where coordinate-independent and physical quantities become infinite. (At the Schwarzschild radius, by contrast, $R_2 = 2\sqrt{3}/R_S^2$, a finite number which is even quite small for large masses.)

Riemann tensor We have seen one curvature invariant, the Ricci scalar R used in the Einstein–Hilbert action. This quantity vanishes for the Schwarzschild metric, as it does for all vacuum geometries according to Einstein's equation with $T_{ab} = 0$. One can construct another invariant from the Riemann tensor, whose components are not all required to vanish. An invariant is constructed by pairing up all indices so as to cancel coordinate changes. One ungainly possibility, $\mathbf{Sum}_{a,b,c,d,e,f,g,h} R_{abc}{}^{d} R_{efg}{}^{h} g^{ae} g^{bf} g^{cg} g_{dh}$ with the Riemann tensor $R_{abc}{}^{d}$, is tedious to compute but provides the nonzero curvature invariant R_2.

Tidal forces Space-time curvature raises the tides; it makes the gravitational force change if we move just a bit. Any extended object then experiences forces differing much on its sides, tearing it apart. By analogy with the forces pulling on Earth and the Moon during their joint orbiting, these forces are called tidal forces. If curvature is infinite, so are tidal forces. Even an object as small as an elementary particle would be torn apart. No matter, as we know it, could withstand the forces active near $r = 0$ in a black hole.

As harmless as the horizon may seem for an infalling observer, it encapsulates a dangerous trap. Not only would observers or any matter be drawn to smaller r, they would be torn apart when they come close to $r = 0$. However, if matter cannot persist in this region, what is $r = 0$? If it were the center of a black hole or of the collapsed matter inside it, it should allow matter to exist. Matter, however, is vanquished at $r = 0$ or near it, and so is even space and time. No coordinate transformation can save the space-time metric from becoming infinite, and so geometry ceases to exist just as matter does. Without a metric, space and time come to an end at $r = 0$, producing tidal forces so strong that not even empty space can withstand them.

Black hole singularity There is a singularity inside a black hole, within the black hole horizon, but it is not its center. A center in the common sense of the word would be a point in space of nearly equal distance from a surrounding surface. Black holes do not have a material surface, but for the present purposes, the horizon is a good substitute. We would then be looking for a spatial point of nearly equal distance from the horizon.

The singularity at $r = 0$ is of equal distance from the horizon at $r = R_S$ constant, but it is not a point in space. Inside the horizon, the radial coordinate r turns timelike; it becomes a time coordinate with directedness toward smaller r. A constant value of r is then a constant value of time, a moment in time rather than a point in space. The black hole singularity is not the center of the black hole; it is its end because the geometry of space-time breaks down when $r = 0$ is reached.

When we fall in a black hole, we do not see the singularity threatening before us. Everything, including light, must move to smaller r within the horizon, and no message can reach us from $r = 0$ as long as we are still at nonzero r. We do notice the ever-increasing tidal forces, but we do not see what awaits. We would be torn apart quite soon, even before reaching $r = 0$, but the end, namely, the singularity, will be reached only at $r = 0$. For all we would know, the singularity does not yet exist on our approach to $r = 0$ because it will form at *time* $r = 0$, in our future when we are still moving to smaller r. The singularity forms only in the moment when we hit it.

Collapse A black hole forms when the surface of a collapsing star falls below its Schwarzschild radius. As the surface crosses the horizon, nothing special happens as seen on the star; only distant observers will notice incessant dimming and redshifting. The star then continues collapsing; in fact, it has to, with all its matter required to move toward the future of smaller r. All matter will reach $r = 0$; it will be torn apart by tidal forces and disappear together with space and time. The star has collapsed into a singularity, and leaves nothing behind except a black hole covered by its horizon.

Figure 3.12 shows the space-time diagram of gravitational collapse. A collapsing star has a shrinking surface, following a timelike worldline toward smaller r. An observer outside sees the star's light as periodic signals. As the star collapses, its surface falls into regions of stronger gravitational force. Light is redshifted and processes of other regular signals are delayed more and more: it takes longer and longer for the light rays to climb out of the gravitational grasp. In the figure, intersections of outgoing light rays with the star surface show that signals must be emitted ever faster so as to reach the distant observer with constant spacing. Signals emitted at regular intervals, such as jets of rotating stars, appear ever more delayed to the distant observer.

Gravitational redshift, by the formula $f_{\text{obs}}/f_{\text{emitted}} = \sqrt{g_{tt}(r_{\text{surface}})/g_{tt}(r_{\text{obs}})} = \sqrt{(1 - R_S/r_{\text{surface}})/(1 - R_S/r_{\text{obs}})}$, shifts all frequencies to zero at $r_{\text{surface}} = R_S$. Zero frequency means an unchanging wave, which cannot be detected. The brightness of the star drops to zero as its surface approaches the Schwarzschild radius.

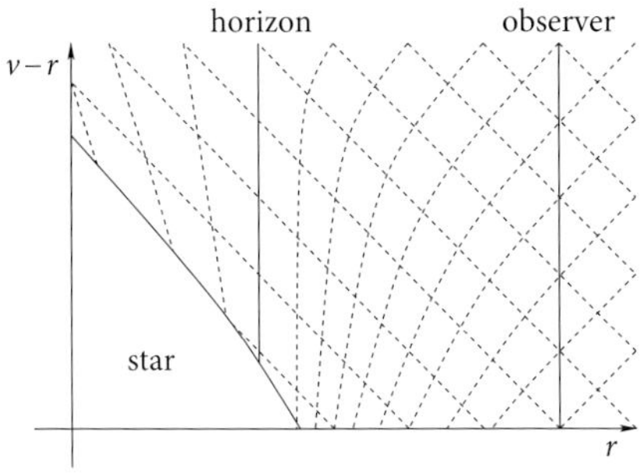

Figure 3.12 A star collapses, its surface falling below the Schwarzschild radius. An outside observer sees the star only as long as light rays can escape from the surface, but the signals appear ever more delayed. When the surface radius reaches R_S, a horizon forms and light is trapped.

To analyze the approach to $r = R_S$ as seen from afar, we connect events on the collapsing surface with the distant observer by outgoing light rays, obeying $v - 2(r + R_S \log|r/R_S - 1|) = C$ constant. For some fixed constant C, we evaluate this equation at coordinates (v_{obs}, r_{obs}) for the observer, and at $(v_{emitted}, r_{emitted})$ for an emission event on the surface: At r_{obs}, the logarithm is much smaller than the linear term in r, and we have $C \approx v_{obs} - 2r_{obs} \approx ct_{obs} - r_{obs}$. At $r_{emitted} \approx R_S$, we have $C \approx -2R_S \log|r_{emitted}/R_S - 1|$. If we combine the two equations by equating the Cs and exponentiate to remove the logarithm, we have $r_{emitted}/R_S - 1 \approx \exp[-(ct_{obs} - r_{obs})/(2R_S)]$.

For an observer stationary at $r = r_{obs}$, the approach of the surface to $r_{emitted} = R_S$ appears to happen by the exponential function: the distance $r_{emitted} - R_S$ is proportional to $\exp[-ct_{obs}/(2R_S)]$, where t_{obs} in the near-Minkowski region far away from the Schwarzschild radius is close to the observer's proper time. The exponential function $\exp(-x)$ (Figure 1.2 reflected) decreases quickly as long as it is larger than one, then approaches zero more slowly but never crosses it. In proper time, the distant observer first sees the star collapse fast until the surface comes near the Schwarzschild radius, and then all processes, including the collapse itself, seem delayed. The observer never sees the horizon or the star surface reach the Schwarzschild radius. Just the immense redshift and dimming can be identified as the sign of a star collapsing into a black hole. What is visible of the collapse happens in a short time: for a star of about the mass of the Sun, the dimming time of $2R_S/c$ equals about 20 millionth of a second.

Black outlook Falling into a black hole is not only bleak, one cannot even be warned of it. As long as an observer is still outside of the horizon, rockets could in principle be used to escape to larger distances. The thrust would have to be immense and difficult to provide with current technology. (Then again, current technology would not even get us near a collapsing star.) However, at least we would not have to violate physical laws in order to escape. Once inside the horizon, the story is different. Even the strongest thrust would not bring us back to larger distances. The radial displacement is now time for us. According to the laws of relativity, we must move toward smaller r.

A daring adventurer might fly close to the horizon of a black hole, hover there for some observing time, and then try to fly back. However, how would such an observer make sure to remain outside the horizon? For someone falling near the horizon, the region does not appear special. Tidal forces are not strong, and space-time appears normal. Since the horizon is not a material surface, it is difficult to measure its position so as to stay at safe distance. Our observer could only hope for external help, waiting for a warning from control center far away before the horizon is reached.

The control center will notice when the horizon is approached, by all the red-shifting and dimming of light signals. If the infalling observer emits regular signals, one can see outside when they become dangerously delayed, indicating that the horizon looms close. Alas, all warnings would be in vain: Signals from the infalling observer move outwards along the light rays in Figure 3.12, reaching the outside observer. The warning signal would have to be sent in to smaller r along ingoing light rays, the 45°-lines of constant v. When the signal delay has become large enough to be noticed by the outside observer, however, the infalling colleague would already be so close to the horizon that the warning signal arrives too late.

These considerations may seem academic because we do not have space ships that could bring us near a collapsing star. However, if a star is massive, its Schwarzschild radius is large and could engulf us even if we are far away. The Milky Way contains a supermassive black hole in its center, of millions of solar masses. While the distance to us is much larger than the corresponding Schwarzschild radius, future collapse of further parts of the Milky Way or a collision with another galaxy could enlarge the central black hole even more. If the horizon would come to lie outside of our own, apparently safe position at the fringes of the Milky Way, we would be part of the black hole. And, no friendly aliens could warn us of our fate.

Limits Space-time ends at a singularity. Even if we don't worry about matter being torn apart, vacuum space-time cannot exist in regions beyond the singularity. There is no space and time after $r = 0$ is reached in a black hole. In an ordinary star, there is nothing beyond $r = 0$ because this is the place of the center; when we cross $r = 0$, we come out by the other side from the center. In a black hole, however, $r = 0$ is a moment in time. When we reach $r = 0$, we cannot go any further, and we cannot go back to larger r because this is prohibited inside of the horizon where r is time. The singularity is a limit to space-time, and thereby a limit to anything we can say about nature within general relativity. The theory, as successful as it may be in describing other phenomena, cannot be complete. We must keep looking for an improved theory, one that can avoid the infinities of singularities.

General relativity's lessons about black holes show other puzzles. A black hole is black because nothing can escape, not even light. Light is an electromagnetic wave, an interplay of electric and magnetic forces. Electromagnetic waves not only provide messengers such as light but also transmit the electric and magnetic forces between charged objects. Similarly, there are gravitational waves that transmit the gravitational force. These waves have a much weaker influence on matter and are more difficult to detect. They are massless objects like light, just related to a different force. They travel at the same speed as light does, the celerity, and move along lightlike geodesics.

If gravity's wavelike force carriers move along lightlike curves, never leaving the region within the horizon, how can it be possible for us to feel the transmitted forces from outside? The force around a supermassive black hole can be enormous; the one in the center of the Milky Way holds the whole galaxy together. However, how can the force come out through the horizon if nothing else can, even its own carriers? These question marks indicate that we must keep looking for a more complete theory that can describe all of space-time, its curvature and the traveling ripples of gravitational waves.

3.3 Cosmology

If space and time can end at singularities and die, they must have a life to relinquish, a life of their own. Space and time are dynamical objects that evolve and change not only when we switch to a different observer, but also when they are watched by one and the same unmoving (but patient) observer.

Space and time outside of a static star or black hole do not change; the metric components have no time dependence. However, once we cross the horizon at $r = R_S$, the Schwarzschild radius, everything is drawn into the singularity at $r = 0$. By the tricks played by the black hole on space and time, their roles then switch and it is the radial distance r, not t, that must change toward the future, as determined by light cones. If r tells the progress of time, geometrical quantities, such as the area $4\pi r^2$ of a sphere at r, become time dependent. As r shrinks to the future, the area of any sphere at time r must shrink as well. Geometry has become dynamical. Just as it is impossible to stop our time and aging processes, it is impossible to prevent a sphere from shrinking to small surface area inside of a black hole. Once time $r = 0$ is reached, all surfaces collapse to zero size; no geometry can survive the singularity.

When we consider a single star or black hole, we assume that space-time far away from the surface or the horizon is nearly empty, except for presumed observers whose measurements we describe. At large r, space-time is then Minkowskian as in empty space where nothing changes. In the real universe, however, stars are not alone and isolated but band together in galaxies, and they often entertain a bunch of planets around them. For every star in the universe, which we could put in the center of our coordinate system, there are other massive objects moving around it in some way. A massive object deforms and curves space-time; when it moves, it tries to carry these deformations along. Space-time reacts to the moving masses by changing its curvature, obeying Einstein's equation all along, with the changing sources of stress-energy. Like a finger drawn through a fresh chunk of cookie dough, a mass drawn through space-time leaves its imprints along its path.

Expansion Space and time are dynamical; they react to masses moving. There is a lot of stuff in the universe, arranged not particularly uniformly as a look at the night sky reveals. Some regions seem almost empty; others abound with stars. Sometimes, we can see the band of the Milky Way, our own galaxy in one of whose fringes we live together with the other stars we see around us. With stronger telescopes, we can see other galaxies, diffuse regions of light formerly called nebulae. Although it has been suspected several times by a range of people (including, perhaps first, Galileo Galilei, the philosopher Immanuel Kant and the writer Edgar Allen Poe[1)]), nebulae were identified as galactic worlds of their own only when cosmological distance measurements became available. In 1925, Edwin Hubble found the last piece of evidence to solve the question.

The Milky Way is surrounded by galaxies such as Andromeda or the Magellanic clouds. They move around one another according the gravitational forces they exert, a huge system of masses bound together by gravity. Often, hundreds of galaxies can be present in such galaxy clusters, each galaxy having hundreds of billions of stars and a size of many quadrillion kilometers. Beyond this scale, however, the gravitational force is no longer strong enough to influence the motion of other masses much. Individual galaxy clusters mind their own business. However, they still react to the overall matter distribution in the universe, not by being pulled by gravity but by following the motion of space-time. If we look at distant stars and galaxies, the light they emit, as noticed by the astronomer Vesto Slipher, is the more redshifted the farther away the stars are from us. With distance measurements developed further, Edwin Hubble demonstrated that the redshift is proportional to the distance.

Redshift has different origins. Shifting frequencies could be a consequence of gravitational redshift, depending on the gravitational potentials where the light is emitted and observed. Such redshift depends on the mass and type of the star, not on the distance from us. Velocity is another origin of redshift, as per the relativistic Doppler effect. If the shift is to the red side of the spectrum and increases with the distance, stars should move away from us at a speed proportional to the distance.

1) "Of these latter the most interesting was the great 'nebulae' in the constellation Orion: – but this, with innumerable other miscalled 'nebulae,' when viewed through the magnificent modern telescopes, has become resolved into a simple collection of stars. [...] Telescopic observation, guided by the laws of perspective, enables us to understand that the perceptible Universe exists as a cluster of clusters, irregularly disposed. The 'clusters' of which this Universal 'cluster of clusters' consists, are merely what we have been in the practice of designating 'nebulae' – and, of these 'nebulae,' one is of paramount interest to mankind. I allude to the Galaxy, or Milky Way. [...] The Galaxy, let me repeat, is but one of the clusters which I have been describing – but one of the mis-called 'nebulae' revealed to us – by the telescope alone, sometimes – as faint hazy spots in various quarters of the sky." Edgar Allen Poe: *Eureka – A Prose Poem.*

Why would such an escape motion happen, away from us? The times when we deemed ourselves the center of the universe are long gone, so it is disconcerting to find ourselves put on the spot again.

There is a more humble explanation for the escape velocities, and the proportionality to the distance is a hint. The more space there is between us and a star, the larger the velocity. What if it is space itself, not a force acting on the star, that produces an apparent motion? If space stretches out in time, that is, if the universe expands, each piece along a line (a geodesic) from us to the star will grow by the same amount in a given time. The more space there is, the larger the total distance increase, and the larger the velocity we notice by redshift of a star. By determining the proportionality of redshift and distance, Edwin Hubble showed that space-time is dynamical and expands.

Origin How can space expand and increase distances without making objects move? If we want to explain Hubble's observations humbly and innocently, without assuming that we are the cause for the stars' escape, the redshifts we see must be due to the expansion of space and nothing else. If all of space expands uniformly, redshift effects are the same around any point, seen the same from all alien planets which have produced astronomers. The redshift depends only on the distance between a star and a place of observation; it does not require a center of the universe from where everything expands.

One often compares expanding space with a balloon blown to larger sizes. If we draw stars or circles on the rubber before blowing air in, we see the distance between all circles increase as the balloon expands. No center is required from where the expansion starts. Rather, the whole balloon is the origin of expansion, for the distances were smallest when the balloon was in its collapsed, air-deprived state.

Similarly, the universe expands without a center. The analogy is even better than the irregular distribution of stars we see in the sky around us would indicate. By now, millions of galaxies have been identified and mapped on the sky, including their distance from us. We obtain a good picture of the matter distribution around us, on scales much larger than we could see with the naked eye. When all these galaxies are considered, the matter distribution they form is almost constant. It is uniform, much like the regular rubber of a balloon filled with air.

If the analogy is better than we could initially expect, we should perhaps draw it even further. For the balloon, we know where everything comes from: a slushy, empty piece of rubber. The universe is expanding without a center, as shown by observations of redshift. Where did all this come from? What do we get if, in our mind, we turn the expansion around and try to see what universe we have in time reverse, collapsing to some original state? It is clear that the density of matter must have been much larger at earlier times, when the same masses were placed in smaller space. To find out what the origin could have been, we must see what densities can be reached, and what physical laws we need to apply. When stars come close to one another, forming a near-homogeneous and hot state, it is no longer just the gravitational force that plays a role. If matter in the whole universe is so dense as it is in a single star, more complicated physics, for instance of nuclear reactions, must be taken into account.

There are forces other than gravity that control the behavior and motion of matter, and they could change the expansion rate. If such forces are strong enough to stop the collapse, our time-reversed expansion of the universe, and provide equilibrium, we could find the analog of the lush piece of rubber of an empty balloon, waiting to be blown up to more glorious sizes. We could find nothing less than the origin of the universe, a state in which all forces balanced one another. Then, what made the universe leave its state of equilibrium? Who blew the air? What made it expand, dilute, and finally be as we see it around us today?

FLRW space-time The history of the expanding universe must be told in the language of general relativity. There is no other framework that could formulate the dynamical nature of space-time, making distances increase without actual motion in space. Distances are the domain of line elements, and so a mathematical description of expansion, and perhaps the origin of the universe, must use a suitable line element for the cosmos.

We have one hint to find the cosmic geometry. On large scales, including billions of galaxies, the universe looks homogeneous in space. Independent cosmological observations, foremost of left-over electromagnetic radiation from the denser phase of the universe at earlier times, called the cosmic microwave background, show the same symmetry. Metric coefficients should not depend on the spatial coordinates, but do so on time for expansion. The mass distribution on larger scales also looks the same in all directions. The universe is isotropic, direction independent.

We are looking for a line element with coefficients depending only on time, and with no distinction of any spatial coordinate or a direction. Any such line element looks like

$$\Delta s^2 = -N(t)^2 (c\Delta t)^2 + a(t)^2 \left[(\Delta x)^2 + (\Delta y)^2 + (\Delta z)^2\right] \tag{3.10}$$

with two functions of time, $N(t)$ and $a(t)$, if we also assume flat space. Space-times of this geometry are called Friedmann–Lemaître–Robertson–Walker space-times for several independent first users of the line element, or FLRW space-times for short.

Let us consider observers stationed at fixed values of the spatial coordinates, obeying the equations $\Delta x/\Delta T = 0$, $\Delta y/\Delta T = 0$, $\Delta z/\Delta T = 0$ with respect to their proper time T. These observers do not move in space because their positions are fixed. Yet, the distance between two such observers, separated by some values of Δx, Δy and Δz, changes. The geometrical distance is measured by the line element, and its value $(\Delta s^2)_{t\,\mathrm{constant}} = a(t)^2 \left[(\Delta x)^2 + (\Delta y)^2 + (\Delta z)^2\right]$ between two points at the same time t depends on what time is assumed. The distance changes in time just because the metric changes, or because the universe is expanding. With the FLRW line element we are able to describe the expanding universe.

The line element also demonstrates that space does not need to expand into anything. Expansion processes of matter on Earth, such as our sample balloon, always happen by motion of a more compact region to larger size, expanding into its surroundings. In such cases, we have actual motion of the matter pieces in space, pushed out of the initial region by some force. The universe also expands, not by motion of points (changing x, y and z), but rather by the motion of geometry or space-time. Distances increase because the metric, our standard for length measurements, is growing. With a ruler, we would measure an increasing distance between two points not because the points are moving apart, but because the scale on the ruler must be continually adapted to the expanding geometry. Nothing material moves in this expansion process, and there need not be anything for the universe to move into.

In fact, the FLRW line element does not require limits on x, y and z, just as Minkowski space-time is valid for unrestricted ranges of coordinates. The total size of the universe at any given time might well be infinite, and yet it can expand in the sense that distances between fixed points increase. This expansion is what we determine for the universe by redshift observations, independent of its total size.

Proper time Two scales may change in an expanding universe, described by the functions $N(t)$ and $a(t)$ in the FLRW line element. The second function is relevant for distance measurements in space, for the actual expansion of the universe. The first function, $N(t)$, should implement changes of the time scale, or time dilation; it is called the lapse function. So far, t is some time coordinate, not yet identified with any observer who would experience change by the progress of this time. Time experienced by an observer is proper time along the observer's worldline. To determine its relation to t, we consider specific observers.

When we do cosmological observations, we always move, rotating with Earth around its axis, orbiting with the planet around the Sun. The Sun moves around the Milky Way, and the Milky Way in its own galaxy cluster. The combined motion affects the progress of our proper time, but it is known and can be factored out of observations to see the pure expansion of the universe. Our viewpoint can be well described by the observers we already introduced: those at constant values of their position coordinates. These observers do not move in space; they follow only the expansion of the universe, their distances increasing even in the absence of motion. Such observers are called co-moving because their sole motion is the one in cooperation with the whole universe, an expansion that nothing can defy.

For co-moving observers, we have $\Delta x = 0$, $\Delta y = 0$ and $\Delta z = 0$ along their worldlines. Their proper time, as determined by the line element, is $\Delta T = \sqrt{-\Delta s^2}/c = N(t)\Delta t$. We can replace the old, nonspecific time coordinate t in the FLRW line element, together with the function $N(t)$, by simple proper-time intervals:

$$\Delta s^2 = -(c\Delta T)^2 + a(T)^2\left[(\Delta x)^2 + (\Delta y)^2 + (\Delta z)^2\right] . \tag{3.11}$$

Only one function $a(T)$ remains to tell the story of the expanding universe.

Scale factor The history of the universe is a history of changing scales. An isotropic universe has only one degree of freedom, the function $a(T)$. In contrast to the lapse function $N(t)$, which we can remove by properly referring to proper time of co-moving observers, $a(T)$ must have observable information; otherwise, our line element would just be a complicated form of Minkowski space-time. There is indeed something measurable about $a(T)$ because its increase as a function of T means that distances between co-moving observers grow. The function determines the total scale of the universe, and is called the scale factor.

However, $a(T)$ is not fully observable; it is not invariant. We may change our spatial coordinates to bx, by and bz with a constant b, and the scale factor changes to $a'(T) = a(T)/b$ so as to keep the line element Δs^2 invariant. The value of $a(T)$ at one instant T_0 cannot be measured, because we could choose $b = a(T_0)$ and have $a'(T_0) = 1$. Any coordinate system is as good as any other one, and so there can be no measurable information in the scale factor at an instant of time, if one can always find coordinates in which the scale factor is just one, or any other value.

Unlike the lapse function $N(t)$, however, we cannot remove the scale factor completely. We cannot define new coordinate intervals $a(T)\Delta x$ and so on because we then refer to different coordinates. The product depends on the changes of more than one variable, of T and of x. It is not possible to write it as a single coordinate change $\Delta x'$, and so we cannot eliminate the scale factor altogether.

The scale factor is not fully observable and not fully disposable either. What is observable refers to values of the scale factor at different times, just what we are talking about with the expansion of the universe. We could look at quantities such as the change of rate $\Delta a/\Delta T$, the analog of a velocity. However, if $a(T)$ can be multiplied with a constant just by using new coordinates, its rate of change is multiplied with the same constant. If we have chosen the constant to be such that $a(T_0) = 1$ at some time T_0, however, the rate of change $\Delta a/\Delta T$ at T_0 can no longer be modified. More generally, without rescaling $a(T)$, we can look at the ratio $H = \Delta a/(a\Delta T)$, in which the multiplicative factors obtained under coordinate changes cancel. The parameter H is of prime interest for cosmology, as the relative rate of expansion of the universe; it can be extracted from redshift observations as undertaken by Edwin Hubble. In his honor, H is called the Hubble parameter.

There are additional invariant quantities, analogs of acceleration or the change of the expansion rate. The expansion of the universe indeed appears to be accelerating, with growing Hubble parameter. The rate of change $\Delta H/\Delta T$ is another observable quantity of current interest in cosmology.

Friedmann equation Space-time evolves according to the matter it contains. Curvature around a massive object, for instance, a black hole, is one example, and the expansion of the universe is another. In general, the line element is determined by Einstein's equation, the stress-energy contributions depending on what matter is present. For a line element of FLRW form, only the scale factor is free, and so Einstein's equation reduces to an equation for the scale factor, or rather the Hubble parameter as the invariant quantity. This equation is the Friedmann equation

$$H^2 - \Lambda = \frac{8\pi G}{3c^2} d \tag{3.12}$$

with the energy density d of matter in the universe and the cosmological constant Λ. (In (3.12), space is assumed flat and Euclidean. If it has positive curvature, of the form of the three-sphere, $H^2 + 1/a^2$ appears instead of H^2, with a not just the scale, but the radius of the three-sphere. Or, H^2 can be replaced by $H^2 - 1/a^2$ for saddle-like, hyperbolic space with negative curvature.)

If we know what kind of matter we have and how it dilutes as the universe expands, we can trace back the changes of $a(T)$ at all times and determine the history of the universe. However, finding the correct dilution behavior of matter is not always obvious. Most of the matter we know dilutes by being expanded in the universe. Its total energy or mass remains constant while the volume it is placed in, proportional to $a(T)^3$, increases. The energy density therefore decreases according to a functional behavior of $d = C/a^3$ with some constant C. The scale factor is not completely observable or invariant, and so C cannot be invariant either. Yet, the dependence of d on a does characterize the dilution behavior, which is enough to determine the history of $a(T)$.

There are energy forms that do not have the same dilution behavior as ordinary matter. Electromagnetic waves also contribute with their energy, and they can be a significant, even dominant fraction at earlier times when the universe was much hotter than it is now. The energy of electromagnetic waves is being diluted as the universe expands, but also the frequency changes according to the same redshift effects that have revealed the expansion of the universe. Redshift affects the wave length, a linear distance scale. If we combine volume dilution with redshift, a density behavior of $d = D/a^4$ with another constant D is obtained. The expansion rate of the universe changes accordingly.

At higher densities and temperatures, we may not even know how to describe matter early in the universe. Particle accelerators can produce higher and higher energies by colliding atomic nuclei or elementary particles, but we may never be able to reach the densities required to understand the complete history of the universe. What we can do instead is an analysis as complete as possible based on what general relativity tells us, with only weak assumptions on the form of matter. Just as in black holes, there may be space-time effects that, while caused by collapsing matter, are insensitive to specific matter properties. Underlining its elegance, general relativity indeed reveals several consequences based on the evolution of space-time alone.

Light cone Space-time effects are determined by the form of light cones. In an FLRW space-time, light rays obeying the equation $\Delta s^2 = 0$ of lightlike worldlines take the form $c\Delta T/\Delta x = \pm a(T)$ for light moving along the x-direction, to the right or the left depending on the plus-minus sign. The slope of such curves in a space-time diagram is time dependent via the scale factor, and need not always be $45°$.

If we choose coordinates such that $a(T_0) = 1$ at time T_0 when light rays are sent out or received, we have light cones with $45°$-slope at their tips. As we move away from the tip, the slope changes. If $a(T)$ is increasing, light rays look steeper in the future: T increases more than x does and we obtain light cones as illustrated in Fig-

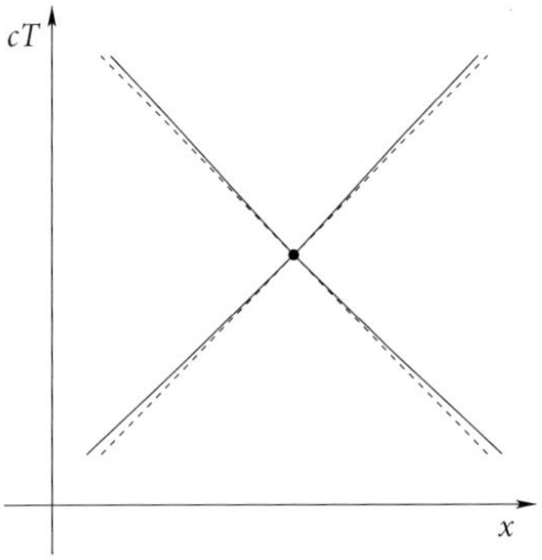

Figure 3.13 Light cones in an expanding universe change their slope away from the tip, bending outwards toward the past and inwards to the future. The dashed lines show the light cone expected in Minkowski space-time, with light rays at 45°.

ure 3.13. Even so, the speed of light does not change, because we can always choose coordinates so that light rays are at 45° near some event, such as the tip of the light cone. In a small-enough region, space-time always looks just like Minkowski space-time and the speed of light is c. We see changes to the space-time geometry only if we travel far, just as we have to travel far on Earth to see its spherical shape. The speed of light does not change in an expanding universe; the light cone changes its slope because light can surf along with the expansion of the universe, not required to do all the motion on its own.

Cosmological observations use messengers such as light or other electromagnetic waves, or perhaps gravitational waves. All these messengers travel at the speed of light; they move along our past light cone toward us. What we see of the universe is only what happens on the light cone, or on the thick lines in Figure 3.13 pointing downwards from the tip. We do not see much of the interior of the past light cone, except for what we directly experience on Earth. Anything that requires wavelike messengers, at the same speed of light, can only show us parts of our past light cone.

The part of the universe filled by light rays reaching us from the past is a rather small fraction of our whole past space-time, from which we try to reconstruct the history of the universe. As we move on in time, continuing to do our observations, the region swept out by past light rays will become larger, but we will always miss much of the interior, as well as the rest of the universe outside of the past light cone. If we assume homogeneity, the geometry is the same at all points in space; knowing it on the past light cone allows us to reconstruct the full space-time geometry. The assumption, however, is a heavy one. As observational data about the universe keep accumulating, we may be forced to change the geometry we assume for space-time on large scales.

Singularity The universe began not in peace, but in a hell of a state. This is at least the story told by general relativity, whose equations we use to follow the expansion of the universe backwards, trying to decipher its origin. Seen backwards, the scale factor decreases as time is wound back, and the density increases.

For ordinary matter humbly called "dust" by cosmologists, notwithstanding the fact that this "dust" includes all matter not only in stars and planets but also in ourselves, the energy density $d = C/a^3$ changes by dilution. The Friedmann equation, with $\Lambda = 0$, then amounts to $\Delta a/\Delta T = \sqrt{8\pi G C/(3ac^2)}$, solved by $a(T)^3 = 6\pi G C[(T - T_{BB})/c]^2$ with a constant T_{BB} related to the value of $a(T_0)$ at some time T_0. At time T_{BB}, the scale factor vanishes, and so do metric components. We are in danger of encountering another singularity, a real danger. Physical quantities, the density or related ones such as temperature, diverge when $a = 0$, and also tidal forces, computed from curvature invariants, become infinite. We cannot trace back the history of the universe as far as we like. The singularity means that the space-time description loses its meaning after a finite time of backward evolution, when time T_{BB} is reached.

Cosmologists, as early as Georges Lemaître, have often attempted to interpret the singularity of FLRW space-times as the beginning of the universe, starting with a bang as big as a singularity. We speak of the big-bang singularity, and T_{BB} is the time of its coming. However, the interpretation as a beginning is unjustified, for it is, first of all, the space-time description of general relativity, one of our theories about nature, that breaks down. If a theoretical description breaks down, interpreting this as the beginning or end of the world constitutes an immense act of cosmic hubris. One should rather admit that one of our best theories reaches its limits and can no longer be applied. As beautiful as general relativity may be, we must look for a more complete (and probably even more complicated) extension that can describe in meaningful terms what general relativity can display only as a singularity.

The theory, after all, is honest about its failure by showing us its own limits by encountering singularities. This degree of honesty is rare in physical theories, and it is a sad fact, perhaps one of human nature, that it is not always matched to the same degree by theorists. Overinterpretations of a singularity (or of some alternative scenarios which always keep cropping up) as the beginning of the universe do not serve the purpose of improving our understanding of nature. They rather mislead research interests and prevent important questions from being addressed. In the example of the big-bang singularity, the failure of the theory is simply accepted and left as part of our world view. We stop looking ahead to understand what the singularity could mean physically, described by a theory whose equations do not break down, and where the singular-looking region itself could have come from. Science, to a large degree, is pushing boundaries. If boundaries are accepted, one stops doing science.

Matter Singularities in general relativity are ubiquitous. If they only occurred in FLRW models, and only for the dust solution we saw, they would not present any danger. Dust is not the main contribution to a hot and dense universe at early times. We should at least include a significant source of electromagnetic radiation, whose energy density dilutes like $d = D/a^4$. With such an energy density, we still encounter a big-bang singularity. Many other dilution behaviors can be tried, and although they change some aspects of the expansion, they do not eliminate singularities.

We do not know how matter behaves at high densities, and therefore must use a rather general description of matter in order to estimate the singular danger. If we look at the Friedmann equation $[\Delta a/(a\Delta T)]^2 = 8\pi G d/(3c^2)$ without using a specific form of the energy density, we cannot solve it for $a(T)$. Instead of inserting different kinds of specific $d(a)$, we can use other equations to tell us how d must change in time. The combined set of equations, one for the change of $a(T)$ depending on d, and another one for the change of $d(T)$ could, with some luck, be solved in tandem.

The energy density is a quantity often used in the subfield of physics called thermodynamics, studying the behavior of matter at different temperatures. One of the laws of thermodynamics, called the first law for its importance, relates the change of energy to work done on a system. From the energy in some region, we can obtain the energy density by dividing by the region's volume, and the work done on a system is related to the change of its volume. By compressing an object, we increase its energy working against internal pressure. The work is proportional to the pressure and the change of volume: $W = -p\Delta V$ with a negative sign because compression, decreasing the volume, increases the object's energy by our work. The first law of thermodynamics equates the energy change to the work, $\Delta E = -p\Delta V$.

Continuity equation Changes of volume and energy occur over time intervals ΔT. From the first law, we obtain an equation for the rate of change, $\Delta E/\Delta T = -p\Delta V/\Delta T$. The volume of a region in a homogeneous universe of FLRW form is $V = a^3 V_0$ with some constant V_0, and it contains an energy $E = da^3 V_0$. We write $V_0 \Delta(da^3)/\Delta T = -V_0 p \Delta(a^3)/\Delta T$ or, since the factors of d and a all change individually,

$$\frac{\Delta d}{\Delta T} + 3\frac{\Delta a}{a\Delta T}(d + p) = 0\,. \tag{3.13}$$

Thermodynamics provides a new equation for the rate of change of $d(T)$, called the continuity equation. We can try to solve it in combination with the Friedmann equation, giving us $a(T)$ and $d(T)$ at the same time. However, now we need to know how the pressure $p(T)$ changes, which again requires a new equation for its rate of change. We could go on and derive equation after equation, but we would never come to an end. It is impossible to derive all properties of matter in the universe from general equations. Something else must be put in as information that completes and closes the equations. Ultimately, the matter behavior is determined by the motion and interactions of its atoms and molecules. However, these equations, combined with those of general relativity, would be much too complicated to solve even for computers. Instead, we have to work with general assumptions about matter, such as the relationship between pressure and energy density, to see how generic singularities are.

Singularity theorem Equations can tell us what we need to know about matter. Both the Friedmann and the continuity equation contain the Hubble parameter $H = \Delta a/(a\Delta T)$ and the energy density. The Hubble parameter squared is proportional to the energy density, and the rate of change of the energy density appears in the continuity equation. If we compute the rate of change of the Hubble parameter, using the Friedmann equation, we can combine it with the continuity equation and produce a new equation, called the Raychaudhuri equation, which contains the key to general properties of singularities.

Raychaudhuri equation We compute the rate of change $\Delta(H^2)/\Delta T = (8\pi G/3c^2)\Delta d/\Delta T$ using $\Delta(H^2)/\Delta T = 2H\Delta H/\Delta T$, and then insert $\Delta d/\Delta T$ according to the continuity equation. After dividing the whole equation by $2H$, the result is

$$\frac{\Delta H}{\Delta T} = \frac{8\pi G}{6c^2 H}\frac{\Delta d}{\Delta T} = -\frac{4\pi G}{c^2}(d+p) = -\frac{4\pi G}{3c^2}(d+3p) - H^2 + \Lambda \,. \quad (3.14)$$

The last step, using the Friedmann equation in $d + p = (d + 3p + 2d)/3$, may seem unnecessary, but it allows us to refer to $d + 3p$ instead of $d + p$, which turns out to obey a general inequality: all the matter we know from experiments satisfies $d + 3p > 0$.

The pressure in matter is related to the motion of its constituents, withstanding further compression. Energy density, in relativity, contains a large contribution from the mass of atoms and molecules even when they are not moving, outweighing not only the kinetic energy of motion but also pressure. Unlike pressure, energy is always positive. Even for electromagnetic waves, which do not have mass, the pressure does not reach the energy density; it always satisfies $p = \frac{1}{3}d$.

Electromagnetic waves With $p = \frac{1}{3}d$, the continuity equation takes the form $\Delta d/\Delta T + 4(\Delta a/\Delta T)d/a = 0$. Dividing by $\Delta a/\Delta T$, we have $\Delta d/\Delta a = -4d/a$, an equation solved by $d(a) = D/a^4$. The relationship $p = \frac{1}{3}d$ therefore provides the correct dilution behavior of electromagnetic waves, including the redshift factor.

In all the standard matter cases, pressure and energy density satisfy the inequality $d + 3p > 0$, called the strong energy condition. The Raychaudhuri equation can then be written as a compact inequality, $\Delta H/\Delta T < H^2$ if $\Lambda = 0$, which in turn implies $\Delta(1/H)/\Delta T > 1$. By adding up the changes of $1/H$ between two times T and T_0, we have $1/H(T_0) - 1/H(T) > T_0 - T$. If T_0 is the current time, the universe is expanding and we have $1/H(T_0) > 0$. At some earlier time $T_1 = T_0 - 1/H(T_0) < T_0$, the inequality gives a Hubble parameter $-1/H(T_1) > T_0 - T_1 - 1/H(T_0) = 0$. At this earlier time, $1/H(T_1)$ was negative; there must be some time between T_1 and T_0 when $1/H$ changed its sign from negative to positive and therefore was zero;

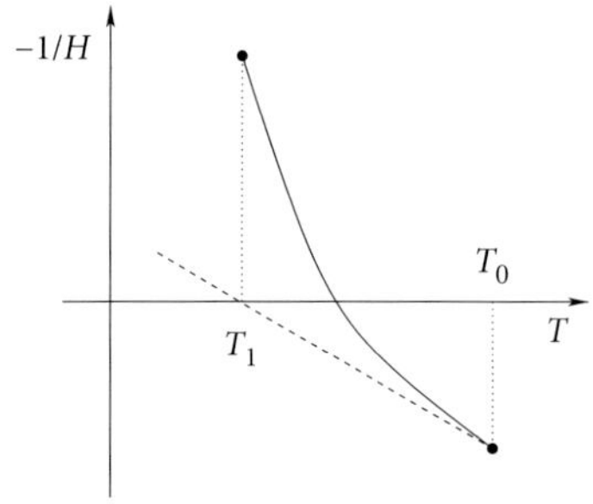

Figure 3.14 If $-1/H$ is larger than $T_0 - T - 1/H(T_0)$, a negative $-1/H(T_0)$ implies that $-1/H(T_1)$ is positive for $T_1 = T_0 - 1/H(T_0)$. Between T_1 and T_0, $1/H$ must vanish. The dashed line shows the linear behavior $T_0 - T - 1/H(T_0)$ obtained if an equality were satisfied for $1/H$.

see Figure 3.14. But H is then infinite, and so is the energy density according to the Friedmann equation.

If $\Lambda > 0$, the arguments no longer apply but can be extended. Positive Λ produces a term in (3.14) similar to a constant energy density $d = 3c^2\Lambda/(8\pi G) > 0$ with negative pressure $p = -d$. Negative pressure can make the expansion rate grow: $\Delta H/\Delta T > 0$, as an otherwise unknown form of dark energy whose presence supernova observations seem to indicate.

Singularity Attraction is fatal. Singularities in FLRW space-times cannot be avoided as long as matter satisfies the strong energy condition. No force in matter, irrespective of its behavior, can withstand the strong pull of gravity once matter has collapsed to small size and high density. This singularity was present in solutions found right after the equations of general relativity had been written down, in functions $a(T)$ computed by Alexander Friedmann. Similar singularities occurred in the black-hole solutions found by Karl Schwarzschild in 1916. Singularities always seemed to form, but the solutions known were symmetric, either isotropic and homogeneous or spherical.

By the symmetry assumed, collapsing matter could only fall toward one center, a single point in space. Singularities are then not surprising; after some time, much of the infalling matter would occupy just one point, making the density infinite. In realistic models, however, we would not have spherical collapse. Matter particles may come close at some time during the collapse and form high density, but without exact symmetry, they would miss each other and do not all reach the same point. After high densities are reached, matter would move further by inertia and disperse again, or it might form a compact and dense stable object without infinite density.

Without assuming at least some form of space-time symmetry, Einstein's equation is difficult to solve. Not much is known about its general solutions. However, one of the general statements refers to singularities, and it tells us that singularities are not just a consequence of symmetry. They occur much more generally, and have to be reckoned with as a generic feature of general relativity. Last doubts were removed when strict mathematical theorems were proven by Roger Penrose, Stephen Hawking and others, that did not require any symmetry.

For general solutions, the properties of singularities that can be inferred seem less daunting. For instance, density and temperature need not always become infinite, and even tidal forces may remain finite. But, in all situations covered by the theorems, the motion of some observers will end after a finite time; there will always be observers, for instance, those falling into a black hole, whose measurements can cover only a limited part of space-time. Space-time retains its limits, and so does general relativity.

As often with mathematical theorems, there are some assumptions that can be violated so as to evade the consequences. General singularity theorems do not require symmetries, but there are some weaker assumptions, for example, about initial values. If the universe is expanding at one time, at all points in space, with a rate of at least some positive number, then there was a singularity in the past. However, if the universe at that time, while still expanding, can have expansion rates arbitrarily close to zero at some points, it may remain non-singular. Explicit nonsingular solutions based on this escape route have been constructed by José Senovilla in 1990. The singularity issue remains delicate. Avoiding the consequences always seems to come at the expense of specific initial values that could not be explained by any means, other than the aim of trying to avoid singularities. General relativity does not provide a mechanism for space-time to evolve without borders. We must keep looking for such a mechanism, or try to extend the theory such that space-time can be nonsingular with new laws of nature.

Cosmic microwave background We do not see singularities. They are covered by a horizon in black holes, and at the big bang they are surrounded by dense and hot matter. We cannot see through hot matter because electromagnetic waves scatter much and cannot travel freely. They are continually absorbed and remitted by matter atoms, distracting them from their original direction. For the same reason, we cannot see the hot interior of the Sun, even though it contains mainly hydrogen and helium, transparent gases. We see only the outer regions of the Sun cool enough for light to travel toward us.

In the universe, electromagnetic waves began to travel freely after the universe had cooled down for scattering processes to end. The temperature was about the same as at the surface of the Sun, roughly 4000°C. This is a large temperature, but cooling down to that value from infinite temperature still takes some time. If we do not take infinity as the starting value where the theory, after all, breaks down, but some other large number, the cooling process takes about 380 000 years, following the expansion of the universe with most of its energy stored in electromagnetic radiation.

The universe at that time became transparent, and it was still as hot and bright as the surface of the Sun. It was as hot and dense everywhere, uniformly filled with electromagnetic waves. Since then, the universe has expanded much for another 14 billion years or so, and cooled down even more. Its old radiation has become diluted and redshifted, but it is still detectable. Its frequency range is now in the microwave spectrum, earning it the name cosmic microwave background, and its temperature a cool 2.7 K. Its properties such as temperature variations over the whole sky have been determined precisely by various detectors and satellites, showing just tiny variations of temperature, about a millionth of the average temperature.

Observations of the cosmic microwave background confirm the near homogeneity of the universe on large scales, as shown also by maps of millions of galaxies. Using the FLRW form of space-time geometry does seem justified, but also here we have to keep in mind that we see only our past light cone, a rather small part of all of space-time.

Horizon problem Uniformization simplifies. Homogeneity makes solving mathematical equations easier, but it comes with several conceptual problems. First off, we would like to explain why the universe is homogeneous, rather than building this property in our theory by assumption. Even if we are not as ambitious as trying to find such an explanation would be, there is a problem when uniformity of the cosmic microwave background is combined with the big bang singularity.

Between the times of high densities and the time when the universe became transparent, about 380 000 years passed. As long as this period may seem, it is little time even for light to traverse vast distances in the whole universe. The values computed for the travel time in various universe models, assuming different forms of dilution behavior, show that it is impossible for light or any other signal to bridge the regions that give rise to the cosmic microwave radiation from different directions. Waves reaching us from opposite directions in the sky come from places so far apart that no signal could travel between them in the period after high density, until the time of transparency, as illustrated in Figure 3.15. However, if no contact could have been established, how can it be that about the same intensities are seen for waves from both regions?

Contact between different regions could be established if the past light cones near the big-bang singularity would bend outwards more strongly than during ordinary expansion of the universe, as in Figure 3.16. Such a rapid expansion with acceleration is called inflation, but it is difficult to achieve: The strong energy con-

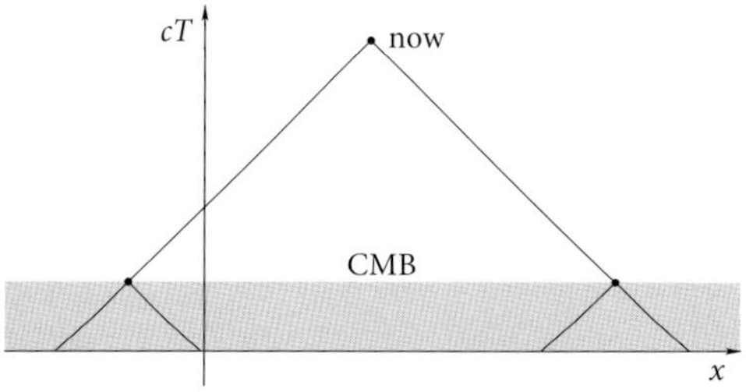

Figure 3.15 Horizon problem: In the early universe, there was not enough time for signals to establish contact between regions seen as opposite points on the sky, and yet the intensity of cosmic microwave background radiation is nearly the same. Past light cones from different regions do not always overlap; the regions of contact are bounded by horizons. The gray area indicates times when the universe was too hot and dense to be transparent.

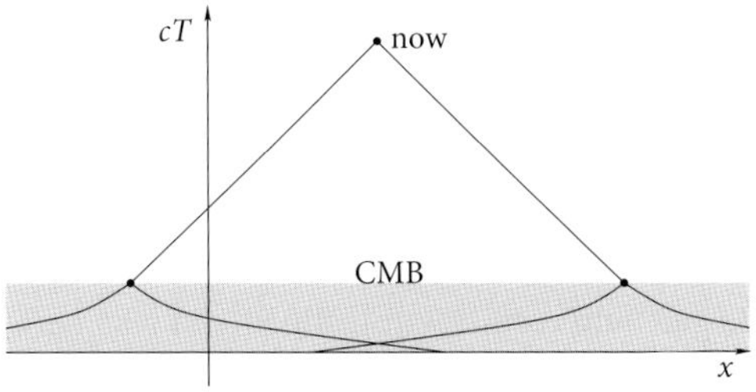

Figure 3.16 Contact between different regions from where we receive the cosmic microwave background could have been established if the universe were expanding rapidly, even in an accelerated manner at very early times, a process called inflation.

dition that gave us the singularity problem also means that the acceleration of $a(T)$, as per the Raychaudhuri equation, cannot be positive, unless $\Lambda > 0$ and large, outweighing $-4\pi G d/(3c^2)$. One has to work hard to construct matter ingredients that could give rise to accelerated expansion.

Nevertheless, the general assumption of inflation, or accelerated early expansion, has astounding consequences. It not only explains why the cosmic microwave background is nearly uniform, it predicts much more precise properties of the statistical distribution of temperature fluctuations, in excellent agreement with observations. Moreover, accelerated expansion does not seem uncommon, for it has been determined also for rather recent times based on the redshifts of supernovae explosions at known distances from us. The origin of the new acceleration, unknown but called dark energy, is most likely different from the one of inflation, but the phenomenon does show that acceleration can happen. There must be some matter or energy constituents that violate the strong energy condition. Such matter has not yet been seen, but if it exists, its properties may depend on quantum physics.

Toward the end of their meeting, the titans grew solemn. They all knew that one task, the foremost one on which all else depended, still lay on the table, the daunting task – daunting even for a titan – of holding the heavens in place. It would mean a lifetime, eternity spent unrecognized, supporting the infinity of space for all others to enjoy. The titans had left this most important task till the very end. Now their eyes turned to the one among them who had remained silent while the spoils were split: Mathlas, a generous giant, a gentle titan, incorruptible and selfless on the verge of self-denial. He never complained and just did what had to be done, not what wanted to be done.
Mathlas saw the other titans' glances turned toward him. He remained silent, nodded and did what had to be done, not what wanted to be done. He stood up, slowly, lowered his eyes and looked down at the tips of his toes. He spread his legs for better support, reached out with strong, muscular hands, grasped all there is at infinity, and pulled it inward, to a single point. Ever since have the heavens been known as the heavenly sphere, or 3-sphere for short.

4
Quantum Physics

Scientists observe nature and try to predict new phenomena from what is known. The notion of observations, even if they are not actively done, is the primary concept; it determines what kind of predictions and theories are possible. Relativity takes seriously the notion that physics can at best describe what a single observer, or perhaps a community of communicating observers, detect and measure. Observers positioned differently or moving in various ways may not measure the same numbers even if they observe the same situation, but their measurement results can be translated into one another. Physics is not solipsistic; on the contrary, it is important to compare what different observers see in order to confirm results by independent means. By considering a large class of observations, done in different ways and from different viewpoints, a great deal of information about nature can be gained. However, even if one uses the collective data of all possible observers, there cannot be an all-knowing entity in physics with access to everything that happens in the universe at a given moment in time. At any moment, we can only know what we have experienced (and remember) or has been communicated to us. Since the number of observations we can make in the finite span of life is limited, and the speed of communication cannot exceed the speed of light, it is not possible for us to gain information about everything that happens at some time.

One may compare the transition from Newtonian physics, where, at least in principle, an all-knowing observer might exist, to relativity theory by the literary transition from an epos to a novel told by a participating hero. An epos is written from the perspective of a watchful author aware of a multitude of simultaneous actions. A novel told by a participating hero can only refer to what the hero knows, even if the author has the whole story plotted out. The prevailing form of Newtonian physics is an epos in which the physicist describes the motion and collisions of different bodies. The heroes of Newtonian physics are lifeless objects such as the planets, stars or galaxies moving and interacting in epic fashion. The heroes of relativity are observers, who experience the motion of objects and communicate with other observers as the story unfolds.

Quantum physics takes the storytelling one step further, to a modernist novel of psychological dimension. In quantum physics, we can make statements only about what an observer is able to know by asking questions and receiving (and interpreting) answers. The knowledge gained in this way may not always be complete, not

The Universe: A View from Classical and Quantum Gravity, First Edition. Martin Bojowald.

even regarding what the observer can access in a given amount of time. Worse yet, the answers received asking different questions or posing them differently may not be consistent with one another. Answers may not be firm; often they are uncertain or vague. Additionally, the state of someone (or something) questioned may change when a question is asked, depending on how it is asked.

Quantum physics is strange and counterintuitive. It has been said that one understands quantum mechanics only when one understands that one does not understand it. If this sounds contradictory to us, well, we just haven't understood quantum mechanics. There is no paradox, no dialectic, and no actual quantum mechanics required to understand this sentence (a superposition? Schrödinger's cat?). It just means that quantum mechanics is not open to intuitive grasp. Quantum physics must deal with the psychological darkness of a confused observer, desperate to find out what happens, unable to read minds. In the history of physics, the succession of Newtonian physics, relativity, and quantum physics resembles the transition from the absoluteness of an epos to the psychologically correct treatment of individual subjects.

4.1 Waves

Relativistic physics distinguishes light from massive objects only by their speed and the corresponding motion through space-time. Light is special because its velocity is the same for all observers, with light rays and the light cone they form as invariant geometrical objects. For motion at the speed of light, time does not progress, owing to extreme time dilation, in stark contrast to the aging of massive objects. However, an observer watching a beam of light travel from a star to a telescope does not see a situation much different from the motion of some high-energy particle with velocity almost the speed of light. Light has characteristic properties such as its color or frequency, but this feature just appears as a specific way of quantifying different types of electromagnetic waves, as we distinguish particles by their energies and masses. If we do not change the viewpoint of an observer, we do not notice the special properties of light related to space-time geometry. Despite its rather different nature as a form of the electromagnetic field, light does not seem too different from the other fundamental players in physics.

Interference A general distinction of physical phenomena is their classification as waves on one hand, and particles on the other. We visualize particles as compact, almost pointlike objects with sharp boundaries, near-impenetrable for other objects they interact with. Thanks to boundaries, particles retain their identity after they collide with other ones or interact in different ways, unless the energy of interactions is large enough to trigger reactions that change the particles involved. Even if particles can change their identities, they acquire their new passports so quickly that, most of the time, they are free of identity crises.

Waves have no sharp boundary, but rather level off slowly outside some central region. They are spatially extended excitations of some medium, such as the surface of a lake or the electromagnetic field, and cannot exist without it. Waves may overlap and form a new, combined wave. The combination of waves gives rise to a typical phenomenon, interference, which allows us to test whether wave or particle properties are predominant in a given situation. As excitations of a medium, combined waves can either support each other in the excitation or counteract. A water wave on a lake pulls the surface periodically up and down. If two waves move in the same region and overlap, the relation of their arrival times determines whether they pull the surface up at the same place and time, or whether one wave pulls down what the other wave is trying to push up. The surface is either rippled even more than a single wave would do, or may stay nearly level when the waves compensate their actions. Arrival times depend on the position, and so one sees a characteristic pattern of places where the waves pull together, constructive interference with even higher waves, next to places where competition, or destructive interference, denies the waves large surface elevation.

Whether waves interact constructively or destructively depends on the relative positions of their maxima and minima, their hills and troughs. Hills and troughs are separated by a distance called the wavelength, a quantity that is determined by the frequency of the wave and its speed. If the excitation of the wave changes periodically with frequency f and the wave travels at speed c, the wave, along its way, lays out a periodic pattern of excitations separated by a distance of $l = c/f$: It takes a time $T = 1/f$ for the excitation to reach a given value at two successive instances, during which time the wave can travel the distance cT. The interference pattern of two overlapping waves becomes visible at distances of the size of the wavelength.

If we try to observe interference in order to test whether a phenomenon is wave-like, we must be able to detect variations of the corresponding excitations over distances of one or a few wavelengths. Such observations are easy for water waves, but can be difficult for other periodic phenomena. With the typical frequencies of visible light and its speed, we obtain wavelengths of less than a millionth of a meter. Nowadays, microscopy with this resolution is routine, but early on in the history of physics, clarifying the nature of light was an important and stimulating problem. In 1803, Thomas Young was able to detect interference fringes in the overlap region of two light rays, coming from the same source, but traveling two different routes through a "double slit." The first confirmation of light as a wave phenomenon was thereby obtained, even before its relation to the electromagnetic field was known.

Particles and waves Particles and waves are two styles of interaction. Particles confront opposition head-on when something encroaches on their spheres of influence. Waves interact politically; depending on the circumstances, they may form coalitions or annihilate opponents (and themselves) without mercy. The behavior realized could depend on the interaction process looked at, and on the resolution of our observations: not anything behind the scenes may fall directly in our eyes. A thin cloud of dust can penetrate another cloud without much perturbation, behaving as a whole more like a wave than a hard-nosed particle, even though it is formed by dust grains, not by excitations of a medium such as the surrounding air. Dust grains do not show interference, and so we are justified in denying them the status of waves.

One can find more ambiguous examples. Swarms of birds often show wavelike patterns when they are attacked by a bird of prey. Birds in the swarm near the place where the attack would most likely happen flee the fastest into the safety of the swarm, but birds in there must first make way. With a finite reaction time, a wavelike excitation builds in the swarm, with birds periodically retreating into parts that opened up, then being slowed down by their neighbors. If we imagine two swarms, each attacked by a bird of prey flying into each other, free spaces in one swarm will more likely be filled by birds from the other swarm than denser regions. The waves overlap, forming a pattern that can resemble interference.

As these examples demonstrate, one must look closely before one can decide whether the constituents are waves or particles. The answer also depends on how one views the phenomena. The birds as constituents of their swarm are particle-like, with sharp boundaries and clear spheres of influence. Their motion relative to one another, on the other hand, can be viewed as an excitation of a uniform distribution of birds in the swarm, and it behaves wavelike. The same phenomenon can be described in different ways, as particle-like in some respects and wavelike in others. Before we decide on which viewpoint to take, we must know much about the underlying processes.

Photoelectric effect Light interacts in wavelike ways, as shown first by Thomas Young's observations, at least when a single beam is separated into two parts, then made to overlap again. However, what happens if we look at a more individualistic situation, like a single bird in the swarm instead of the relative positions of all birds? Does light always interact in wavelike ways with itself or with matter when it is absorbed or reflected? Is it possible to look at light so closely that we see single constituents, individual particles like the birds of a swarm? These constituents could be tiny; we cannot rule out their existence just because it may be difficult to observe them. By looking at the interactions of light at different energies or over different distances and failing to see particle-like aspects, we can only conclude that light constituents are not realized in the energy and distance ranges probed.

A good way of testing properties of light is to analyze its interaction with matter. We are not interested in seeing the whole swarm, but rather its individuals. If we vary the intensity and the frequency of light, the two main characteristics of a wave, we may get into a range where only a few constituents, if they exist, play a role. There is a class of experiments with effects happening in small portions. The individual portions are not seen for light itself, but rather for matter it interacts with. If we shine a beam of light on the surface of a piece of metal surrounded by vacuum, it may happen that single electrons are ejected from the metal. One can detect the electrons by the marks they leave on a scintillator screen (much like an old-fashioned TV set) opposite the metal surface, and one can count how many electrons are ejected and what their energies are, depending on the light source used. A single electron is a rather small portion of energy. It allows us to

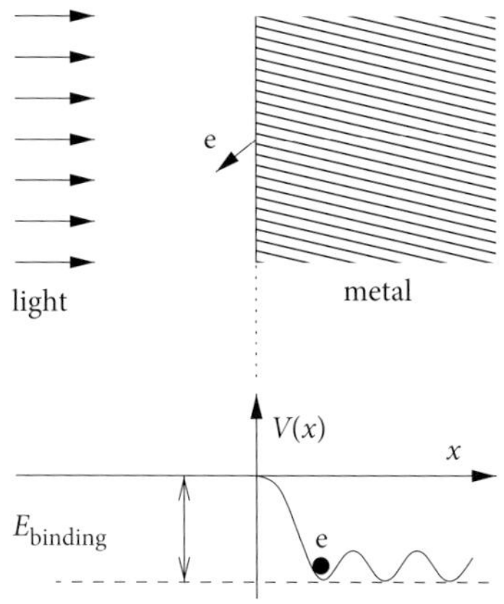

Figure 4.1 The photoelectric effect: Light shines through vacuum on a metal surface, ejecting electrons. A classical model would have metal electrons bound in a periodic potential. After light transfers more than the binding energy $E_{binding}$ on one of them, it can leave the metal. The energy of a classical wave comes from its intensity, or the amplitude of its oscillations.

test whether it rises from the metal by interactions with some constituents of light, or with the whole wave.

If the whole wave makes electrons leave the metal, the number and energies of electrons should depend on the light source in a specific way. An electron is held in the metal by electric forces pulling it near the nuclei of atoms forming the metal; see Figure 4.1. Light, as a periodic excitation of the electromagnetic field, also exerts a force on the electrons near the metal surface. If the force near the maximum of the wave is stronger than the force from nuclei, the electron can be pulled out of the metal, enter the vacuum region and then move further away. The maximum electric field exerted by the wave is its amplitude A. If its pull is larger than the typical distance of the nearest electrons from the surface, the electron will escape. The amplitude is independent of the frequency of the wave; it is rather related to the intensity of light. We expect that there should be some threshold intensity for light to be able to yank electrons from a piece of metal, irrespective of the frequency. Above the threshold, the energy of escaped electrons should be larger for higher light intensity.

The experiment, performed first by Heinrich Hertz in 1887, presents a different outcome. The energy does depend on the frequency in a visible way, of linear form as in Figure 4.2. Instead of a threshold intensity, there is a limiting frequency below which no electrons appear. While the electron energy stays the same for larger intensity, more electrons, all of the same energy if the frequency is unchanged, leave the metal.

These observations are incompatible with wavelike properties of light in the interaction process. As noticed by Albert Einstein in 1905, they can be explained if light has particle properties, with constituents called photons of the following qualities: The energy of a single photon in a light beam of frequency f has energy

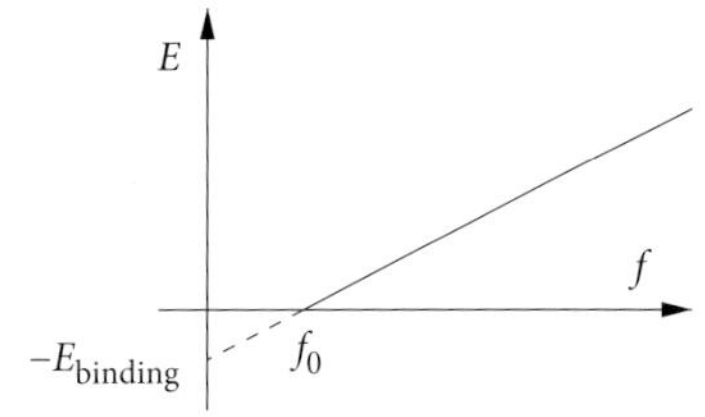

Figure 4.2 Measurement results for the photoelectric effect: The energies E of emitted electrons are in linear relationship with the frequency f of light used, with a minimum f_0 required to emit any electrons.

$E_{\text{photon}} = hf$ with a constant h. A photon interacts with an electron in the metal like two particles do, by duels. The electron, hit by a photon, can absorb the photon's energy or at least a significant part of it. The added energy allows it to escape the binding forces holding it in the metal. If some binding energy E_{binding} is required for an electron to escape, it will be left with an energy of $E_{\text{electron}} = E_{\text{photon}} - E_{\text{binding}} = hf - E_{\text{binding}}$ in the vacuum region. The electron energy indeed depends on the frequency of light in a linear fashion. Moreover, the photon energy must be at least the binding energy, and no electrons are ejected if light of a frequency less than $f_0 = E_{\text{binding}}/h$ is used. If we increase the intensity, on the other hand, we will have more photons, all of the same frequency-dependent energy, to push electrons out of the metal, also in agreement with observations.

The constituent picture shows that light of a given frequency f is quantized: It is impossible to decompose it in energy portions less than the photon energy hf.

Particle waves Light has particle constituents, and particles have wave properties. If one looks closely, one can see that not only light, under the right circumstances, interacts with matter by collisions with photons. One can also find that matter, such as electron beams, sometimes interact like waves. If we send two electron beams through a double slit, repeating Young's experiment with matter, we see interference over small distances. For an electron, we normally do not measure a frequency, but rather an energy. The interference pattern is compatible with a relationship $E_{\text{electron}} = hf$ between the measured energy and a frequency f that we can assign to it, just as in the case of photons, with same h. Moreover, by scattering electron beams, one can determine that their momentum is related to a wave length l via $p = h/l$.

The interchangeability of particle and wave properties, depending on the type of interactions, is not surprising, as the example of swarms of birds demonstrates. However, Newtonian mechanics, and even relativity, is incompatible with objects described sometimes as a wave, and sometimes as a particle. A particle is located in a small region or at a single point and moves along some curve; it is characterized by its position and velocity at different instances of time. A wave is extended and moves while changing its excitations, raising them at the front and letting them level off in the back; it is described by a function that tells us what the excitation is at all positions in space at different instances of time. One may take a particle viewpoint as an approximation of a wave excited in some small region, as we do when we speak of a light beam or a ray instead of an extended wave. However, such an approximate view cannot explain how a wave sometimes interacts with matter by particle-like collisions.

Just as we had to modify Newtonian physics in order to make it compatible with the universality of the speed of light, we must modify classical physics, in Newtonian or relativistic form, in order to allow objects to have both wave and particle properties. We need to develop a formalism that shows us when a wave and when a particle is realized. The mathematical description of a wave is more general than that of a particle. The size of the region in which the wave is substantially excited can be changed, while we cannot extend particles. In order to find our generalization of classical physics, we should begin with a particle and try to describe it as a wave.

Energy Particle motion is quantified by energy. We associate energy with motion so much that we even use it to describe the excitability of people. Energy is important for technical processes, to tell how much movement some source of energy can generate. Mathematically, energy is a conserved quantity, so that potential landscapes can be used to visualize how an object moves. As a function of position and momentum, the energy is all we need to formulate motion at an exact level. The energy function is the constitution of mechanics; its form determines all other laws of motion.

An object of mass m has energy $E(x, p) = p^2/2m + V(x)$ when it is at a position x where the potential energy is $V(x)$, with momentum $p = m\Delta x/\Delta t$. We have seen how general properties of this function affect the form of motion in classical mechanics. For instance, the kinetic energy is always positive, so that a given value of E may allow the object to move only in some bounded regions where $V(x)$ is less than E. We have also seen that Newton's laws of motion, especially the second law $\Delta p/\Delta t = -\Delta V/\Delta x$, ensure that energy does not change in time even as x and p change for a moving particle, provided the potential $V(x)$ is not modified as time goes on.

All three equations in the preceding paragraph are interlinked. One tells us how the momentum is related to the change of position, another shows how the momentum changes with a given potential energy, and we have an equation that combines momentum and the potential energy to the total energy. The total energy then turns out to be constant as a consequence of the other equations. A constant is a simpler quantity than a function of time, and it has much information if it initially appears as a combination of time-dependent functions such as $x(t)$ and $p(t)$.

There is much power in the notion of formulating change by the constants of life. Unchanging character traits can explain much about decisions, successes and failures. In mechanics, we endow objects with certain traits if we decide on their initial values, for instance of position and momentum which then determine what conserved quantities they have. The constant energy trait allows us to see whether the object is able to surpass a hill in the potential landscape. Referring to the energy is an immense shortcut compared to solving the differential equations specified for $\Delta x/\Delta t$ and $\Delta p/\Delta t$.

We begin with the energy expression and see it as the basic one, as more fundamental than Newton's equations for the change of x and p. Even for questions about the specific rate of change of one of these numbers, we can recover the value from the energy function. Newton's laws of motion can be derived from the combination of kinetic and potential energy added up in $E(x, p)$. We can write Newton's second law as $\Delta p/\Delta t = -\Delta_x E/\Delta x$ if on the right-hand side we take $\Delta_x E$ as the change of E at a constant value of p, so that the kinetic energy does not contribute. We obtain a reformulation which is rather simple but convenient because it resembles another equation that we can use in lieu of the momentum equation $p = m\Delta x/\Delta t$. We write this equation as $\Delta x/\Delta t = p/m = \Delta_p E/\Delta p$, using $\Delta(p^2) = (p + \Delta p)^2 - p^2 = 2p\Delta p + (\Delta p)^2 \approx 2p\Delta p$ for small Δp in the rate of change. Newton's equations of motion are then written by direct reference to the total energy,

$$\frac{\Delta x}{\Delta t} = \frac{\Delta_p E}{\Delta p}, \quad \frac{\Delta p}{\Delta t} = -\frac{\Delta_x E}{\Delta x}. \tag{4.1}$$

These equations are called Hamilton's equations of motion.

Poisson bracket Hamilton's equations look symmetric regarding the roles of x and p, except that one equation has a negative sign. One can make the appearance more equal by writing

$$\frac{\Delta x}{\Delta t} = \{x, E\} \quad \text{and} \quad \frac{\Delta p}{\Delta t} = \{p, E\} \tag{4.2}$$

with the Poisson bracket defined as

$$\{f(x, p), g(x, p)\} = \frac{\Delta_x f}{\Delta x}\frac{\Delta_p g}{\Delta p} - \frac{\Delta_p f}{\Delta p}\frac{\Delta_x g}{\Delta x} \tag{4.3}$$

for any two functions f and g of x and p. To compare this general form with the equations of motion for x and p, we take $g = E$ and $f = x$ or $f = p$, noting that $\Delta_x x/\Delta x = 1$ and $\Delta_p x/\Delta p = 0$. The Poisson bracket not only allows us to write the equations of motion in more symmetric form, it also generalizes them to rates of change of arbitrary functions of x and p: $\Delta f/\Delta t = \{f, E\}$.

Average A particle is as dynamic as the energy it has. If we are to describe motion in terms of extended waves so that narrow shapes reproduce Newtonian motion, the energy should play an important role. Having reformulated Newton's equations of motion in terms of a single function $E(x, p)$, we can now generalize them to waves.

In the relationship between particles and waves, the best guess for a position associated to a narrow wave is some kind of center of the region in which the wave is excited much. Most waves have hills and troughs and not just one unique maximum that could be identified with a particle's position. However, even for complicated wave forms, there is a general mathematical procedure to compute a central value, the average. The distribution of some characteristic, such as the height, in a population does not always show a clear maximum that could be associated with its typical value. Instead, one commonly calculates the average and views it as the most typical example. If there are N_{H} individuals of height H, rounded to some precision so as to produce a finite number of possible heights, the average is obtained as

$$\overline{H} = \frac{\mathbf{Sum}_{\mathrm{H}}(H\, N_{\mathrm{H}})}{\mathbf{Sum}_{\mathrm{H}}(N_{\mathrm{H}})} \,. \tag{4.4}$$

If all individuals have the same height H_0, there is only one nonzero term in the sums, and $\overline{H} = H_0$. If the heights vary, the more common ones with larger N_{H} have the strongest influence on $\overline{H}$. For a narrow sample, the average is near the center of the distribution, where it has its maximum. In more complex cases, there may be several peaks in the distribution which compete for dominance in the average. Two peaks, one for short and one for tall individuals, for instance, would average out to some medium size, even if there are not many medium-sized people in the population. In some cases, the average may even yield a height that no individual represents. The average therefore should not be seen as the most likely value realized in the population. It represents what its name and formula indicate: if we measure the height for many people then $\overline{H}$ is the average of all values obtained.

A wave, at a given moment in time, can be seen as a distribution over positions in space, corresponding to the distribution N_{H} of the number of individuals of some height H. The excitations of a wave can go in two different directions, such as up and down for a water wave. Negative values would represent one of the directions, but the number of individuals cannot be negative. Some waves may take only positive values (or perhaps zero), but to be safe, we compare the height distribution not with the value of the wave at a point, but with the value squared.

It is more likely to find a particle at some place x if the wave takes a large value at x, producing a large value N_x for its square. If we associate the position of the particle with the average $\overline{x}$, computed by (4.4), regions with large excitations of the wave have the strongest influence on the position. With several peaks in the wave, there may be competing effects from different places and the average might be in a region where the wave is low. If the wave has only one narrow peak, the average faithfully reflects this position, but in more complex shapes the average may not be the most likely position we would find in the wave. In a statistical sense, nonetheless, the average is the best measure we can take for a general distribution of some quantity.

Variance Statistical statements are not precise, but there are precise laws of statistics. The average, for instance, does not reflect the whole population. Still, it is precisely defined and provides an exact number to quantify some aspects. If we wanted to reflect all characteristics of the population in exact detail, doing justice to every single individual, we could only consider the whole set of individuals and their characteristics. The flood of data would be overwhelming; we would most likely miss the forest for the trees. A crisp briefing, providing the average and perhaps a small number of some other quantities, gives us a better and faster picture.

In addition to the average, we may be interested in how much individual values spread around it. There could be one narrow peak around the average, or a wide one, or two narrow but separated peaks. All those distributions could produce the same average, but still differ in other respects. A statistical quantity to capture such properties of the distribution is the variance. We start with the average $\overline{x}$ and calculate the deviations $\Delta x = x - \overline{x}$ for all individuals. The average of these deviations always vanishes and is of no interest:

$$\overline{\Delta x} = \frac{\mathbf{Sum}_x \left[(x - \overline{x}) N_x \right]}{\mathbf{Sum}_x N_x} = \frac{\mathbf{Sum}_x (x N_x)}{\mathbf{Sum}_x N_x} - \overline{x} = 0 \, .$$

In this calculation, $\overline{x}$ is the same for all x in the sum, and can therefore be factored out.

Instead of averaging Δx, it is more interesting to average its square. The square is never negative and cancellations that made $\overline{\Delta x}$ vanish no longer happen. We obtain a parameter to quantify deviations from the expectation value, called the variance

$$\mathbf{Var}_x = \sqrt{\overline{(\Delta x)^2}} = \sqrt{\frac{\mathbf{Sum}_x \left[(x - \overline{x})^2 N_x \right]}{\mathbf{Sum}_x N_x}} \, . \tag{4.5}$$

The variance is zero if all individuals have the same value of x, in which case only $N_{\overline{x}}$ is nonzero. It grows as the wave spreads out, or when it has several separated peaks.

We can go on and introduce an infinite number of parameters to describe the distribution, modeled on the mathematical expression for the variance: Instead of taking squares of the deviation, we take all positive integer powers. Unless the integer is one, an interesting and nonzero number is obtained. In formula, we have the equation

$$\mathbf{Mom}_{x,a} = \overline{(\Delta x)^a} = \frac{\mathbf{Sum}_x \left[(x - \overline{x})^a N_x \right]}{\mathbf{Sum}_x N_x} \tag{4.6}$$

for moments of order a. Some of the moments have special names, such as the variance (squared) for $a = 2$, or skewness for $a = 3$. For arbitrary a, however, it would be tedious to come up with an infinite number of individual names, other than a itself.

We have arrived at an infinite number of parameters, the moments combined with the average, by which we can describe an arbitrary distribution. There seems to be no advantage because we are just trading the infinite number of individual measures for an infinite number of moments. We express the original information in statistical terms, but there would be the same information overflow if we were to work with all the moments. The great benefit of the moments is that a small number of them often captures the main information contained in the population, even if a finite number would not give us an exact description. If we were to pick a small number of individuals and take their characteristics as a picture of the whole population, we could be badly mistaken. By statistical methods, we rearrange and organize the information, starting with the average value and amending the information it provides with measures for average deviations from the average. A small number of these deviation parameters, in most cases, is all we need to have a good view of the whole population.

Particular moments Moments are our best chance to relate particle properties to waves. At any given time, a particle is characterized by a small number of parameters, its position in space and the momentum. A wave at a time, by contrast, is specified by an infinite number of parameters, the excitations at all positions and also their rates of change. Statistics gives us methods to compress a large, perhaps infinite number of quantities to a small amount. For every particle, we assume that there is some wave or a distribution. At a fundamental level, the wave is the dynamical, moving object. It shows interference as seen in high-resolution experiments, but certain measurements will be sensitive only to the average or some moments. In the latter case, the wave behaves like a particle placed at the average value of position.

For a particle, position and momentum are independent quantities. When we throw a ball, the motion of our hand determines the position and velocity, or momentum, at the time of release. For the same position, many different momenta are possible, depending on the strength of our arm. The average position therefore cannot be sufficient to describe all motion, and moments present unrelated information. We need a distribution $N_{x,p}$ in two independent variables x and p, with two independent averages $\overline{x}$ and $\overline{p}$ to be identified with the particle's position and momentum at any given time.

If there are two independent averages, there must be different classes of moments. We can compute deviations $\Delta x = x - \overline{x}$ of positions from their average, and $\Delta p = p - \overline{p}$ for momenta. From independent deviations, we construct independent moments

$$\mathbf{Mom}_{x,a,p,b} = \frac{\mathbf{Sum}_{x,p}\left[(\Delta x)^a(\Delta p)^b N_{x,p}\right]}{\mathbf{Sum}_{x,p} N_{x,p}} . \tag{4.7}$$

For quadratic moments, we have the position variance $\mathbf{Var}_x = \sqrt{\overline{(\Delta x)^2}}$, the momentum variance $\mathbf{Var}_p = \sqrt{\overline{(\Delta p)^2}}$ and the covariance $\mathbf{CoVar} = \overline{(\Delta x)(\Delta p)}$.

Momentous dynamics As a wave moves, the average position and the moments change. There are equations of motion for these quantities just as we have Newton's second law for the position of a particle in classical mechanics. The particle's position changes according to the forces it encounters as it moves. A wave, an extended object, can change in more complicated ways. Its motion is sensitive not only to the force at the average position, but also to the forces that all its parts encounter. It is sensitive to a force field, that is, a different value of the force at every point in space, or the whole position-dependent potential from which we calculate the force. A water wave in shallow waters, for instance, changes its motion according to the depth of the water underneath, a position dependent function to which the wave reacts.

With several forces acting on all the parts of the wave, there may be different accelerations that move some parts of the wave away from others. The whole wave spreads out and gets wider, and accordingly the moments, foremost the variance, increase. Tails of the wave reach out to separate places, encountering new regions of the force field. During all these processes, the average of the wave cannot remain unmoved. Even if the tails are tiny, their motion changes the average, even if there is no force acting at the place pointed at by the average. To capture the whole motion of the wave, we should not work with equations only for the average position. We need a more complicated set of equations, collaborative equations that refer to the variance and other moments in order to determine the motion of the average.

A point particle moves according to equations controlled by the energy $E = p^2/2m + V(x)$. If wave effects are negligible, for instance, when the wave is narrow, the same quantity should determine how the average moves: $E_{\text{average}} = \overline{p}^2/2m + V(\overline{x})$, with $\Delta\overline{x}/\Delta t = \Delta_{\overline{p}} E_{\text{average}}/\Delta\overline{p}$ and $\Delta\overline{p}/\Delta t = -\Delta_{\overline{x}} E_{\text{average}}/\Delta\overline{x}$. However, some of the moments are nonzero even for a wave just slightly extended; they might affect the changes of averages. Our E_{average}, in which we replaced the general position x and momentum p with the average, cannot be equal to E. To be exact, we should write $x = \overline{x} + \Delta x$ and $p = \overline{p} + \Delta p$ as we did when we introduced the moments. The energy $E = (\overline{p} + \Delta p)^2/2m + V(\overline{x} + \Delta x)$ then depends on the positions and momenta of all parts of the wave. The average energy $\overline{E}$ refers not only to the average position and momentum, but also to moments via averages of products of Δx and Δp. Hamiltonian equations for $\overline{x}$ and $\overline{p}$ then depend on some of the moments, and the moments themselves evolve according to their own equations.

Average energy By taking the square in the kinetic energy and approximating the potential by a polynomial $V(x) = \mathbf{Sum}_n V_n x^n$ of some possibly high order, we compute

$$\overline{E} = \frac{\overline{p}^2}{2m} + V(\overline{x}) + \frac{\mathbf{Var}_p^2}{2m} + \mathbf{Sum}_n\left(V_n \mathbf{Sum}_{i=2}^n \frac{n!}{i!(n-i)!}\overline{x}^{n-i}\mathbf{Mom}_{x,i}\right) \quad (4.8)$$

with the factorial $n! = 1\cdot 2\cdots(n-1)\cdot n$. The first two terms add up to the classical energy E_{average} evaluated for the average position and momentum; the remaining terms are corrections due to wave effects. They are larger the more spread-out the wave is. Hamilton's equation for $\overline{x}$ retains its classical form, but the equation for $\overline{p}$ acquires new terms from the x-moments:

$$\frac{\Delta\overline{p}}{\Delta t} = -\frac{\Delta V(\overline{x})}{\Delta\overline{x}} - \mathbf{Sum}_n\left(V_n \mathbf{Sum}_{i=2}^n \frac{n!}{i!(n-i-1)!}\overline{x}^{n-i-1}\mathbf{Mom}_{x,i}\right). \quad (4.9)$$

Momentous dynamics The change of the moments is also controlled by $\overline{E}$. Hamilton equations give

$$\frac{\Delta \mathbf{Var}_x^2}{\Delta t} = \overline{\{\mathbf{Var}_x^2, E\}} = \overline{\frac{\Delta_x (\Delta x)^2}{\Delta x} \frac{\Delta_p E}{\Delta p}} = 2\overline{\Delta x \cdot \frac{p}{m}} .$$

If we write $p = \overline{p} + \Delta p$, the $\overline{p}$-term does not contribute to the total average because $\overline{\Delta x} = 0$, and from Δp, we have

$$\frac{\Delta \mathbf{Var}_x^2}{\Delta t} = \frac{2}{m} \overline{(\Delta x)(\Delta p)} = \frac{2}{m} \mathbf{CoVar} . \tag{4.10}$$

By similar calculations, we obtain

$$\frac{\Delta \mathbf{CoVar}}{\Delta t} = \frac{1}{m} \mathbf{Var}_p^2 - 2V_2 \mathbf{Var}_x^2 - \ldots \tag{4.11}$$

with other terms depending on higher orders in the polynomial expression for the potential. Finally, and also with several other terms,

$$\frac{\Delta \mathbf{Var}_p^2}{\Delta t} = -4V_2 \mathbf{CoVar} - \ldots \tag{4.12}$$

These equations, together with equations for all the other moments, provide a complete, but horribly complicated set of laws for the motion of the whole wave.

Spreading Waves have a tendency to spread out. Unlike a particle, which one may think of as a solid ball or a single point, a wave has several parts that move at different speeds. In special instants, the parts could be close and form a narrow wave. However, if they move at different speeds, their closeness will soon be lost. They may split up into separate waves, or still be part of the same wavy excitation, much more spread out.

A wave spreads out even if no forces act on it. Without forces, or with a vanishing potential, the average energy of the wave is $\overline{E} = \overline{p}^2/2m + \mathbf{Var}_p^2/2m$. The position and momentum average obey the same equations as the classical quantities: $\Delta\overline{x}/\Delta t = \overline{p}/m$ and $\Delta\overline{p}/\Delta t = 0$. The kinetic energy makes the average position move just as classically, but without a force or a potential, the momentum remains unchanged. The average position just keeps moving if it is pushed on at one time, or it stays at rest if not pushed.

In addition to the kinetic energy for the average position, the average energy has a contribution from the variance of momentum. It is characteristic for a wave, and its presence changes wave properties such as the spreading. The variance term does not affect the motion of averages, but it implies equations for the moments. The momentum variance itself is constant, $\Delta\mathbf{Var}_p^2/\Delta t = 0$, but the position variance changes according to $\Delta\mathbf{Var}_x^2/\Delta t = 2\mathbf{CoVar}/m$, depending on the covariance. The latter, in turn, is subject to its own equation of change: $\Delta\mathbf{CoVar}/\Delta t =$

$\mathbf{Var}_p^2/m$. If the (constant) momentum variance is not zero, it changes the covariance, which in turn makes the position variance change. With a constant rate of change, the covariance increases linearly in time, $\mathbf{CoVar}(t) = Ct + \mathbf{CoVar}(0)$ with $C = \mathbf{Var}_p^2/m$ constant. The rate of change of the position variance grows linearly in time, and the position variance itself increases quadratically: $\mathbf{Var}_x^2(t) = \mathbf{Var}_p^2 t^2/m^2 + 2\mathbf{CoVar}(0)t/m + \mathbf{Var}_x^2(0)$.

Taking a square root, the position variance $\mathbf{Var}_x$ of a particle not subject to forces is almost linear in time. The wave keeps spreading out and will eventually occupy a wide region. Particle aspects, realized to a good degree if the wave is narrow, disappear as it becomes more difficult to assign a sharp boundary to the spread-out wave. We should be able to observe sharply defined particles only in rare circumstances, just when all parts of the wave are located close to a central point. In our everyday experience, however, we do not worry that an object, for instance, a car we are driving, becomes wider and wider and at some point loses its confining walls in the unsteady wavy thing it becomes. The reason for the classical behavior of objects we usually deal with is the rate of change of $\mathbf{Var}_x$, inversely proportional to the mass. The momentum variance of a narrow wave is much smaller than the momentum itself, and therefore the position variance increases much more slowly than the position if the particle is moving. For heavy objects, we do not notice the spreading because it takes too long. Lighter objects, for instance, the individual particles we deal with in atomic and particle physics, are much more sensitive to the spreading or other wave phenomena; these particles do show the aspects of quantum physics. However, at a precise level, quantum physics does apply to all objects in the universe, no matter how heavy.

A wave not subject to any force keeps changing, even if the particle it should resemble does not move at all. Absolute rest is more difficult to achieve in quantum physics than in classical mechanics. It is possible only if all forces are balanced. In our example, we have assumed the potential to vanish as there are no forces in the classical understanding. However, the average energy of a wave has an additional contribution on top of the kinetic energy, even if there is no potential. The form of the wave, characterized by the momentum variance, enters a new potential that makes other moments change. As a truly quantum phenomenon, the variance force cannot be balanced by any classical force. We could at best try to balance it with other forces of quantum physical nature, namely, other contributions from the shape of the wave.

Harmonic oscillator Harmony improves collaboration. When all parts of a wave change harmoniously, the whole wave will keep together, not spread out. A particle without forces is not harmonic; it has constant momentum variance (squared) added to the average energy, making the position variance change in time. We cannot eliminate the momentum variance from the average energy because it is a faithful fellow of kinetic energy. However, we can try to arrange a system of forces that adds extra variance to the average energy, so that changes of position and momentum variances compensate each other.

A hint is found in the equations for the particle without forces. The rate of change of the position variance does not directly depend on the momentum variance, but rather on the covariance. The change of the covariance, in turn, is proportional to the position variance. If there is an extra term that cancels the position variance in the rate of change of the covariance, the covariance will remain constant. If it vanishes at one time, it will always be zero and then leave the variances undisturbed, provided that there are no other terms in the variance equations that we bring in for our new system.

Position and momentum behave in a rather symmetric way in the Hamiltonian formulation of equations of motion, except for the occasional minus sign. A particle without forces breaks the symmetry because it has only kinetic energy, depending on the momentum, but no position-dependent potential. We can make the system more symmetric regarding the roles of position and momentum if we assume a potential quadratic in x, giving rise to a linear force. In classical mechanics, such a system is realized to a good degree for a swinging mass or one suspended on a spring, provided the extensions are small, and for several forces in atomic physics, the model is a good approximation. Such a system is called a harmonic oscillator; it has potential energy $V(x) = \frac{1}{2}kx^2$ with the spring constant k.

> The average energy now takes the form $\overline{E} = \overline{p}^2/2m + k\overline{x}^2/2 + \mathbf{Var}_p^2/2m + k\mathbf{Var}_x^2/2$. Equations of motion for the average position and momentum, $\Delta\overline{x}/\Delta t = \overline{p}/m$ and $\Delta\overline{p}/\Delta t = -k\overline{x}$, do not depend on the moments; they are identical in form to the classical equations of the same system. These two equations are coupled: The rate of change of $\overline{x}$ depends on $\overline{p}$, and the rate of change of $\overline{p}$ depends on $\overline{x}$. We obtain a decoupled equation just for $\overline{x}$ if we compute the rate of change of its rate of change: the acceleration $\Delta^2\overline{x}/(\Delta t)^2 = m^{-1}\Delta\overline{p}/\Delta t = -k\overline{x}/m$. Any such function, whose negative acceleration is proportional to the original function, must be either a sine or a cosine. From the time dependence $\overline{x}(t) = A\sin(\bigcirc\sqrt{k/m}\ t/2\pi) + B\cos(\bigcirc\sqrt{k/m}\ t/2\pi)$ with constants A and B depending on the initial position and velocity, we see the oscillating nature, of frequency $f = \sqrt{k/m}/2\pi$. With an active force, the average position is not constant; the particle cannot be at rest unless it always stays at $\overline{x} = 0$ where the force vanishes.

The moments behave according to $\Delta\mathbf{Var}_x^2/\Delta t = 2\mathbf{Var}_p^2/m$ as before, but now we have $\Delta\mathbf{CoVar}/\Delta t = \mathbf{Var}_p^2/m - k\mathbf{Var}_x^2$ for the covariance and $\Delta\mathbf{Var}_p^2/\Delta t = k\mathbf{CoVar}$ for the momentum variance. The two variances change in general, but we can keep both of them constant if the covariance vanishes: $\mathbf{CoVar} = 0$. The covariance itself must obey an equation for its rate of change; it can remain a constant zero only if the position and momentum variances match so that $\mathbf{Var}_p^2/m = k\mathbf{Var}_x^2$. If this is the case, neither the variances nor the covariance change, and the wave does not spread out as the average position keeps oscillating. If we have such a nonspreading wave, the average energy is $\overline{E} = \overline{p}^2/2m + k\overline{x}^2/2 + k\mathbf{Var}_x^2$. The constant position variance raises the energy compared to the classical value.

Uncertainty relation Harmonic oscillations allow waves to move without changing shape. Only the average position swings along the classical motion, and the average momentum wanes and waxes. All along, the width of the wave, expressed by the position or the momentum variance, remains constant. The wave always stays in shape, no parts of it bulge out. All this is quite different from a sedentary particle not subject to any force, whose girth widens as time goes on.

Not all harmonic waves behave in this way; after all, we had to assume the constancy of variances. However, the fact that waves forever in shape are possible if there is a harmonic force acting on the particle distinguishes this system among all others of quantum physics. If there is no force, for instance, we have already seen that constant position variance is possible only if the momentum variance vanishes.

However, the momentum variance of a wave cannot vanish, and also the position variance must be nonzero: The position variance would vanish when the wave is excited at just one point, like a single water molecule swinging up and down in the ocean. Such a skinny wave would not only be invisible, it would be impossible. Physical waves arise because some medium is excited from its equilibrium state. The forces that are balanced in equilibrium then become active and make the medium swing back, but it overshoots and pushes the wave on as long as friction still leaves some of its energy. A single molecule pulled out of the ocean would be no wave. It would fall back without making a splash, or be carried away with the wind. A wave can form only when enough water is elevated above the surface, or pulled down beneath the surface, so that it is still connected with the rest of the water body. Cohesion then pulls along other parts of the water, and the wave moves.

A wave must have a nonzero width and curves, and its position variance cannot be zero. The momentum variance is not easy to visualize, but it is related to an intuitive wave property by one experimental hint. Scattering experiments with electrons show that a particle of momentum p behaves like a wave with wave length $l = h/p$, the distance between two successive wave hills. For vanishing momentum variance, the wave would have constant wave length, the same value everywhere. However, if the wave length does not change as we move along a snapshot of the wave at one time, the wave pattern must keep repeating itself, everywhere in space. Such a wave is impossible to build, not because there would be too few displacements from equilibrium, but too many. A wave cannot continue the same pattern everywhere in space; such a wave would not only cost an infinite amount of energy, it would also require an infinite amount of time to build.

Waves with vanishing position or momentum variances cannot exist. What our failed constructions show goes even further: We can try to reduce the momentum variance only at the expense of a larger position variance. Starting with some wave, we could try to make its pattern repeat itself more by suitably exciting the wave in regions where it was not very strong. The wave length would become more constant, but by extending the wave into regions it had not yet reached we increase the position variance. Conversely, if we try to decrease the position variance by shrinking the wave to be supported in a smaller region, we deprive it from space in which it could repeat its pattern with unchanged wave length; we increase the momentum variance.

Nonvanishing variances imply uncertainties of the values we can assign to position or momentum. The averages are good first guesses, but the wave is excited also at places different from the position average, and some of its parts move with momenta different from the momentum average. The wave does not allow us to specify its exact position and momentum, and the variances, as measures for the spread of the wave around its average values, quantify uncertainty. The uncertainty cannot vanish, and it is not possible for the variance of both position and momentum to be small. One of the two variances can be reduced only at the expense of increasing the other. This uncertainty relation takes its mathematical form in the inequality

$$\mathbf{Var}_x \mathbf{Var}_p \geq \frac{h}{4\pi} \tag{4.13}$$

with the same positive constant h seen before. For short, we define $\hbar = h/2\pi$ (h-bar).

We will later see why the inequality takes this form, and how the value $h/4\pi = \frac{1}{2}\hbar$ is determined. For now, we note that it reflects all the properties we have seen by our considerations of waves. Neither variance can vanish because the left-hand side would then be zero, smaller than $\frac{1}{2}\hbar$. Starting with some values of the variances, it is possible to reduce one of them while keeping the other constant, or even to decrease both of them. However, once the product of variances is small, reaching the bound of $\frac{1}{2}\hbar$ allowed by the uncertainty relation, any further reduction of one variance can be accomplished only if the other is enlarged.

Zero-point fluctuations A wave must always move. If we would arrest the shape of a wave for all times, it would cease to be a wave and become something of a mountainscape. Motion is an essential part of what makes a wave a wave, and absolute rest is impossible. We have seen this general property realized in our example of a particle free of forces. Even if $\overline{x}$ stays at rest, the wave spreads out and changes. Absolute rest would have been possible if the momentum variance had vanished, but we now know that such a value is not allowed by the uncertainty relation. A wave must always keep on changing.

The uncertainty relation is one important property of waves related to their incessant motion. Even if we could make the average momentum vanish, so that a classical particle would not move unless it is pushed on by forces, the momentum variance of a wave had to remain nonzero. The wave keeps on changing even if its average stays fixed. We can see the new option for waves to move, compared with particles, in the average energy. Without forces, the average energy is $\overline{E} = \overline{p}^2/2m + \mathbf{Var}_p^2/2m$. If $\overline{p}$ is zero, there is still a form of kinetic energy left, referring to $\mathbf{Var}_p$ instead of $\overline{p}$.

Harmonic oscillations, in some way, are the opposite of a free particle. They allow waves with constant shape but changing averages. Again, the uncertainty relation implies that it is not possible for the variances to be zero, but there is no such restriction for the covariance. A vanishing **CoVar** for waves of the harmonic oscillator, we recall from our earlier investigation, is the condition for unchanging variances, which then obey the equation $\mathbf{Var}_p^2/m = k\mathbf{Var}_x^2$. If we use this equation in the uncertainty relation, we obtain a minimal possible value for the variances: $\mathbf{Var}_{x,\min}^2 = \frac{1}{2}\hbar/\sqrt{mk}$ and $\mathbf{Var}_{p,\min}^2 = \frac{1}{2}\hbar\sqrt{mk}$. The variances contribute to the average energy of the wave, $\overline{E} = \overline{p}^2/2m + k\overline{x}^2/2 + \mathbf{Var}_p^2/2m + k\mathbf{Var}_x^2/2$. For the minimum variances, the energy is the classical expression evaluated for the averages, plus a constant contribution of $E_{\text{zero}} = \frac{1}{2}\hbar\sqrt{k/m} = \frac{1}{2}hf$ with the oscillation frequency $f = \sqrt{k/m}/2\pi$.

If the average position stays fixed at $\overline{x} = 0$, an equilibrium point where the force vanishes, the wave still has energy. It is called the zero-point energy because it is left even if we manage to slow down all particles to absolute rest, for instance, by freezing them at fixed position when the temperature is at absolute zero. It is the minimum amount of energy that must be kept in the wave because a wave cannot stop changing. When the wave has attained a state in which the energy is as small as possible, it has reached its ground state, the state to which it would always return after any excitement, after all expendable energy has been used up. The ground state of the harmonic oscillator with frequency f has an energy of $E_0 = \frac{1}{2}hf$.

Quantum dynamics Free and harmonic systems display different behaviors of waves. They allow us to derive some new features of wave dynamics compared to particle dynamics, but they are a rather restricted class of models. In both cases, the average values of position and momentum obey equations of motion that are independent of the variances and the covariance, or any other moments. In fact, the equations for averages in these cases are identical to the classical equations, namely, Newton's second law for the position and momentum of a particle. If quantum physics would merely enrich classical mechanics by new variables and their equations of motion, not much would change. The description would just be uneconomical; with all the moments, we would introduce an infinite number of parameters to obtain unchanged equations for the average values.

Quantum physics, however, is different from classical physics, sometimes radically so. We have seen one new feature in the uncertainty relation, a consequence of the transition from particle mechanics to wave mechanics. Even if the averages obey the same equations as the classical variables and follow the same motion, we could not determine the precise position or momentum because they are subject to uncertainties. Sharp values for these important physical quantities are not defined; they do not exist. The best approximations are the averages, but they remain approximations and are affected by uncertainties. Measuring position and momentum therefore shows crucial new features in quantum physics compared to classical mechanics, even if position and momentum sometimes change in time just as they do classically.

In addition to uncertainties, there are important deviations of quantum dynamics from classical dynamics in systems not free or harmonic. If the potential is not a quadratic polynomial, the average energy (4.8) has extra terms which depend on the moments but, unlike in the free and harmonic cases, also on the average position and momentum. For a free particle and the harmonic oscillator, wave properties could be seen in the average energy only by a zero-energy contribution independent of average values.

For other potentials, there are product terms of averages and moments, for instance, a product $\overline{x}\,\mathbf{Var}_x$ of the average position and the position variance if the potential contains a third power of x. We can view such a term as a new kind of potential depending not just on the average, but also on the variance. The appearance as a product means that the average interacts with the variance: In order to know the potential energy for the average position, we must know the position variance and vice versa. In physics, objects move in such a way that they minimize the total energy. In order to minimize a product term of two time-dependent variables, both variables must move in concerted ways. The rate of change of one variables, the average position say, depends on the value and the time dependence of the other, the position variance. We can no longer solve equations for average values and for variances in separation; they are subject to one coupled set of equations because they all describe the motion of a single wave. Additional variance-dependent terms appear in the equations of motion for averages, providing an important source of quantum corrections to the classical dynamics.

Wavelike motion implies new contributions to the energy and the potential, affecting the motion if the wave nature is much pronounced. New forces arise, without classical analog because they depend on variances which do not exist for a classical particle. Many properties of matter, for instance, aspects of its long-term stability, rely on such new quantum forces, as do some effects in the universe at early times, when the density was large. In order to grasp the consequences, we need to look in more detail at the waves themselves, rather than the average and moment parameters they possess.

4.2 States

Quantum mechanics treats particles as waves, with the pointlike hard-sphere nature of our usual imagination seen as an approximation. A wave, in contrast to a particle, is extended without sharp boundaries. The values it takes at different places change in time in related but unequal ways; when the wave rises at one place, it shrinks somewhere else. Waves are the result of a deviation from some equilibrium configuration, kept in motion by an interplay of forces that try to restore rest. Without deviation from a stationary state, without motion, a wave does not exist, unlike a particle which can find its equilibrium at some force-free resting place without losing its existence.

Motion is an integral part of waves. In quantum physics, new contributions to the energy come about by remnant, zero-point motion which cannot be stopped. Even objects that appear unmoved are brimming with microscopic motion, much as an adversary's poker-face is infused with blood. A wave must always change, or else it disappears, and so does the object it describes. Quantum physics means move or die.

Wave function What kind of wave are we dealing with? The wave phenomena we know from experience begin with excitations of some medium, like the water surface of a lake. If the level surface is disturbed, water moves the surface in wavelike form, propagating the disturbance. If a wave has disappeared by breaking on the shore or by mere friction, the water surface remains. What is the analog of the surface or the whole lake for the waves we have in quantum mechanics? This question, as per current knowledge, has no answer. The waves of quantum mechanics are not physical. They are waves in the sense that they propagate, always keep moving, and are extended around their averages without sharp boundaries, but of sizes determined by variances rather than physical surfaces. And, they do not begin with excitations of a medium.

The waves of quantum mechanics begin with the existence of a particle, either from the beginning of time or in a transmutation process from one particle to another. The medium being excited is not physical, but rather an abstract notion that we could, for lack of better words and only in negative terms, call nonexistence. The equilibrium state of quantum physics, when no wave is excited, is absolute emptiness, total nonexistence. If there is a wave, there is something, for instance, a particle, whose range of positions is described by the wave, a wave in the lake of existence. The values of the wave at different places show how much nonexistence is excited into existence. Unlike the binary quality of existence, which can either be realized or not, a wave can take on a continuous range of values. If we want to know what kind of waves we are dealing with in quantum mechanics, we must extend the dichotomy of existence and nonexistence to a whole range.

It might be difficult to imagine an object half existing and half nonexisting, but the extended range of existence values does have a meaningful interpretation: probability of existence. If the wave gives us a value of one half in some region in space, we need not make sense of half-existing particles. We rather interpret the value by saying that there is a 50%-change of encountering the particle in this region. If we look for the particle or measure its position, we will find the particle in our region in half the cases, and outside the region in the other half. This probabilistic interpretation matches well with the statistical concepts of averages and variances we have used to describe waves.

If the wave has disappeared or takes the value zero everywhere, it is impossible to find the particle anywhere. The likelihood to see the particle, even indirectly by forces it might exert on other particles, is zero. For all practical purposes, the particle does not exist. If we want to describe a particle by mathematics, an unquestionable assumption is that the particle exists. For all our equations, we therefore assume that the probability to find the particle just somewhere is 100%, or takes the value one. Any wave in quantum mechanics must fulfill this normalization condition.

The fact that the waves of quantum physics do not have a physical interpretation, but are mathematical, probabilistic objects, just like poll data for a population, is the cause of some difficulties and counterintuitive features. Mathematics, however, is incorruptible and does not require visualization, although a good picture always helps. Many predictions can be extracted from quantum mechanics, and they all agree with measurements. Quantum mechanics is a successful theory, perhaps the most successful one we have if we consider that it applies universally, to all objects small and large. In some cases, we obtain theoretical results about physical effects from the dynamics of averages and moments, as already used. Mathematical solutions for averages and moments then have a good physical interpretation: they literally correspond to averages and moments such as variances of the distribution of measurement results. In other cases, it is more convenient to work directly with the wave function, telling us the value of the wave everywhere in space. Devoid of a physical interpretation, the wave function is less intuitive than the averages and moments. It is the main carrier of counterintuitive features of quantum mechanics, and for the thorniness of some of these issues, it is denoted by the cactus symbol Ψ. (Actually, the Greek letter Psi.)

Waves Archetypical functions describing waves are the sine and cosine. Plotting these functions (Figure 1.3) shows their regular variations with hills and valleys separated by the wavelength l. The two functions differ only in that they are shifted against each other by a quarter wavelength; still, it is often convenient to treat both twins as individuals. In functional form, a wave with wavelength l is obtained by $\Psi(x) = \sin(\bigcirc x/l)$ or $\Psi(x) = \cos(\bigcirc x/l)$, or more generally, a combination $\Psi(x) = a\sin(\bigcirc x/l) + b\cos(\bigcirc x/l)$. Depending on the values of a and b, such a function has its first zero not at zero (like the sine) or at $\frac{1}{4}\bigcirc$ (like the cosine), but somewhere in between. With trigonometric identities, one can rewrite the wave as $\Psi(x) = A\sin(\bigcirc x/l - d)$ to make it clear that the first zero is found at $x = dl/\bigcirc$.

The functions so far only depend on the position x, not on time. They are waves arrested at one moment of time, and while they show spatial variations, they do not display oscillations in time. Oscillations are measured in terms of a frequency f such that the whole process repeats itself after a period $T = 1/f$. For the time dependence of a wave of frequency f and wavelength l, we write $\Psi(x,t) = a\sin(\bigcirc x/l - \bigcirc ft) + b\cos(\bigcirc x/l - \bigcirc ft)$. Variations in space and time are not independent; they are coordinated in such a way that the pattern of the wave, or the positions of individual hills and valleys, moves according to the equation $x/l - ft = C$ constant. Writing this linear equation as $x(t) = Cl + flt$ shows that the wave moves with velocity $v = fl$.

Wavelengths, frequencies and wave velocities are classical properties, often observed for a water wave. As a new aspect of the wave function Ψ in quantum mechanics, we do not directly measure the wave but rather particle-like properties associated with it, the momentum or energy. A classical wave has energy and momentum, well known, for instance, for light as an electromagnetic wave with its capacity to warm, but there are various notions of energy for different waves. We must find out how the energy of particles in quantum mechanics is related to the wave function, using as guidance the classic results of the photoelectric effect and related experiments.

Light in quantum mechanics is made from photons as smallest energy packets. In the photoelectric effect, light of frequency f carries photons of energy $E = hf$ each. A wave function therefore determines the particle energy by its time dependence, for instance, the prefactor of t in a sine or cosine wave. Electron scattering experiments show that the momentum of a particle in quantum mechanics is obtained from the wavelength l of the wave function, $p = h/l$. For the momentum, spatial variations of the wave are crucial.

Operators Energy and momentum are related to changes in time and space. If we know that a wave is a sine or cosine, it is possible to look at the function and read off coefficients of t and x to determine E and p. For other functions, such as saw curves, the procedure might not be obvious. We should look for direct mathematical operations to compute variations of any wave function. We know a good candidate to determine the size of variations in space and time: derivatives. The energy E should be related to $\Delta\Psi/\Delta t$, and momentum p to $\Delta\Psi/\Delta x$, both with a factor of $h/2\pi = \hbar$.

If we compute the rates of change for a sine wave, we indeed obtain the frequency and wavelength as prefactors: $\Delta \sin(\bigcirc x/l - \bigcirc ft)/\Delta t = -2\pi f \cos(\bigcirc x/l - \bigcirc ft)$ and $\Delta \sin(\bigcirc x/l - \bigcirc ft)/\Delta x = (2\pi/l)\cos(\bigcirc x/l - \bigcirc ft)$. However, the sine has changed to a cosine because the zero of the sine corresponds to the maximum rate of change, while the maximum of the sine has a vanishing rate of change. We could ignore the functions and just look up prefactors to tell us the frequency and wavelength. However, there might be complications for other waves if we do not know how to separate the interesting quantities from other factors that might appear.

There is one even simpler wavelike function, though related to sine and cosine. Its properties are more illuminating than the waves we used so far because the function itself is not modified when we compute its rate of change, up to a factor that shows us the frequency or wavelength. We have seen one function unfazed by computing the rate of change: the exponential function, Figure 1.2. But it is ever-increasing, not oscillating like a wave. We obtain an unfazed wave by combining sine and cosine.

Imaginary unit For a simple wave of the form $\Psi(x,t) = a\sin[\bigcirc(x/l - ft)] + b\cos[\bigcirc(x/l - ft)]$ with suitable coefficients a and b, we have the rate of change $\Delta\Psi/\Delta t = 2\pi f\{b\sin[\bigcirc(x/l - ft)] - a\cos[\bigcirc(x/l - ft)]\}$. It is proportional to the original wave function, $\Delta\Psi/\Delta t = -2\pi i f\Psi$, only if there is a number i such that $a = bi$ and $b = -ai$. Taken together, the two equations imply $a = -ai^2$, or for nonzero a, $i^2 = -1$. The square of a real number is always positive and cannot equal -1. Instead, the number i with the property $i^2 = -1$, called the imaginary unit, is an example for a complex number, an element of a larger set than the real numbers. A combination of sine and cosine with $b = 1$ and $a = i$ has a rate of change proportional to itself, just like the exponential function but with an oscillating shape. We write $\cos(\bigcirc x) + i \sin(\bigcirc x) = \exp(2\pi i x)$ to indicate the similarity.

Waves of the form $\Psi(x,t) = \cos[\bigcirc(x/l - ft)] + i \sin[\bigcirc(x/l - ft)] = \exp[2\pi i (x/l - ft)]$ are of special interest in quantum mechanics. They allow us to compute variations in space and time in a direct way: $\Delta\Psi(x,t)/\Delta t = -2\pi i f \Psi(x,t)$ and $\Delta\Psi(x,t)/\Delta x = (2\pi i/l)\Psi(x,t)$. We compute energy $E = hf$ and momentum $p = h/l$ as the coefficients of $i\hbar\Delta\Psi/\Delta t$ and $-i\hbar\Delta\Psi/\Delta x$, respectively, with $\hbar = h/2\pi$. Distinguishing the computations of energy and momentum from the actual waves, we associate the operations $\hat{E} = i\hbar\Delta/\Delta t$ to the energy and $\hat{p} = -i\hbar\Delta/\Delta x$ to the momentum, identified as official mathematical operators by their hats.

Complex numbers Measurement results, such as the direction of a hand, are always real numbers. If we use complex numbers in physics so as to have simpler or more elegant mathematical descriptions, we must ensure that physical predictions we derive are all real. We have chosen a complex wave function as an example of a wave in order to have a simple form for the rate of change. Since the wave function itself is not observable in quantum mechanics, we are free to use complex values for it as long as there is a clear procedure that guarantees physical properties to be real.

A complex number is a combination $z = a + ib$ of two real numbers a and b with the imaginary unit i. For a complex number in this form, a is called the real part of z and b the imaginary part. We calculate with complex numbers just as we do with real numbers, using the usual rules for addition and multiplication.[1] The sum of two complex numbers is obtained by adding the real parts and imaginary parts, respectively; we add all terms in the two complex numbers and factor out i in two of them. To multiply two complex numbers $z_1 = a_1 + ib_1$ and $z_2 = a_2 + ib_2$, we begin ignoring the complex nature and obtain $z_1 z_2 = a_1 a_2 + i(a_1 b_2 + a_2 b_1) + i^2 b_1 b_2$. The imaginary unit i was defined such that $i^2 = -1$, departing from what we know for real numbers with their never-negative squares, and we write the product $z_1 z_2 = a_1 a_2 - b_1 b_2 + i(a_1 b_2 + a_2 b_1)$ to read off the real and imaginary parts.

The product of two complex numbers shows us how we can assign to a complex number a quantity guaranteed to be real. In the square of a complex number, multiplying it with itself, the imaginary part in general is nonzero; the result is another complex number. If we want to make the imaginary part $a_1 b_2 + a_2 b_1$ of the product of two complex numbers $z_1 = a_1 + ib_1$ and $z_2 = a_2 + ib_2$ vanish, we could choose $a_2 = a_1$ and $b_2 = -b_1$. Then, $z_1 z_2 = a_1^2 + b_1^2$, always a real number. We call the specific number z_2 obtained in this way the complex conjugate of z_1 and write it as $z_2 = z_1^* = a_1 - ib_1$. The square root of the product of z_1 with its complex

1) Treating complex numbers by the rules of real numbers does not allow us to manipulate the square root $i = \sqrt{-1}$ in any way other than by $i^2 = -1$. Otherwise, we could derive the contradiction $1/i = 1/\sqrt{-1} \overset{\text{No!}}{=} \sqrt{1/(-1)} = \sqrt{-1} = i$, which would imply $i^2 = 1$ if we multiplied the equation with i. The correct calculation is $1/i = i/i^2 = i/(-1) = -i$.

conjugate is called the norm of z_1: $|z_1| = \sqrt{z_1^* z_1} = \sqrt{a_1^2 + b_1^2}$, interpreted as the magnitude of the complex number.

With this procedure applied to the complex wave function $\Psi(x,t)$, we compute $d(x,t) = |\Psi(x,t)|^2$ for all x and t and obtain a real function $d(x,t)$. This function, being real, can be related to physical predictions, and by its definition, it is related to the wave movement as well. However, unlike for classical waves, physics and wave motion are not described by the same object. If we have two waves moving toward each other and overlapping, it is the moving wave functions Ψ_1 and Ψ_2 that are added, or superposed, to obtain the combined wave. In classical physics, we would see consequences of the combination from the sum. In quantum mechanics, on the other hand, we compute the function d for the superposition, $d = |\Psi_1 + \Psi_2|^2 = (\Psi_1^* + \Psi_2^*)(\Psi_1 + \Psi_2) = |\Psi_1|^2 + |\Psi_2|^2 + \Psi_1^*\Psi_2 + \Psi_1\Psi_2^*$. Compared with the sum of d-functions of Ψ_1 and Ψ_2, there are additional terms which often have implications for physics.

Physical interpretations require the use of the real function d instead of the wave function Ψ. At the beginning of our endeavors in quantum mechanics, we have seen that physical predictions are made in a statistical fashion, with a nonnegative function that tells us likelihoods of individual measurement results. These likelihoods are used to compute averages, variances and other moments to be compared with the distribution of measurements. We cannot use Ψ for this purpose because it is not real, let alone nonnegative. However, $d = |\Psi|^2$ is not only guaranteed to be real, it never takes negative values. It satisfies all requirements for the calculation of probabilities.

Interaction What do the wave functions of quantum mechanics look like? We can never see them because they are not measurable, but we can derive and picture their form by mathematics. We have seen the examples of sine, cosine, and exponential waves with their constant momenta and energies. More interesting physics always occurs when there are interaction processes in which momentum or energy is exchanged between different objects, or when a force acts to change the momentum by accelerating or decelerating an object. When this happens in quantum mechanics, the wavelength and frequency of the wave function must change, for they tell us what the momentum and energy are. However, how does the form of the wave function depend on the force? What wave equation must we solve to derive the wave function?

We begin our search for a wave equation by looking back to the harmonic oscillator with average energy $\overline{E} = \overline{p}^2/2m + \frac{1}{2}k\overline{x}^2 + \mathbf{Var}_p^2/2m + \frac{1}{2}k\mathbf{Var}_x^2$. We found that the averages of position and momentum satisfy the classical equations $\Delta\overline{x}/\Delta t = \overline{p}/m$ and $\Delta\overline{p}/\Delta t = -k\overline{x}$, and that there are states with unchanging variances provided the covariance vanishes and we have $\mathbf{Var}_p^2 = mk\mathbf{Var}_x^2$. The minimum value allowed for $\mathbf{Var}_x^2$ by the uncertainty relation is then $\mathbf{Var}_{x,\min}^2 = \frac{1}{2}\hbar/\sqrt{mk}$. With this value and the corresponding one for $\mathbf{Var}_{p,\min}^2 = \frac{1}{2}\hbar\sqrt{mk}$, we have minimal average energy: the zero-point energy. The state in which the energy is minimal is called the ground state.

We obtain more information about other, excited states, if we consider a harmonic oscillator interacting with some other energy form such as light. We can excite a classical oscillator by moving it in periodic fashion, with the strongest effect when we move it with its own frequency $f = \sqrt{k/m}/2\pi$, at resonance. The resonance effect is well known from experience and can often be seen when tall trees or even buildings are pushed on by the wind and start swinging with a characteristic frequency, always the same for a given building no matter how the wind is blowing. The resonance frequency is the frequency by which the object would swing on its own, determined by the inertia of its mass resisting a change of motion, and the restoring elastic forces that try to bring the object back to its average position.

Quantum mechanics must explain our observations, and so it should show resonance effects. We can excite the harmonic oscillator from its ground state by pushing it on periodically with its own frequency f. The oscillator could be an electron bound by a force $F = -kx$, with light of frequency $f = \sqrt{k/m}/2\pi$ shining on it. Light as an electromagnetic wave moves the charged electron periodically, exciting the oscillator.

With light shining on the electron, the combined expression for the energy changes. The electric field of the light wave implies an extra energy contribution of the electron. We must work to move the electron against the electric field, and the electron gains energy when it is accelerated by the field. The work is given by the product of the distance traveled with the force to be overcome. For the electric force, with a field e assumed to be constant in space, we have to work an amount $q\overline{x}e$ in order to move a charge q by an amount $\overline{x}$. In our oscillator excited by light, both $\overline{x}$ and e depend on time, $\overline{x}(t) = \overline{x}(0)\cos(\bigcirc f t) + [\overline{p}(0)/(2\pi m f)]\sin(\bigcirc f t)$ according to the equations of motion, and $e(t) = a\cos(\bigcirc f t) + b\sin(\bigcirc f t)$ for a harmonic light wave of frequency f. The product of these quantities must be added to the average energy to describe the motion of the oscillator in the combined system with light.

To see what energies are involved, we write the electric field in terms of exponential waves: $e(t) = A_-\exp(2\pi i f t) + A_+\exp(-2\pi i f t)$ with $A_\pm = \frac{1}{2}(a \pm ib)$. We write the average position as $\overline{x}(t) = a_+\exp(2\pi i f t) + a_-\exp(-2\pi i f t)$ with $a_\pm = \frac{1}{2}[\overline{x}(0) \mp i\overline{p}(0)/(2\pi m f)]$. The new energy contribution is then $\overline{E}_{\rm int} = q\overline{x}(t)e(t) = A_-a_- + A_+a_+ + A_+a_-\exp(-4\pi i f t) + A_-a_+\exp(4\pi i f t)$. The last two terms vanish on average over time, while the first two give constant contributions depending on the amplitude of light and the motion of the electron. (We can see the importance of resonance: Had we not used the same frequency for the light wave and the oscillator, all terms in the product of $\overline{x}(t)$ and $e(t)$ would be oscillating and average out over time.)

For long time intervals, there are two important terms in the combined average energy of light and a harmonic oscillator of the same frequency: $\overline{E}_{\rm int} = A_-a_- + A_+a_+$. In energy expressions, products of quantities corresponding to two different physical objects always describe interactions. For instance, the product of two masses in the gravitational potential indicates that they together determine the gravitational force between the objects. If we increase the energy by working against a force, for instance raising a mass up a certain amount, we have to add the product of the force and the distance to the energy expression. The force and the distance do not strictly interact, but together they determine the work. In our energy contribution from light, we have two products $A_- a_-$ and $A_+ a_+$, with light amplitudes $A_\pm$ and oscillator initial values depending on $a_\pm$. Light and the oscillator interact: if the amplitude of light changes, the oscillator must change its motion, in such a way that the total energy, by the new interaction terms, changes according to the work done.

Discrete energy Nothing is for free in physics. If we want to change some state, we must supply energy from another source so that the total value of energy remains unchanged. To excite a harmonic oscillator to higher energies above its ground state, energy from the light wave it is coupled to can be used.

At the resonance frequency of the oscillator, our light wave has two components, one with amplitude A_+ and one with amplitude A_-. We compute the energy of each contribution by considering the time dependence, or acting with the operator $\hat{E} = i\hbar\Delta/\Delta t$. Using $i^2 = -1$, the contribution with amplitude A_+ then has energy $E_+ = hf$, and the one with amplitude A_- has energy $E_- = -hf$.

It is not possible for a photon to have negative energy in an absolute sense; here, we are not describing the whole wave and its energy but rather the changes it suffers from interacting with the oscillator. Negative energy means work done by the oscillator, whose energy decreases by emitting a photon. An amount hf is subtracted from the oscillator energy, calming it down as measured by a_-. If light does positive work, as indicated by the A_+-term, the oscillator is excited to a more energetic state by a_+.

The complex numbers $a_\pm$ cannot directly correspond to the amplitude of harmonic oscillations. We rather have to look for their real and imaginary parts in order to find initial values of $\overline{x}(0)$ and $\overline{p}(0)$ according to the definition $a_\pm = \frac{1}{2}[\overline{x}(0) \mp i\overline{p}(0)/(2\pi m f)]$. After an interaction with light, a_+ and a_- change; if we reset the clock, new oscillations start with initial values $\overline{x}(0) = a_+ + a_-$ and $\overline{p}(0) = 2\pi i m f(a_+ - a_-)$.

In quantum mechanics, we need not only initial values for averages but also values of all the moments in order to see how the wave function behaves. In particular, the variances appear in the average energy; they increase when the energy grows. We have already computed the energy $E_0 = \frac{1}{2}hf$ of the ground state, using the minimal values of variances allowed by the uncertainty relation. Every absorbed photon increases the energy by an amount of hf, and we obtain a whole range of discrete energies $E_n = (n + \frac{1}{2})hf$ with an integer $n \geq 0$. No other energy can be realized because absorption processes would average out to zero over time. Only for brief time intervals is it possible to deviate from the discrete energies, another example of an uncertainty relation: If the time uncertainty is small, energy uncertainties $\mathbf{Var}_E$ can be large. However, high levels of energy cannot be maintained for long times.

The ground-state energy was obtained with a minimal position variance $\mathbf{Var}^2_{x,\min} = \frac{1}{2}\hbar/\sqrt{mk} = \frac{1}{2}\hbar/(2\pi m f)$, with energy $E_0 = k\mathbf{Var}^2_{x,\min}$. In general, the energy also depends on the momentum variance, but for stationary states the two variances are related to each other by $\mathbf{Var}_p^2/m = k\mathbf{Var}_x^2$. When the energy is raised to $E_n > E_0$, the position variance must increase to $\mathbf{Var}_{x,n}^2 = (n+\frac{1}{2})\hbar/(2\pi m f)$, and the momentum variance to $\mathbf{Var}_{p,n}^2 = 2\pi(n+\frac{1}{2})m\hbar f$. The higher the energy in an excited state, the more the wave function spreads out.

Black-body radiation Darkness can only be as dark as heat radiation allows. If we poke a whole through the wall of an otherwise enclosed cave, we should not expect to see much light, except when the whole cave is so hot that it radiates like a lamp. Only the temperature should influence the form of radiation, providing a controlled setting useful for analyzing and standardizing types of waves.

The cave walls contain atoms which oscillate when hot. If they are harmonic, with a characteristic frequency f, they occupy energies $E_n = (n+\frac{1}{2})hf$, with a selection of n depending on how they are all excited, in turn depending on the total energy and temperature. The larger n, the more difficult it is to excite an oscillator to energy E_n, but high temperature makes it easier. The likelihood of realizing energy E_n is a function that decreases with larger n and increases with the temperature T. In thermodynamics, one assumes that the likelihood function is exponential, $p_T(n) = C\exp[-E_n/(k_B T)]$ with a constant k_B named after Ludwig Boltzmann, and another constant C to normalize the probabilities. The resulting radiation spectrum is shown in Figure 4.3.

Planck's formula The probability of finding some energy is one, $\mathbf{Sum}_n p_T(n) = 1$. The sum $\mathbf{Sum}_n \exp[-nhf/(k_B T)]$ evaluates to $1/\{1-\exp[-hf/(k_B T)]\}$, and therefore $C = \exp[hf/(2k_B T)] - \exp[-hf/(2k_B T)] = 2\sinh[hf/(2k_B T)]$. We can compute the average energy $\overline{E}_T = -\frac{1}{2}hf\mathbf{Sum}_n E_n p_T(n)$ from the normalization sum: if $s(q) = \mathbf{Sum}_n q^n$, $q\Delta s/\Delta q = \mathbf{Sum}_n nq^n$. Using $q = \exp[-hf/(k_B T)]$, we find $\overline{E}_T = hf/\{\exp[hf/(k_B T)]-1\}$. In the real world, oscillators can swing in three directions: we add up three harmonic energy contributions, $E_{n_1,n_2,n_3} = (n_1+n_2+n_3+\frac{3}{2})hf$. The larger $n = n_1+n_2+n_3$, the more options there are to realize an energy value E_n. For the average energy, this amounts to an extra factor of f^2: an energy value lies on a sphere of radius nf in the space of all E_{n_1,n_2,n_3}; if we account for the growing area by the f^2-factor, we can sum over all integer n as before. We obtain Planck's formula $I_T(f) = hf^3/\{\exp[hf/(k_B T)]-1\}$ for the intensity of waves of frequency f, emitted in an enclosed cave at temperature T.

Ladder operators If a wave changes, a mathematical operation cannot be far. We have introduced operators to compute the energy and momentum of a wave, and changes of waves from absorption or emission of energy have their own mathematical operators.

The amplitudes $a_\pm$ tell us how the initial values of oscillations must change if we reset the clock after an exchange of energy. In quantum mechanics, the oscillator is described by a wave function, which requires many more initial values than an oscillating particle. For a complete description, we need an infinite number of

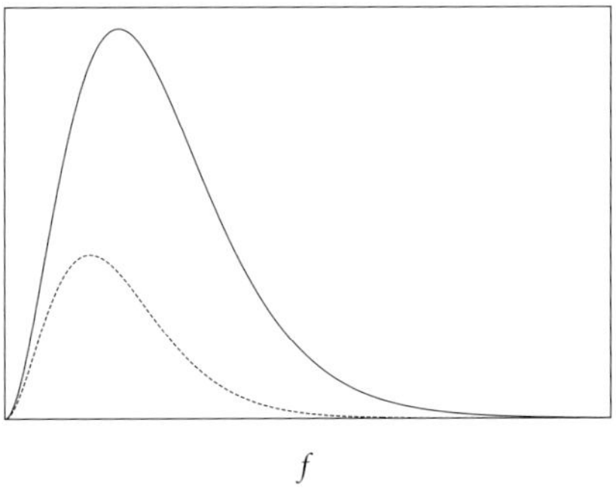

Figure 4.3 Planck's formula for the frequency-dependent intensity of radiation emitted by matter at temperature T (solid) and $\frac{3}{4}T$ (dashed). The total intensity, given by the area under the curve, and the frequency at which most energy is emitted decrease for smaller temperature.

initial values, described either by the values the wave function takes at all points in space or by the set of all moments.

We have already seen some changes in the moments, in particular, in the variances of excited states. For the calculation, we used the fact that the average energy depends on the variances, combined with the known discrete values of the oscillator energy. Other moments cannot be computed in the same way because they just don't affect the energy of a harmonic oscillator. Instead, we can use the amplitudes $a_\pm$ which do change after each exchange of energy. If we turn the complex numbers $a_\pm$ of classical oscillations or of the average oscillations of a wave into operators, we can derive all the wave properties and their changes when the energy is increased or decreased.

In terms of average position and momentum, the interaction amplitudes are $a_\pm = \frac{1}{2}[\overline{x} \mp i\overline{p}/(2\pi m f)]$. We already know that the average momentum is replaced by an operator $\hat{p} = -i\hbar\Delta/\Delta x$ to compute it for a wave function $\Psi(x,t)$. Instead of using the average position $\overline{x}$, we can, when dealing with a whole wave, use the position x at which we evaluate the wave function. If energy is exchanged, the wave function changes from $\Psi(x,t)$ to $\hat{a}_\pm\Psi(x,t)$ with the ladder operators $\hat{a}_\pm = \frac{1}{2}\{x \mp [\hbar/(2\pi m f)]\Delta/\Delta x\}$.

To increase the energy from the ground state to an excited state of level n, the oscillator must absorb n photons of energy hf. We act n times with the operator $\hat{a}_+$ starting with the ground state Ψ_0: $\Psi_n(x,t) = (\hat{a}_+^n/N_n)\Psi_0(x,t)$. The numbers N_n must be chosen so that $\Psi_n(x,t)$, like any admissible wave function, gives total probability one for finding the particle just anywhere, at all times: $\mathbf{Sum}_x|\Psi_n(x,t)|^2\Delta L(x) = 1$ (summoning over small x-intervals of sizes $\Delta L(x)$). We will compute these numbers, called normalization factors, later. To lower the energy, we act with the operator a_-. If we view the discrete energy values and the corresponding states and wave functions as the rungs of a ladder, we climb up with $\hat{a}_+$ and down with $\hat{a}_-$. These ladder operators tell us everything about the states.

Moments Averages on the ladder provide us with moments of states. We use the complex wave function $\Psi(x,t)$ to compute probabilities $d(x,t) = |\Psi(x,t)|^2$, which in turn are used for the averages in the definition of moments.

The average position in an excited state $\Psi_n(x,t)$ is $\overline{x}_n = \mathbf{Sum}_x \Psi_n(x,t)^* x \Psi_n(x,t)\Delta L(x)$. We bracket the position (operator) x between the two factors of the wave function because we know how to act with an operator on a single wave function. Similarly, we write the momentum average as $\overline{p}_n = -i\hbar \mathbf{Sum}_x \Psi_n(x,t)^* [\Delta\Psi_n(x,t)/\Delta x]\Delta L(x)$. As a shorthand, we often use the notation $\overline{x}_n = \langle\Psi_n|\hat{x}|\Psi_n\rangle$ and $\overline{p}_n = \langle\Psi_n|\hat{p}|\Psi_n\rangle$, bracketing each state and operator factor. Writing Ψ_n twice may seem redundant, but we sometimes need to compute such expressions with different states inserted, for which we can then use the same notation. (For instance, if $\Psi = (\Psi_1 + \Psi_2)/\sqrt{2}$, $\langle\Psi|\hat{x}|\Psi\rangle = \frac{1}{2}(\langle\Psi_1|\hat{x}|\Psi_1\rangle + \langle\Psi_2|\hat{x}|\Psi_2\rangle + \langle\Psi_1|\hat{x}|\Psi_2\rangle + \langle\Psi_2|\hat{x}|\Psi_1\rangle)$.)

Expressed in terms of the ground state, we have $|\Psi_n\rangle = (\hat{a}_+^n/N_n)|\Psi_0\rangle$. The mirror image $\langle\Psi_n|$ in the average formulas comes from the complex conjugate in the probabilities, and we must take the conjugate of $\hat{a}_+$,which has an i-factor in its momentum term, if we express $\langle\Psi_n|$ in terms of $\langle\Psi_0|$: $\langle\Psi_n| = \langle\Psi_0|\hat{a}_-^n/N_n$. Also the operators $\hat{x}$ and $\hat{p}$ in the averages can be expressed by the ladder operators, via linear combinations. Written for the averages of $\hat{a}_\pm$, we have $\overline{a}_{\pm,n} = \langle\Psi_0|\hat{a}_-^n\hat{a}_\pm\hat{a}_+^n|\Psi_0\rangle/N_n^2$. If we know the moments of the ground state, we can compute moments in excited states by iteration using the ladder operators.

Ground state The ground state has the lowest energy; it is the lowest rung of the ladder. If we act with $\hat{a}_-$ on Ψ_0, we cannot go to any other state further down the ladder. It is impossible to lower the energy below the ground-state level. We can avoid reaching a lower-energy state only by requiring that $\hat{a}_-\Psi_0$ be no state at all, or that it does not exist and just vanishes as a function; $\hat{a}_-$ annihilates Ψ_0, sending it to nonexistence.

If $\hat{a}_-$ annihilates the ground state, we know that the average and all moments of $\hat{a}_-$ vanish in the ground state: $\langle\Psi_0|\hat{a}_-^n|\Psi_0\rangle = 0$ for $n > 0$. In position and momentum averages of excited states, we have operators $\hat{a}_-^n\hat{a}_\pm\hat{a}_+^n$ where we act with $\hat{a}_+$ on $|\Psi_0\rangle$. These averages are nonzero, despite some factors of $\hat{a}_-$: Operators, unlike numbers, do not always commute. However, there are relationships that tell us what to do when we want to change their order, allowing us to compute moments in terms of those in the ground state.

Commutator For the ladder operators acting on some state Ψ, we have

$$\hat{a}_-\hat{a}_+\Psi = \frac{1}{4}\left(x + \frac{\hbar}{2\pi m f}\frac{\Delta}{\Delta x}\right)\left(x - \frac{\hbar}{2\pi m f}\frac{\Delta}{\Delta x}\right)\Psi$$

$$= \frac{1}{4}\left(x^2\Psi - \frac{\hbar}{2\pi m f}x\frac{\Delta\Psi}{\Delta x} + \frac{\hbar}{2\pi m f}\frac{\Delta(x\Psi)}{\Delta x} - \left(\frac{\hbar}{2\pi m f}\right)^2\frac{\Delta^2\Psi}{(\Delta x)^2}\right)$$

$$\hat{a}_+\hat{a}_-\Psi = \frac{1}{4}\left(x^2\Psi + \frac{\hbar}{2\pi m f}x\frac{\Delta\Psi}{\Delta x} - \frac{\hbar}{2\pi m f}\frac{\Delta(x\Psi)}{\Delta x} - \left(\frac{\hbar}{2\pi m f}\right)^2\frac{\Delta^2\Psi}{(\Delta x)^2}\right).$$

The difference of the two orderings is $(\hat{a}_-\hat{a}_+ - \hat{a}_+\hat{a}_-)\Psi = (\hbar/4\pi m f)[\Delta(x\Psi)/\Delta x - x\Delta\Psi/\Delta x] = (\hbar/4\pi m f)\Psi$. We can commute the two operators provided that we add a constant $\hbar/(4\pi m f)$, called the commutator $[\hat{a}_-, \hat{a}_+]$.

For averages, we can bring all operators $\hat{a}_-$ to the right until they act on Ψ_0, giving zero. The constant contributions we must add, however, stay nonzero and tell us what values we have in excited states. Any such term, for the averages $\overline{x}$ and $\overline{p}$, has at least one ladder operator left because in averages there is an odd number (one) of these operators to begin with, and each commutation eliminates two ladder operators. In each term, we will be computing averages with two different states, the ground state and an excited state an odd number of rungs higher. They average each other out, and the result is zero: $\langle\Psi_0|\hat{a}_-^n|\Psi_0\rangle = 0$ and $\langle\Psi_0|\hat{a}_+^n|\Psi_0\rangle = 0$ for $n > 0$ because $\hat{a}_-|\Psi_0\rangle = 0 = \langle\Psi_0|\hat{a}_+$. Averages of the ladder operators, or of position and momentum, vanish in all excited states. Only moments, beginning with the variances, differ.

Norm We compute variances as averages of $\hat{x}^2 = (\hat{a}_+ + \hat{a}_-)^2 = \hat{a}_+^2 + \hat{a}_+\hat{a}_- + \hat{a}_-\hat{a}_+ + \hat{a}_-^2$. The first and last contribution bring us to two different states, again averaging each other out. The two middle terms, however, do not change the overall level of the ladder, and provide nonzero results. If we count how many times we must commute the ladder operators in $\langle\Psi_0|\hat{a}_-^n(\hat{a}_+\hat{a}_-)\hat{a}_+^n|\Psi_0\rangle$ in order to have at least one $\hat{a}_-$ hit the ground state, we know what values the variances have. It is more convenient to proceed by iteration and relate the moments in state Ψ_n to moments in state Ψ_{n-1}. We just switch the ladder operators in the middle to obtain $\langle\Psi_n|a_+a_-|\Psi_n\rangle = N_{n+1}^2/N_n^2 - \hbar/(4\pi m f)$ with $N_n^2 = \langle\Psi_0|\hat{a}_-^n\hat{a}_+^n|\Psi_0\rangle$.

Variance We still need to compute the ratio of normalization factors. Again, we proceed by iteration and move one of the leftmost $\hat{a}_-$ all the way through to the right, commuting it with $n+1$ operators $\hat{a}_+$ for N_{n+1}^2. The first commutator gives

$$N_{n+1}^2 = \langle\Psi_0|\hat{a}_-^{n+1}\hat{a}_+^{n+1}|\Psi_0\rangle$$
$$= \frac{\hbar}{4\pi m f}\langle\Psi_0|\hat{a}_-^n\hat{a}_+^n|\Psi_0\rangle + \langle\Psi_0|\hat{a}_-^n\hat{a}_+\hat{a}_-\hat{a}_+^n|\Psi_0\rangle\,.$$

Commuting the $\hat{a}_-$ to the right $n+1$ times, we obtain $n+1$ commutator constants $\hbar/(4\pi m f)$, and we are left with $N_{n+1}^2 = (n+1)N_n^2\hbar/(4\pi m f)$. Then, $\langle\Psi_n|a_+a_-|\Psi_n\rangle = n\hbar/(4\pi m f)$. With this, $\mathbf{Var}_{x,n}^2 = \langle\Psi_n|a_+a_- + a_-a_+|\Psi_n\rangle = 2\langle\Psi_n|a_+a_-|\Psi_n\rangle + \hbar/(4\pi m f) = (n+\frac{1}{2})\hbar/(2\pi m f)$.

Wave function A wave can be described by its width and shape, or by the profile it takes over space. Width and shape are quantified by moments; the profile is the wave function $\Psi(x,t)$, or its height at all positions and times. We should not think of $\Psi(x,t)$ itself as the height of the wave because it may not be a real number. From the wave function, we obtain the norm squared $|\Psi(x,t)|^2 = \Psi(x,t)^*\Psi(x,t)$, guaranteed to be real. However, if two waves move into each other and overlap, we add the wave functions $\Psi(x,t)$, not the squared norms. From the perspective of superposition, it is therefore more natural to consider values of $\Psi(x,t)$ as the analog of the height of a wave.

We have computed moments of a state in several examples. In principle, it is possible to construct the wave function that gives rise to those values, providing not only an additional visualization of the state but also alternative computation tools. The wave function, unlike moments and averages, is not directly accessible by physical experiments. For one, it takes complex values which are not all real, in both a mathematical and physical sense. Values of the wave function can only be found in indirect ways, for instance, from the moments or from other experiments such as interference that test how different wave functions overlap. However, such reconstruction procedures are complicated, often allowing not more than trial and error.

Computational methods often go the opposite way. Mathematical tools to solve for the evolution of a single wave function $\Psi(x,t)$ are more standard than those for an infinite set of moments. Once the wave function is known, it is easier (although still quite involved) to compute its moments, rather than going the other way around. Even though values of the wave function cannot be measured directly, they do provide valuable information.

Ground state To follow this procedure, we must find equations that the wave function satisfies. Some states obey rather simple conditions, for instance, the ground state of the harmonic oscillator. We have already used the fact that it is annihilated by the lowering operator $\hat{a}_- = \frac{1}{2}\{x + [\hbar/(2\pi m f)]\Delta/\Delta x\}$, representing the complex combination $a_- = \frac{1}{2}[x + ip/(2\pi m f)]$. The ground state therefore has a wave function $\Psi_0(x,t)$ with $\Delta\Psi_0(x,t)/\Delta x = -(2\pi m f/\hbar)x\Psi_0(x,t)$ at all times. We solve this differential equation by rewriting it as $\Psi_0(x,t)^{-1}\Delta\Psi_0(x,t) = -(2\pi m f/\hbar)x\Delta x$ and noting that $\Psi_0^{-1}\Delta\Psi_0 = \Delta\log\Psi_0$ and $x\Delta x = \frac{1}{2}\Delta x^2$ by reverse differentiation. With an arbitrary function $f(t)$ that only depends on time, the ground state satisfies $\Psi_0(x,t) = \exp(-\pi m f x^2/\hbar)f(t)$.

We determined the position dependence of the ground state by using the lowering operator together with the relation of momentum p to the oscillation rate in space. We now find the time dependence by using the relation of energy E to the oscillation rate in time, $\hat{E} = i\hbar\Delta/\Delta t$. We already know the energy of the ground state, a constant $E_0 = \frac{1}{2}hf$. Therefore, $\hat{E}\Psi_0(x,t) = i\hbar\Delta\Psi_0(x,t)/\Delta t = \frac{1}{2}hf\Psi_0(x,t)$, or $\Psi_0(x,t) = g(x)\exp(-i\pi f t)$ with some function $g(x)$ that depends only on x.

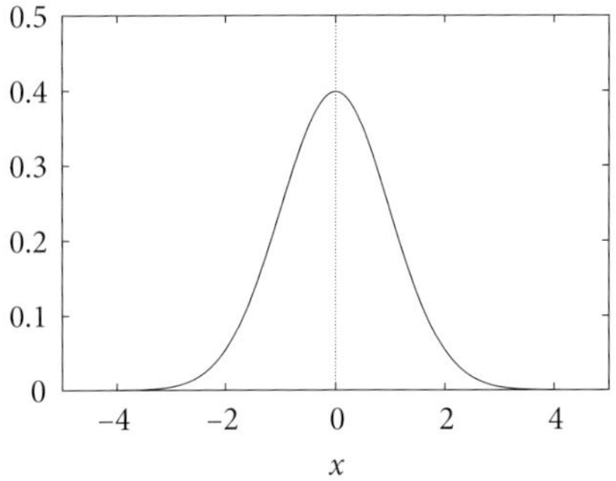

Figure 4.4 The probabilities $|\Psi_0(x,t)|^2 = \exp(-\frac{1}{2}x^2)/\sqrt{2\pi}$ showing the shape of the harmonic-oscillator ground state.

Combining the two solutions we found for the position and time dependence, we have $\Psi_0(x,t) = C\exp[-\pi f(mx^2/\hbar + it)]$ with a constant C that depends neither on x nor on t ($C = \sqrt[4]{mf/\hbar}$ for Ψ_0 normalized). The shape of the wave function, shown in Figure 4.4, and its oscillations have been determined.

Schrödinger equation Energy determines motion, classical and quantum. In classical physics, the dependence of energy on position and momentum implies all (Hamilton's) equations of motion. In quantum mechanics, the relationship between energy and momentum tells us how oscillations in space and time are coordinated.

Energy and momentum of a wave function are computed by operators $\hat{E} = i\hbar\Delta/\Delta t$ and $\hat{p} = -i\hbar\Delta/\Delta x$. If the classical energy for some potential $V(x)$ is $E = p^2/2m + V(x)$, the oscillation rates of a wave function should obey

$$i\hbar\frac{\Delta_t\Psi}{\Delta t} = -\frac{\hbar^2}{2m}\frac{\Delta_x^2\Psi}{\Delta x^2} + V(x)\Psi\,, \tag{4.14}$$

a relationship called the Schrödinger equation. The ground state $\Psi_0(x,t)$ of the harmonic oscillator satisfies this equation with $V(x) = \frac{1}{2}m(2\pi f)^2x^2$. In general, if we know what the wave function $\Psi(x,t_0)$ is at some initial time, such as $t_0 = 0$, the Schrödinger equation tells us how it will change as time goes on. Adding up the changes, that is, solving the differential equation, we find the wave function at all times.

It is often useful to assume a fixed energy E while relating momentum to the rate of change in space. We write

$$E\Psi = -\frac{\hbar^2}{2m}\frac{\Delta^2\Psi}{\Delta x^2} + V(x)\Psi\,, \tag{4.15}$$

the time-independent Schrödinger equation. It is easier to solve than the original equation because we need not take into account time dependence for the class of states considered. For fixed E, the time dependence can only be of the form

$\exp(-iEt/\hbar)$, for the rate of change is $\hat{E} = i\hbar\Delta/\Delta t$. Although the wave function does depend on time (unless $E = 0$), its norm squared is time-independent: $|\exp(-iEt/\hbar)|^2 = \exp(iEt/\hbar)\exp(-iEt/\hbar) = 1$. These states are called stationary.

Time dependence is an important aspect in physics, showing motion and change. One may think that stationary states are too special to be of much interest. However, it turns out that knowing all stationary states, for all possible values of E, gives us complete access to all solutions. If the set of stationary states is $\Psi_n(x,t)$, labeled by some number n, any solution to the Schrödinger equation can be written as $\Psi(x,t) = \mathbf{Sum}_n C_n \Psi_n(x,t)$ with some complex coefficients C_n. The sum is a solution to the Schrödinger equation because the equation is linear: If we know two solutions, their sum is another solution, and so are the solutions multiplied with constants. In order to compute the coefficients C_n for a solution of interest, we have to find how the initial wave function $\Psi(x,t_0)$ can be written as a sum of $C_n\Psi_n(x,t)$. Since the coefficients C_n are time-independent, they will remain the same at later times.

Bound states Although much of physics is about motion and change, sometimes one has to rest. Stationary states describe systems in equilibrium, in which, for some while at least, nothing happens. Learn how a system rests, and you can find out how it moves when more active. If interactions happen as quick bursts, we may approximate the changing system by a succession of equilibrium states, interrupted by brief and rapid bursts of evolution.

The possible energy values of stationary states tell us what energies the system can sustain for extended periods of time. If the state changes, the energy difference must be supplied by or returned to an external source. In a single dose, the energy carrier could be a photon emitted or absorbed. Differences of the stationary energies then tell us what energies the photons can have, or what their frequencies $f = E/h$ are. For the harmonic oscillator, we saw that the energy in a state $\Psi_n = (\hat{a}_+^n/N_n)\Psi_0$ is $E_n = (n+\frac{1}{2})hf$, with integer $n = 0, 1, \ldots$ All energy differences are $E_{n+1} - E_n = hf$; all emitted or absorbed photons have the same frequency as the oscillator, being in perfect resonance.

For other potentials, when energies are not equally spaced, photons of different frequencies can be emitted or absorbed. For instance, the infinite square well is an idealization of a particle trap with forces so large that no particle can escape from a bounded region of size a, no matter what energy it has. If the particle cannot escape, it will never be found outside of the trap. The probability to measure the particle position outside of the trap is zero, and the position average is always in the well, $0 < \overline{x} < a$. The wave function $\Psi(x,t)$, determining probabilities, is zero for $x \leq 0$ and $x \geq a$ at all times. In the well, the wave function depends on the forces acting there. For simplicity, we may assume that there is no force in the well, just the sharp walls.

This idealization of a realistic trap allows us to solve the Schrödinger equation: In the well, the time-independent equation reads $\Delta^2\Psi/\Delta x^2 = -2mE\Psi/\hbar^2$. With the second derivative proportional to the wave function, including a negative sign, the solution is of sine or cosine form. The sine is zero at $x = 0$, the cosine is not.

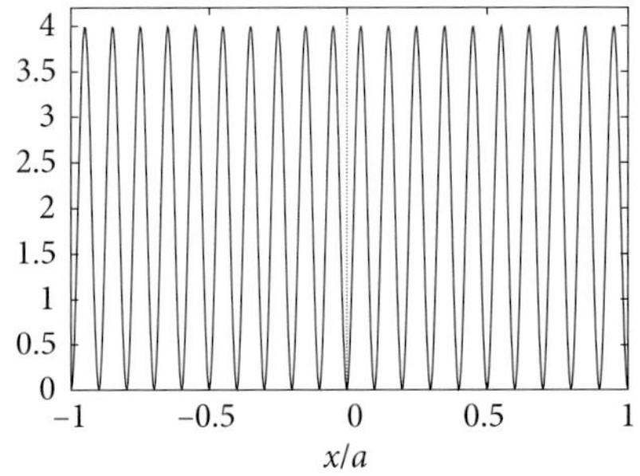

Figure 4.5 The probabilities $|\Psi_{10}(x,0)|^2$ in a state of sharp energy in an infinite square well, with excitation level $n = 10$. The wave function varies much within the well, but provides an almost constant probability for position measurements with low position resolution. For such measurements, the likelihood of finding the particle is uniform within the well.

For wave functions to vanish at one side of the well, we can only allow solutions $\Psi(x,t) = A\sin(\bigcirc\sqrt{2mE}x/h)\exp(-iEt/\hbar)$. They must also vanish at the other side of the well: $\sin(\bigcirc\sqrt{2mE}a/h) = 0$. With the parameters m, a and $\hbar$ fixed by particle properties, the setup of the trap, or as a constant of nature, the allowed energies E obey $\sqrt{2mE}a/h = \frac{1}{2}n$ with nonzero integer n. An example of a wave function with $n = 10$ is shown in Figure 4.5. Negative n do not give us wave functions independent of those for positive n, and therefore we obtain a set of energies

$$E_n = \frac{n^2h^2}{8ma^2} \quad \text{for} \quad n = 1, 2, \ldots \tag{4.16}$$

The set of energies is discrete, but not equally spaced. Differences of neighboring energy levels increase as $(n+1)^2 - n^2 = 2n+1$ as n grows. Photons with frequencies $f = (E_{n+1} - E_n)/h = h(2n+1)/(8ma^2)$, proportional to an odd integer, can be emitted or absorbed.

The frequencies, not just the energies of photons, are proportional to h. If h is much smaller than other quantities, as in many classical systems, photons of almost any frequency can be emitted or absorbed. There is no resonance phenomenon as with the harmonic oscillator. A classical particle in the trap would move freely, undisturbed by any force, until it reaches the walls. Instantly reflected back, it then continues its motion with the same velocity, just in the opposite direction. It bounces back and forth, retaining the value of its kinetic energy. If it is hit by a photon, it can absorb all the energy and move faster. There is no influence by a force that would make it react selectively to varying photon frequencies.

Free particle Nothing in physics is bound forever. We would need infinite forces or infinite jumps in potentials to keep an object confined to a bounded region. Such strong forces, like anything infinite, cannot exist in nature, and for a given setup of a trap it is always possible to escape if only the energy is large enough.

If we tear down the walls of the infinite square well, the particle is set free. We have already solved for some of the moments of its wave function: the average follows the classical trajectory while the position variance keeps increasing. We have the same wave function $\sin(\bigcirc kx/2\pi)$ and $\cos(\bigcirc kx/2\pi)$ with $k = \sqrt{2mE}/\hbar$, or wavelength $l = 2\pi/k = h/\sqrt{2mE}$, as inside the well, but they are now free of boundary conditions at the walls. No restriction for k or the energy E follows. Particles not bound to a finite region can take energy values in a continuous range; the energy spectrum is not discrete.

If we want to specify the momentum of the particle, not just its energy, we should switch from $\sin(\bigcirc kx/2\pi)$ and $\cos(\bigcirc kx/2\pi)$ to $\exp(ikx)$ and $\exp(-ikx)$. The former functions can always be expressed as combinations of the latter two, and so we do not lose (or gain) any information. So far, we have defined $k = \sqrt{2mE}/\hbar$ as a positive number. If we allow negative values as well, we need just one function $\exp(ikx)$ to cover both cases, a function that corresponds to momentum $p = \hbar k$ via $\hat{p}\exp(ikx) = -i\hbar\Delta\exp(ikx)/\Delta x = \hbar k\exp(ikx)$. The sign of k determines whether the particle moves to the right or the left, in both cases with energy $E = \hbar^2 k^2/2m = h^2/(2ml^2)$.

Combined with time dependence according to E, the wave function is $\Psi_l(x,t) = C\exp[2\pi i(x/l - ft)]$ with $f = E/h = h/(2ml^2)$. Again wave functions with positive l (or $k = 2\pi/l$) move to the right: a landmark on the wave, a wavemark like a hill or a trough, has a constant value of $x/l - ft = D$. Motion by $x = flt + Dl$ has positive velocity fl for positive l and negative velocity for negative l.

Group velocity A group does not move like an individual. Evaluating the wave velocity derived for a single wavemark, we obtain $fl = h/(2ml)$. Identifying $h/l = \hbar k$ with the momentum, we have velocity $p/2m$, unlike the classical relation $v = p/m$ between velocity and momentum. The violation of the velocity-momentum relation is independent of h; it occurs even under circumstances in which we expect classical physics to be valid. We cannot blame the mismatch on the smallness of h compared to other parameters involved. We must find an alternative explanation.

Wave functions $\Psi_l(x,t)$ show the wavelike nature by their oscillations. However, the probabilities $d(x,t) = |\Psi_l(x,t)|^2$ do not: $|\Psi_l(x,t)|^2 = 1$ is constant and does not oscillate. This observation is not altogether surprising. We obtained these wave functions as solutions of the time-independent Schrödinger equation, as stationary states. For a wave whose probabilities vary in space and time, corresponding to visible motion, we must superimpose several components with different values of l or E, writing $\Psi(x,t) = \mathbf{Sum}_l C(l)\Psi_l(x,t)$. If the function $C(l)$ is nonzero for several l, the time dependence does not disappear if we compute the norm squared.

If several l are involved, the momentum is no longer sharp. We obtain a certain result for the average, $\overline{p} = \mathbf{Sum}_l (h/l)|C(l)|^2$, given that each component with amplitude $C(l)$ and probability $|C(l)|^2$ contributes momentum $\hbar k = h/l$. Different velocities are realized for the contributions to the wave function. If $C(l)$ is large only for a small range of l-values around an average $1/l_0 = \overline{p}/h$, we can better compare the velocities involved. The function $f(l) = E(l)/h = h/(2ml^2)$ in each relevant stationary contribution $\exp[2\pi i(x/l - ft)]$ with large coefficient $C(l)$ then does not depend much on l.

We write $f(l) = f(l_0) + \Delta(1/l) \cdot \Delta f/\Delta(1/l)$ with $\Delta(1/l) = 1/l - 1/l_0$ and

$$\frac{\Delta f}{\Delta(1/l)} = \frac{h}{2m}\frac{(1/l_0 + \Delta(1/l)^2)^2 - 1/l_0^2}{\Delta(1/l)} \approx \frac{h}{m l_0}\,.$$

The stationary contributions split into a factor $\exp\{2\pi i[x/l_0 - f(l_0)t]\}$ independent of l, which can be factored out of the sum in $\Psi(x,t)$ and does not contribute to the norm squared, and $\exp\{2\pi i[x - th/(ml_0)]\Delta(1/l)\}$.

Comparing space and time variations, the correct velocity-momentum relationship $v = \hbar k/m = h/(ml) = \Delta f/\Delta(1/l)$ follows.

Dispersion relation Using $\Delta f/\Delta(1/l)$, we have corrected a factor of 1/2. It is more impressive to look at the relationships of special relativity. We have $E(p) = \sqrt{m^2c^4 + p^2c^2}$, and so $f(l) = \sqrt{m^2c^4/h^2 + c^2/l^2}$. The product $f(l)l = \sqrt{c^2 + l^2m^2c^4/h^2}$ is never smaller than c, and can take any larger value. The derivative $\Delta f/\Delta(1/l) = c/\sqrt{1 + l^2m^2c^2/h^2} = c/\sqrt{1 + m^2c^2/p^2}$ amounts to the correct velocity in $p = mv/\sqrt{1 - v^2/c^2}$.

The group velocity $v_{\text{group}} = \Delta f/\Delta(1/l)$, corresponding to the velocity of a small group of waves, gives the correct relationship with classical physics. It differs from the phase velocity $v_{\text{phase}} = fl$ for the velocity of a wavemark, or the phase. The difference of these two kinds of velocities can be seen not only in quantum mechanics, but also in more familiar forms of waves, even in swarms of birds: The whole swarm flies with a (group) velocity different from the speed that can be reached by a single bird, or the phase velocity. Birds often arrange their position in the swarm so that each one spends about the same fair amount of time at the wind-exposed and dangerous fringes. There is motion within the swarm when birds switch their position, and not all their velocity corresponds to the whole velocity of the swarm. When the swarm is attacked by a bird of prey, closer birds flee first, generating a wave within the swarm that moves with a larger (phase) velocity than the whole group. Values of up to 25 m/s have been observed.[2)]

Scattering Particles unbound move and scatter. If there is no force at all, particles are free and move freely, always with the same, conserved momentum. Forces change the momentum and can turn around the velocities of particles they act on.

Changes of velocities depend on the strength of the force, and so physicists often make use of particles shot with some energy or velocity at a region where an unknown force acts. By analyzing the scattered particles after they have encountered

2) A. Procaccini *et al.* (2011) Animal Behaviour, **82**, 759.

the force, one can draw conclusions about the force itself. The structure of atoms, nuclei, and some elementary particles, for instance, the proton, has been uncovered by such scattering experiments with particle accelerators. In the first famous example, Ernest Rutherford aimed an electron gun at a metal foil, or the atoms it contained, and found that electrons are scattered just as one would expect if they encountered the electric Coulomb potential of a single charge, rather than a uniform charge distribution. Rutherford concluded that most of the positive charge repelling electrons is located in a tiny region, the nucleus.

If a particle encounters a region in which the potential changes, implying a force, its wave function varies according to the Schrödinger equation. It is more difficult to solve differential equations with varying functions in it, such as $V(x)$, compared with constant potentials used for the free particle. As seen for the infinite square well, we may model changing potentials by sharp jumps between different constant values of V. We first solve the wave equation in each region where the potential is constant, and then implement suitable matching conditions at the jumps to ensure that we still have one cohesive wave rather than separate pieces in each region of constant potential. The cohesion condition, mathematically, means that the wave function and its derivative (we need it to compute one more derivative for the Schrödinger equation) each approach the same value from both sides of the jump: they are continuous functions.

We apply this method to a finite barrier as a model for a repulsive force, like one encountered in collisions with a hard ball or an impenetrable particle. We choose a potential that is zero outside a region $|x| < a$ and takes a positive constant value $V(x) = V_0 > 0$ within $|x| < a$. Outside, we have our free wave functions. Inside, the form of the solution is similar, except that we have to replace $k = \sqrt{2mE}/\hbar$ with $\ell = \sqrt{2m(E - V_0)}/\hbar$ to account for the extra term of $V(x)$ in the Schrödinger equation. The constant potential implies a threshold: For energies $E > V_0$, ℓ is a positive real number and the solutions $\exp(i\ell x)$ are oscillating. For energies $0 < E < V_0$, however, $\ell = id$ is imaginary, as the square root of a negative number, and solutions $\exp(i\ell x) = \exp(-dx)$ are exponential. They always decrease (or increase for $d < 0$) depending on the decay rate d. The probability of finding the particle keeps shrinking as we move deeper into the barrier. Indeed, the energy range $V_0 > E$ would not allow the particle to penetrate the barrier in the classical case.

We now have solutions $A\exp(ikx) + B\exp(-ikx)$ in the free region to the left of the potential, $C\exp(i\ell x) + D\exp(-i\ell x)$ in the barrier, and $F\exp(ikx) + G\exp(-ikx)$ to the right. With positive k, a contribution of $\exp(ikx)$ implies a particle moving to the right, while $\exp(-ikx)$ implies motion to the left. If we describe a scattering experiment with particles coming in from one side only, we should set either A or G to zero. We choose $G = 0$, so that particles move in only from the left. Contribution from F are particles moving to the right, transmitted through the barrier; contributions from B are particles moving to the left, reflected at the barrier.

If we relate B and F to the amplitude A of incoming particles, we can compute how likely it is for particles to be reflected or transmitted. These probabilities are quantified by the transmission coefficient $T = |F/A|^2$ and the reflection coefficient

$R = |B/A|^2$. We relate B and F to A by implementing our matching conditions to ensure that we have one complete wave function approaching the same values from the right and the left of the jumps in the potential. Since particles must be either reflected or transmitted if there is no friction to slow down and trap particles in the barrier, $R + T = 1$.

Continuity conditions We impose $A\exp(-ika) + B\exp(ika) = C\exp(-i\ell a) + D\exp(i\ell a)$ and $ik[A\exp(-ika) - B\exp(ika)] = i\ell[C\exp(-i\ell a) - D\exp(i\ell a)]$ for continuity of Ψ and $\Delta\Psi/\Delta x$ at $x = -a$, as well as $C\exp(i\ell a) + D\exp(-i\ell a) = F\exp(ika)$ and $i\ell[C\exp(i\ell a) - D\exp(-i\ell a)] = ikF\exp(ika)$ for continuity at $x = a$. Adding and subtracting the last two equations, after dividing the last one by $i\ell$, we find $C = \frac{1}{2}\exp(i(k-\ell)a)(1 + k/\ell)F$ and $D = \frac{1}{2}\exp[i(k+\ell)a](1 - k/\ell)F$. Similarly, we obtain $A = \frac{1}{2}\exp(ika)[(1 + \ell/k)\exp(-i\ell a)C + (1 - \ell/k)\exp(i\ell a)D] = \frac{1}{4}\{(1 + k/\ell)(1 + \ell/k)\exp[2i(k-\ell)a] + (1 - \ell/k)(1 - k/\ell)\exp[2i(k+\ell)a]\}F$ and another (equally long) expression for B. If these relations are solved, we have a complete and continuous wave function with one free parameter A multiplying the whole function. The transmission coefficient is

$$T(E) = |F/A|^2 = \frac{1}{1 + (V_0^2)/[4E(E - V_0)]\sin\left[2 \bigcirc a\sqrt{2m(E - V_0)}/h\right]^2}\,. \tag{4.17}$$

The reflection coefficient follows from a similar calculation, or from $R = 1 - T$.

Tunneling Unreal objects can enter regions that real ones cannot. The complex spectre of a wave function stays firm even at a barrier larger than the particle energy. With an imaginary wave number $\ell = id$, $d = \sqrt{2m(V_0 - E)}/\hbar$, or momentum $p = i\hbar d$, in the barrier, the wave function decreases but does not drop to zero. Although there is no moving particle in a classical sense, not even one at rest, the probability of finding the particle in the barrier does not vanish. At the side of the barrier opposite from where the particle beam impinges, the wave function, with amplitude $F \neq 0$, oscillates again: Figure 4.6. With $T \neq 0$, as shown in Figure 4.7, a particle is transmitted even if its energy is so low that classical laws would imply reflection.

The wave function makes it possible for particles to penetrate and enter regions not allowed classically, giving rise to a whole new tunneling effect. The barrier must be small and narrow for the effect to be noticeable for macroscopic objects, but microscopic ones can count on it on a regular basis. Electric engineers and hardware designers use the effect in important electronic devices. For instance, a special kind of transistors can be built in which a voltage is used to tune the barrier of another electric current, with fine sensitivity to control how much current is let through.

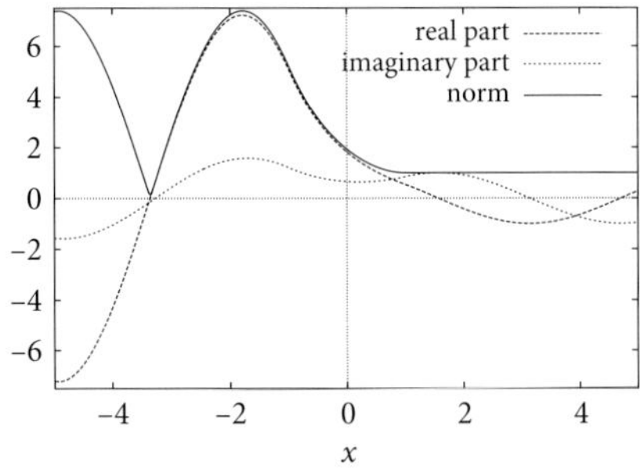

Figure 4.6 The tunneling wave function, oscillating outside of the barrier between $x = -1$ and $x = 1$, decreasing within.

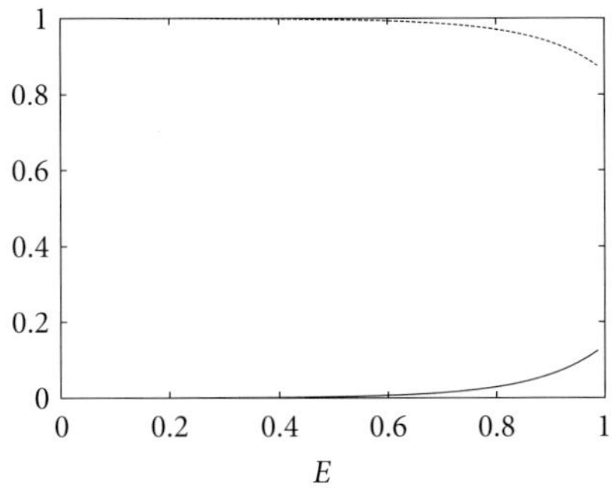

Figure 4.7 The transmission (solid) and reflection coefficients (dashed) for a particle tunneling with energies $E < V_0$ less than the barrier potential V_0. The function $T(E)$ with $E < V_0$ is obtained from (4.17) by writing the sine of an imaginary argument in terms of the exponential function: $\sin(\bigcirc ix/2\pi) = \frac{1}{2}i[\exp(x) - \exp(-x)] = \frac{1}{2}i\sinh(x)$.

Tunneling in electronic devices can be a nuisance, too, for it lets leakage currents flow. There can be no perfect insulator in quantum physics. An insulator is supposed to be a strong barrier for electrons making up the current. A current is operated with a certain maximum amount of energy, and we would think that we just need the insulating barrier to provide a higher potential. However, in quantum physics, tunneling currents would still flow, enlarging leakage.

The eye of the scanning tunneling microscope, as its name suggests, is the tunneling effect. If we have a fine metallic needle almost, but not quite, touching a conducting probe, with some voltage between needle and probe, we can determine how many electrons tunnel through the gap by measuring the current flowing in our circuit. The number of electrons tunneling through depends on the tunneling probability or the transmission coefficient, which in turn depends on the size of the barrier or the distance between needle and probe. Measuring the tunneling current with the needle placed over many points on the surface of the probe, we deduce the surface profile. Fine needles, ideally with a single atom at the tip, resolve the surface with atomic precision.

Bound states To escape bounds, one tunnels. In quantum mechanics, no barrier is too high because the wave function enters regions in which its energy would be less than the threshold required for a classical particle. It is difficult to reach deep into the barrier and the wave function quickly wanes inside. However, it never reaches zero, and what is left of it can enjoy its freedom once the barrier's other end is reached.

We have already seen several results for bound states, such as the discreteness of their energy spectrum. How can bound states exist if wave functions can always enter barriers, as by tunneling? A particle can always escape, no matter how strong the forces, if the barrier is of finite thickness and eventually drops down. In our examples for bound states, the barrier remained, no matter how far away from the central region we went. For the infinite square well, the barrier suddenly becomes infinite and stays so if we move right or left beyond the walls. Here, a particle has no chance of tunneling. The harmonic oscillator's potential, $V(x) = \frac{1}{2}kx^2$, stays finite also when the energy threshold E is exceeded. However, it keeps increasing and never drops back below the energy; the barrier has infinite extension, and the particle will never see freedom.

Both examples are idealized because real forces are finite; there is always some energy by which bounds can be breached. Particles with large energies can escape, but less energetic ones are controlled by the properties of bound states, seen by the examples of the infinite square well and the harmonic oscillator.

The infinite square well, with its sharply delimited walls, provides an intuitive explanation for the discreteness of the energy spectrum and the associated bound states. For a wave function to satisfy the conditions that no particle can be found in regions of infinite potential, secluded by infinite forces, we must be able to fit an integer number of oscillations within the confines of the well. The wave function must vanish at the walls (and beyond) and therefore swing its oscillations back to zero just at the right moment. The oscillation length in space is given by the momentum p, and the momentum of a free particle (between the walls) determines the kinetic energy $E = p^2/2m$. Only some energy values can fit an integer amount of wave lengths between the walls, resulting in the spectrum $E_n = n^2h^2/(8ma^2)$ as in (4.16).

The energies increase as a quadratic function of the integer n, their separation becoming larger and larger. For the harmonic oscillator, the energy spectrum, by comparison, is denser: Here, the energies $E_n = (n + \frac{1}{2})hf$ increase linearly with the integer n, their separation staying constant. The harmonic-oscillator potential never becomes infinite at a finite place, and we do not have sharp conditions for places at which oscillations must return to zero. The oscillation length is not even constant throughout space because the potential keeps changing as we move around. Nevertheless, once the potential at some place is so large that it is above the energy, the wave function turns from oscillations to steep decrease. The decreasing behavior is shared with tunneling, but the particle will never make it out. Matching oscillations in the central region with decrease in the exterior regions again requires the energy to be tuned to some value in a discrete set.

For the harmonic oscillator, we have an additional argument thanks to the resonance phenomenon. Energy is absorbed most likely when the excitation frequency agrees with the intrinsic frequency f of the oscillator. The energy should increase by constant amounts in each absorption process, adding integer multiples of hf. For other potentials, there is no unique frequency, so that the spacing of the energy spectrum may change as we look at larger or smaller energies.

With the requirement of fitting oscillations of an energy-dependent length within the region in which a classical particle could move, we have an intuitive argument to tell us how the energy spectrum in a general potential should look. A classical particle could move anywhere where its energy is larger than the potential, and it would be reflected back when it reached turning points where the energy equals the potential. The points x where $V(x) = E$ delimit the region of motion for a classical particle, and of oscillations of the wave function in quantum mechanics. For given energy E, we must fit an integer number of oscillations into this region. If the potential increases quadratically, the harmonic oscillator tells us that we add a constant amount to the energy in order to reach another value of successful fitting. If the potential increases faster than quadratically, the region of oscillations widens more slowly than in the harmonic case, and we have to add larger and larger amounts of energy to fit the oscillations. The infinite square well is an example for this case, with energies increasing by quadratic dependence on n.

If the potential increases more slowly than a quadratic function, the spacing of energies becomes smaller as we move up on the energy ladder. The potential walls turn away from the central region as the energy gets higher, the turning points of a classical particle move out, and it becomes possible to fit in one more oscillation even if we have not increased the energy as much as in previous steps. The energy spectrum becomes denser. In many cases of realistic examples, the potential flattens out far away from the central region and approaches some finite value, as it is impossible to sustain confining forces for all energies. When the classical turning points move out to infinity within a finite range of energies, we can fit in almost any number of oscillations (Figure 4.8). Discrete energy values pile up; they come to lie denser and denser until they reach the continuum of energies possible for a particle whose energy is above the barrier. Quantum mechanics and discrete ener-

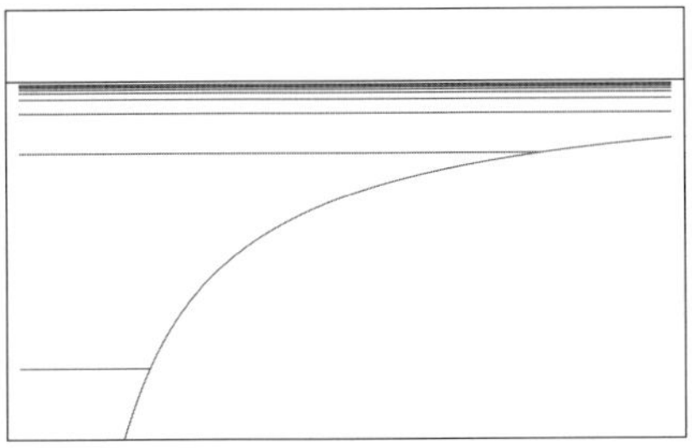

Figure 4.8 The possible discrete energies of bound states in the Coulomb potential for a hydrogen atom. As shown by the horizontal lines, infinitely many values E_n lie in a finite range, crowding at $E = 0$.

gies are most pronounced for small energies near the minimum of the potential, near the ground state where the spectral spacing is large.

Hydrogen Hydrogen is not only the most ubiquitous element in the universe, it is also the simplest example of an atom described by quantum mechanics. A hydrogen atom contains just one electron, bound to a nucleus. The positive charge of the nucleus is provided by a proton, which in most hydrogen atoms is the sole particle in the nucleus. In some cases, the proton can be accompanied by a neutron, forming a deuteron as the nucleus of deuterium, or even by two neutrons in the unstable tritium. The neutrons do not change the electric charge, just the mass of the nucleus.

The two charged particles of tiny extensions are bound to each other by the Coulomb force, inversely proportional to the square of their distance. The nuclei of other atoms are still tiny, even though they contain more neutrons and protons, but the electrons, with their small mass, cannot be confined to a region as small as the nucleus. Electric forces also act between the electrons, making the total description of the force, with several distance parameters, more complicated.

For the Coulomb force $F(r) = -Ce^2/r^2$ with a constant C, the elementary electric charge e, and the distance r between proton and neutron, we have a potential $V(r) = -Ce^2/r$. The light electron around the heavy nucleus can move in all three spatial directions, not just in one as we often assumed so far for simpler models. The most important coordinate is the radial one if we choose our origin at the position of the nucleus, because the force just depends on the distance between the two charged objects. The situation is similar to the classical problem of planetary orbits, where we restricted the main equation to one for the radial distance. However, polar angles may change in a given solution, and the angular motion does affect the radial one; it takes some energy for itself, no longer available for the changing radius. For planetary systems, we were able to take into account the energy heist from angular motion by amending Newton's potential to an effective potential. In quantum mechanics, we should expect a similar role of angular momentum in systems with spherical potentials.

We have already expressed linear momentum in quantum mechanics by the spatial oscillations of a wave, associating it with an operator $\hat{p} = -i\hbar\Delta/\Delta x$. In three-dimensional space, we obtain all three components of the linear momentum $\vec{p}$ by associating them with the three independent spatial variations. Angular momentum in classical mechanics refers to changing angles in orbiting or rotating motion. For instance, the z-component of angular momentum tells us how fast the longitude **Lo** changes.

In classical physics, $L_z = (2\pi/\bigcirc)mr\Delta\mathbf{Lo}/\Delta t$. In quantum mechanics, we should associate L_z with an operator $\hat{L}_z = -i\hbar\Delta/\Delta(2\pi\mathbf{Lo}/\bigcirc)$. Its values can be seen directly in wave functions with a **Lo**-dependence $\exp(2\pi i\ell\mathbf{Lo}/\bigcirc)$, such that L_z has the value $\hbar\ell$. This result looks just like the relationship between linear momentum and the wave number k, but there is an interesting difference: Although the longitude **Lo** can be any real number, the point it implies at fixed r and **La** does not change if we add an integer multiple of the full circular angle $\bigcirc$

to it. The wave function must take the same values for **Lo** and **Lo** $+ n\bigcirc$ with some integer n. The complex exponential function, related to sine and cosine via $\exp(2\pi i\ell\mathbf{Lo}/\bigcirc) = \cos(\ell\mathbf{Lo}) + i\sin(\ell\mathbf{Lo})$, has the required periodicity provided that ℓ is an integer. Angular momentum L_z takes discrete values: integer multiples of $\hbar$.

The other components of angular momentum are more complicated because they do not correspond to just one changing polar angle **Lo** or **La**. We derive their expressions using the classical relations such as $L_x = yp_z - zp_y$ and inserting the operators we know for linear momentum components. With a transformation of coordinates, we write them as derivatives by **Lo** and **La**. They turn out to take only discrete values, too. If we are interested in the precise values, we encounter a typical problem of quantum mechanics: Operators change wave functions they act on, and the order of different operators sometimes matters for the results. If we start with our angular wave function $\exp(2\pi i\ell\mathbf{Lo}/\bigcirc)$ adapted to $\hat{L}_z$, applying $\hat{L}_z$ will just multiply it with a number, interpreted as the value of L_z for such a wave function. Operators $\hat{L}_x$ or $\hat{L}_y$, however, change the wave function and we will no longer be able to tell what the L_z-value is afterwards. In quantum mechanics, we cannot determine sharp values of all angular-momentum components (or just of two of them) in the same state. Just as position and momentum measurements are limited in their precision by the uncertainty relation, there are similar inequalities for the angular-momentum components.

What we can measure sharply together with L_z is the value of the total angular momentum, $L^2 = L_x^2 + L_y^2 + L_z^2$. It has discrete values that must be larger than $\hbar\ell$ (the L_z-value) and turn out to have the form $\hbar^2 j(j+1)$ with an integer j such that $-j \leq \ell \leq j$. For every value of total angular momentum given by j, there are $2j+1$ different values of ℓ, telling us how we can align the L_z-component of angular momentum. Since the other two components are undetermined, we should think of angular momentum in quantum mechanics not as a sharp vector, but as a fluctuating one, with endpoints on a circle obtained by intersecting a sphere of radius $\hbar\sqrt{j(j+1)}$ with the plane at L_z-value $\hbar\ell$.

In all these considerations, we have distinguished the L_z-component over the others because the usual mathematical conventions relate it in a simple way to the longitude **Lo**. By rotating our coordinate system, we could choose a different component as the simple one with sharp values. Which component is sharp depends on the setup of an experiment measuring one of the components.

As with the classical effective potential for a central force, the value j of total angular momentum L^2 shows that we must look for bound states in the effective potential $V(r) + \hbar^2 j(j+1)/2mr^2$. Unlike our previous examples, we do not have a finite minimum of the potential if angular momentum vanishes, but there is an upper bound because it is always negative. Even without a minimum, there is a ground state of lowest, finite energy because the potential becomes narrow at its depth, making it impossible to fit in even one oscillation of the wave function. Above the ground state, the potential flattens out to approach zero at infinity. As described before, the potential, staying finite as we turn away from the central region, makes it possible to fit in more oscillations even if the energy is not increased much. The spacing of the energy spectrum becomes smaller and smaller, and approaches the continuum near zero energy; Figure 4.8. The exact energy values can only be found by solving the wave equation, or the time-independent Schrödinger equation: $E_n = -Ce^2/n^2$ with a constant C and the electron's charge e, quantized by an integer n that must be larger than the value j of angular momentum in a state of the same energy.

Spectroscopy The measurement of frequencies emitted or absorbed when a system changes between two of its bound states is a powerful tool to probe matter and atoms, or to identify ingredients in an unknown source. Spectroscopy plays an important role in astrophysics, where the ratios of different elements in stars can be determined from the emitted light. One identifies frequencies of different elements by the characteristic pattern they form when they are drawn as lines on the frequency axis. Measurements on Earth tell us what lines we should see; when we recognize the same pattern in star light, we can conclude that the same element must be present. The absolute value of the frequencies, however, is different for stars because they move; frequencies are redshifted by the relativistic Doppler effect. In this way, one measures escape velocities of stars, or the expansion rate of the universe.

Spectroscopy has played a crucial role in unraveling the nature of atoms. Throughout the nineteenth century, physicists showed much zeal in looking at spectra to identify more and more, fainter and fainter lines in emission and absorption. "Fraunhofer first saw eight lines; then there were six hundred, and in our time, a hundred thousand and more have been seen. In fact, growing enormously in number, they are innumerable, yet counted and numbered."[3] recounted the Swedish writer August Strindberg in 1896, impressed by the accelerating accumulation of new data. Quantum mechanics, in the early twentieth century, explained this wealth of information by a few principles, summarized for instance in the Schrödinger equation. Although the equation may look complicated, it is much more elegant than a list of all possible spectral lines for all elements. This compactness of mathematical descriptions for complex-looking phenomena is what physicists identify as beauty in their theories.

A similar pattern was repeated in the twentieth century, when accelerator experiments at increased energies showed more and more peaks in the reaction rates, identified as resonances of particles much like lines in frequency spectra. Again, these resonances were reduced to complicated-looking but much more elegant formulas that describe the forces governing elementary particles, mainly the strong nuclear force.

Atoms An atom is a positively charged nucleus surrounded by electrons. Electrons are much lighter than a nucleus, at least by a factor of 2000, and so the former tend to roam around more than a tardy nucleus. In the simplest case, hydrogen, a single electron is bound by the Coulomb force to the nucleus containing one pro-

3) August Strindberg, *Scientific and philosophical notes*, first published in the journal *L'Initiation*.

ton and, in deuterium and tritium, up to two neutrons. The Coulomb potential is the main contribution to the binding force, used to find energies of bound states.

The Coulomb potential is not a complete description of the force. Relativity changes the motion even at typical speeds of an electron in hydrogen, less than 1% of the speed of light. There are also effects from magnetic interactions of the current, implied by the moving electron, with a spinning nucleus. Moreover, the nucleus is not a single particle but a bound system as well. The uncharged neutrons do not matter much for the electric interactions, but the proton, unlike the electron as it appears in present measurements, is not an elementary particle; it is made of even smaller ones, the quarks. The proton has a radius of about a trillionth of a millimeter, which is small, but still large compared with the current resolution of accelerator experiments: a thousandth of the proton radius, at which the electron still appears pointlike. In the Coulomb potential, we assume a pointlike or at least spherical distribution of the central charge. When this assumption is violated, corrections appear in calculations of the spectrum of hydrogen. Modern atomic spectroscopy can be used to gain information about the size of a proton.

Atoms other than hydrogen have at least two electrons. They interact with each other by the Coulomb force, not just with the nucleus, making the problem of bound states nonspherical and much more complicated to solve: At least three distances must be taken into account, not just one radius parameter. For light elements, helium, lithium and so on, the pattern of states found for hydrogen can be used as an approximation to the energies realized. The larger central charge implies a stronger force toward the nucleus, making the inner core of atoms more compact. Energy levels of bound electrons lie much lower than in hydrogen, requiring more energy to eject one of the inner electrons. If all electrons could occupy the same ground state, ionization energies required to free one of them would be immense. Atoms other than hydrogen would hold their own electrons so tight that they would be loath to react with other ones. The rich structure of chemistry, with atoms of all kinds happy to interact and bind, is vivid testimony to the fact that not all electrons of an atom can be in the ground state.

Pauli's exclusion principle When left alone, all objects strive to minimize their energy. In a hydrogen atom, an electron excited to an energy above the minimum will soon emit a photon with the energy gained, and return to the cozy ground state. In heavier atoms, electrons would come to lie so close to the nucleus that an enormous amount of energy would be required to make them react with other atoms. The next-lightest element, helium, is indeed inert, as the prime example of a noble gas too aloof to intermingle with any other atom. One step further brings us to lithium with three protons and three electrons (and some neutrons). Heavier than helium, it should display even more gravitas, but regarding reactivity and promiscuity, it is no second to hydrogen.

There must be something that prevents most electrons from reaching the ground state in atoms heavier than hydrogen or helium. Some electrons must be left to roam farther from the apron strings of the mother nucleus, striving to hook up with other atoms and form new relations. Classical physics suggests no force other than Coulomb's. It offers nothing to prevent electrons from reaching the ground state, and no other energy barrier can be imagined. Quantum mechanics leads to new forces from the wavelike nature or from its variances, but how can they be strong enough? Why can't we just stack together all electron wave functions with the same energy? We have to look into fundamental properties of wave functions to see what holds most electrons apart, enabling reactions of chemistry.

A single electron, as in a hydrogen atom, has a wave function $\Psi(x, y, z, t)$, telling us how likely it is to find the electron at different places. With two electrons, we have a wave function for each of them, $\Psi_1(x_1, y_1, z_1, t)$ and $\Psi_2(x_2, y_2, z_2, t)$ for the two independent positions at the same time t. Measuring two independent positions implies that the probability to find electron 1 at place (x_1, y_1, z_1) and electron 2 at place (x_2, y_2, z_2) is the product of the individual probabilities. If we find electron 1 half the time in some region 1 no matter what the position of electron 2 is, and if we find electron 2 half the time in some region 2 no matter what the position of electron 1 is, we find electron 1 in region 1 and electron 2 in region 2 a quarter of all times. The combined probability is the product $\frac{1}{4} = \frac{1}{2} \cdot \frac{1}{2}$. Individual probabilities are given by the norms squared of the wave functions, and so we multiply the wave functions for the combined state of independent objects.

One too-classical ingredient crept into the preceding consideration. If we think of the two particles as electron-1-at-position-1 and electron-2-at-position-2, we must be able to identify and label the electrons. For all we know, electrons are pointlike and have no qualifying characteristic to distinguish them from one another, except their possibility to spin in different ways. In classical physics, if we consider the motion of identical-looking objects, we could take a picture of all of them at one time, come up with imaginative names to label their images or paint them with different colors, and then follow their trajectories to identify all objects at later times. In quantum mechanics, we deal with overlapping waves, not with individual objects moving along trajectories. We cannot recognize an electron as a specific one; we can only recognize it as one of many equals. The probability question we posed was incorrect: We cannot ask how likely it is to find electron 1 in region 1 and electron 2 in region 2; we may only ask how likely it is to find one of the two electrons in region 1 and one in region 2.

If it is impossible to distinguish two electrons from each other, we cannot assign individual positions to them. Probabilities, or the wave function, can only tell us how likely it is to find some electron, either electron 1 or electron 2, in a region. If we keep working with individual positions $\vec{r}_1 = (x_1, y_1, z_1)$ and $\vec{r}_2 = (x_2, y_2, z_2)$, we must ensure that the combined probabilities do not allow us to tell what coordinates correspond to the position of which electron. The probability must be symmetric under switching the two positions, described by one probability function $\mathbf{Prob}(\vec{r}_1, \vec{r}_2)$ depending on all coordinates such that $\mathbf{Prob}(\vec{r}_1, \vec{r}_2) = \mathbf{Prob}(\vec{r}_2, \vec{r}_1)$. The labels or places of the two electrons are no longer distinguished from each other; we are just dealing with two unidentified, indistinguishable electrons.

The wave function allows different options with the required symmetry. It must be such that $\Psi(\vec{r}_1, \vec{r}_2)$ has norm squared unchanged when we switch the two positions: $\Psi(\vec{r}_1, \vec{r}_2) = A\Psi(\vec{r}_2, \vec{r}_1)$ with a complex number A such that $|A|^2 = 1$.

There are many such numbers, all of the form $A = \cos(a) + i\sin(a)$ with a real number a. However, if we switch the positions once more, back to their original order, we get the original wave function: $\Psi(\vec{r}_1, \vec{r}_2) = A\Psi(\vec{r}_2, \vec{r}_1) = A^2\Psi(\vec{r}_1, \vec{r}_2)$. The implied equation $A^2 = 1$ is stronger than $|A|^2 = 1$. It has just two solutions, 1 and -1.

For both values, probabilities are symmetric under switching the positions, but the wave function is either symmetric ($A = 1$) or antisymmetric ($A = -1$). In the former case, we can use our earlier product form if we "symmetrize" it as $\Psi_+(\vec{r}_1, \vec{r}_2) = \Psi_1(\vec{r}_1)\Psi_2(\vec{r}_2) + \Psi_2(\vec{r}_1)\Psi_1(\vec{r}_2)$. The two electrons are then indistinguishable by their positions. In the antisymmetric case, we write $\Psi_-(\vec{r}_1, \vec{r}_2) = \Psi_1(\vec{r}_1)\Psi_2(\vec{r}_2) - \Psi_2(\vec{r}_1)\Psi_1(\vec{r}_2)$.

In a symmetric wave function, all electrons can occupy the same state $\Psi_1 = \Psi_2$, for instance, the ground state. The antisymmetric combination of $\Psi_1 = \Psi_2$, however, results in a vanishing wave function, an impossible result if we know that there are two electrons. If electron wave functions are combined antisymmetrically, two of them can never occupy the same state. If one of them is in the ground state, another one must contend itself with a state of the next available energy. By this exclusion principle, named after Wolfgang Pauli, chemistry as we know it becomes possible.

Periodic table The bound-state energies of an atom provide seats of different kinds for its electrons. It is most convenient for an electron to sit in the front row and be near the nucleus, but each coveted spot can be taken by only one electron. Starting with a bare nucleus, a complete ionization of the atom, and adding electron after electron, the best spots will be taken first, just as unreserved theater seats will be filled from the front to the back if everyone wants to be as close to the action as possible. The front row will be filled first, each audience member taking one seat, perhaps starting from the middle and proceeding to both sides. Once the first row is full, the pattern is repeated for the second row, and so on. The theater is filled in periodic fashion, row by row. Atoms are filled in periodic fashion, just as they are arranged in the periodic table.

Energy levels of the hydrogen atom only depend on one integer quantum number n seen also in one-dimensional problems. If slight deviations from the Coulomb potential due to an extension of the nucleus or interactions of all the electrons are included, the levels also depend on the angular momentum j. However, changing j with fixed n implies a much smaller change of energy than changing n. All states with fixed n, called a shell in atomic physics, are like a row of the theater, of similar distance to the center.

The number of seats in a shell depends on the energy. There are two conditions relating the quantum numbers we have seen: n must be larger than j, and ℓ for the L_z-component of angular momentum cannot have magnitude larger than j, but can take negative values or zero. With these ranges, there are $2j + 1$ possible values for ℓ, and $\mathbf{Sum}_{j=0}^{n-1}(2j + 1) = n^2$ values of j and ℓ for given n. Moreover, two electrons can differ by whether they spin clockwise or counterclockwise (two options somewhat analogous to theater seats being to the left or right of the center).

For every n, we therefore have $2n^2$ different states of similar energy, or $2n^2$ places in every shell.

The first shell, for $n = 1$, has two states, one filled in hydrogen and both in helium. With a full shell, helium is inert to chemical reactions; removing or adding an electron is not favorable from the energy viewpoint. The next shell, $n = 2$, is started in lithium, a reactive element with a single electron in the outer shell. For $n = 2$, there are eight different states, all filled when we have ten electrons total, two in the first shell and eight in the second. We have arrived at neon, another noble gas. Continuing in this way, one can explain the whole arrangement of the periodic table. (Some amendments of our simple arguments based on hydrogen are required, especially for heavier elements with many interacting electrons.) This feat is an impressive demonstration of the concept of bound states in quantum mechanics combined with Pauli's exclusion principle.

4.3 Measurements

A particle, or any object, in quantum mechanics does not have a sharp position. The possible positions where it might be seen, and how likely each one is to be found, are determined by the wave function. The more narrow it is, the more certain one will be with position estimates. However, the uncertainty principle prevents us from making the wave as narrow as we like, and narrowing it down even by limited amounts can only be afforded if we increase the uncertainty of momentum.

A good way of making certain statements in uncertain times is to be statistical. The wave function indeed provides us with statistical or probabilistic information by the average values and variances (or other moments) we compute from it. We identify these numbers with the averages or variances (also called standard deviations) of actual measurement results, done many times for a large number of systems prepared in the same state. If we measure the size of a hydrogen atom in its ground state, the results will show some spread but on average agree with the average radius $\bar{r}$ in the ground-state wave function. Even without absolute certainty, the theory allows empirical tests.

Doing multiple measurements before we compute the average or variance, we get a bunch of individual measurement results. Physics is supposed to describe and probe nature by the experiments and observations we do. A good theory should be able to make predictions also for a single measurement result, not just for the average of many ones. Averages are statistical and do not provide sharp statements for single outcomes, just as we cannot tell an individual vote from general election results. However, the more we know about the voting populace or the physical system, the better we can tell how an individual may have voted. In political elections, demography is a crucial aspect. In quantum mechanical expectation values, the wave function is the key player.

If we know a lot about the wave function, we can compute more than the average and variance it implies for measurements. We can use the values it takes at individual points to see how likely it is to measure the particle at that point. As already used in passing when we formalized averages and variances in terms of the wave function, we assign the probability $\mathbf{Prob}(x, t) = |\Psi(x, t)|^2 \Delta L(x)$ to a measurement outcome of position x within some range Δx (at time t), for a particle with wave function $\Psi(x, t)$. The norm squared is a positive real number, as required for probabilities. Also, if we choose the right prefactor of the wave function, we can make sure that the total probability to find the particle anywhere, summing the individual probabilities $\mathbf{Prob}(x, t)$ over all possible x-values, is 100%, or normalized to the number one. Using the norm squared of the wave function, we can find out how likely it is to measure each possible position in an individual measurement.

Observables Experimental physicists are skilled in measuring different quantities, with position only one of them. The position of an object, and the way it changes in time, is one of the most obvious things to observe. It played an important role during the beginnings of modern physics in the times of Galilei and Newton, and continues to do so in current experiments even at high energies. Position and motion is close to our intuition, much closer than concepts such as energy or momentum which had to be developed, "found" in the build-up of theory, over several centuries.

The senses we come equipped with are precise and quantitative when regarding positions of objects we can see. However, even though we perceive heat or can be badly affected by collisions, as some examples of energy "measurements," we have to work much harder to find quantitative measures. To appeal to our intuition, experiments with atoms or elementary particles still show their results by positions and motion, whenever possible. Particle detectors in high-energy accelerators, for instance, draw fountains of particles spraying away from the collision point. Still, of most interest for modern physics are the energies of different particles, or other, more abstract quantities. When positions or trajectories are used, they serve an indirect purpose, to determine the momentum of a charged particle from the deflection it suffers in a magnetic field.

In quantum mechanics, we begin with a wave function $\Psi(x, t)$ that depends on position and time in order to extend the notion of a classical particle at position x. We have derived energy and momentum from oscillations of $\Psi(x, t)$ in time and space, giving us important information, for instance, for spectroscopy. However, position and energy measurements, as two examples, are so far based on rather different footings. For position, we have the whole wave function, telling us the probabilities to find a specific result in every single measurement. For energy, we can find the value only in specific states, and we do not have a direct way of telling individual probabilities. For a complete physical evaluation of the theory, we need a systematic way of determining possible values and individual likelihoods for any given observable we might be interested in.

Stationary states Spectroscopy is a direct way of measuring energies. A single photon emitted from an excited atom shows by its frequency how the energy of the atom changed. The possible energies of an atom are given by the discrete values realized in stationary states, in which the atom can stay at constant energy for a long time. Stationary states are closely related to individual measurements of energy.

If we look at emission or absorption spectra of atomic gases, the bright or dark frequency spots are not sharp "lines," but have some frequency spread. Spectra cannot be direct images of the discrete set of energies of stationary states, although the spots are centered at frequencies that agree with energy differences divided by h. Different factors affect the observed frequencies:

First, atoms in a gas move with average velocities that depend on the temperature. An atom emits or absorbs photons with frequencies according to the energy spectrum, provided the frequencies refer to the rest frame of the atom. Atoms are in relative motion compared with an observer measuring frequencies in a laboratory, and so the frequencies measured are shifted relative to the energy spectrum, depending on the motion of atoms according to the Doppler effect. One can cool down the gas to minimize the velocities and Doppler shifts, but some motion (at least zero-point energy) will always remain, broadening the frequency lines.

Secondly, quantum mechanics implies that energy and the time in which it is realized cannot both be measured to arbitrary precision. Just as with position and momentum, there is an uncertainty relation for energy and time. An excited state has a finite lifetime because a lower-lying state, such as the ground state, is preferred. The atom in its excited state has an energy with some uncertainty $\mathbf{Var}_E$, related to the lifetime $\mathbf{Var}_t$ by $\mathbf{Var}_E \mathbf{Var}_t \geq \hbar/2$. The variation $\mathbf{Var}_E$ implies an additional spreading of frequency lines, which cannot be reduced by changing properties of the atomic sample.

In most cases, the broadening of lines is much smaller than the distance between neighboring ones. Even if the photons emitted or absorbed do not have sharp frequencies, we can say with certainty which two states of the atom were involved in the process. A single measurement, that is, the detection of a single photon, can tell us what the new energy is and how the wave function has changed.

Energy eigenstates Measurements shape wave functions. When some value for the energy of an atom has been measured, for instance, by spectroscopy, the wave function takes a form that belongs to the energy observable. After detecting the photon, we know that the atom is in some state in which its energy E_n is related to the oscillation period of the wave function Ψ_n by $i\hbar\Delta\Psi_n/\Delta t = E_n\Psi_n$. According to the Schrödinger equation, Ψ_n then obeys $-(\hbar^2/2m)\Delta^2\Psi_n/\Delta x^2 + V(x)\Psi_n = E_n\Psi_n$. Possible stationary states Ψ_n and associated energy values E_n belong to the energy observable; no information other than the energy operator and its relation to the potential is needed. With a bit of German heritage,[4] the Ψ_n are called eigenstates and the E_n eigenvalues of energy.

4) It is not easy to provide a unique translation of "eigen." The word can have several different meanings, including "proper" or "own," but also "peculiar." The best way of thinking of the meaning of "energy eigenstate" is perhaps as a state characteristic of a certain energy value.

After we measure the energy, we know what value E_n it takes. The probability of finding the same result in a second measurement right afterwards, leaving no time for further emission or absorption processes, must result in the same value, with certainty because we already know what the energy is. The wave function contains all information about the system, in a statistical way because our knowledge is never complete. However, if an observable has already been measured, we know its value right after the measurement, without qualifying statistics. After measuring an observable, the state must be one that fully belongs to the observable, or an eigenstate with the eigenvalue measured.

Energy eigenstates are nothing but stationary states. The new terminology is useful because it applies to any observable O. If we know the operator $\hat{O}$, we can compute eigenvalues and eigenstates by the equation $\hat{O}\Psi_n = O_n\Psi_n$. In analogy with energy eigenstates, we say that a system takes on an eigenstate Ψ_n of $\hat{O}$ if the observable O has been measured with value O_n. For momentum, for instance, we have the operator $\hat{p} = -i\hbar\Delta/\Delta x$, and eigenstates are $\Psi_k(x) = \exp(ikx)$ with eigenvalues $p_k = \hbar k$. As in this case, eigenvalues do not always form a discrete set.

We know what states we can have after measuring an observable, and what measurement results are possible. The last piece, the actual value that we will measure, remains statistical. If the state is not an eigenstate to begin with, after a measurement different eigenstates can be reached with some probabilities. The usual statistical properties of the wave function appear: A general state can be written as $\Psi(x,t) = \mathbf{Sum}_n C_n \Psi_n(x,t)$ in terms of eigenstates Ψ_n of the observable we measure, with complex numbers C_n that tell us how each eigenstate contributes. After the measurement, we find one state Ψ_n, and the probability by which it appears is $|C_n|^2$.

Collapse of the wave function After being measured, the wave function collapses in submission. It bends to the will of the observable we measure (provided the measurement was successful) and takes on a form dictated by the observable, an eigenstate. The change of the wave function is a consequence of our new information gained about the state. After a measurement, we know with certainty which value we have measured; the wave function must correspond to a sharp value of the observable. It is still a wave, without certain answers for most questions we could ask about it. However, the one quality we have just measured is known, turning the wave function into an eigenstate.

One often calls the transition from some form of the wave function to an eigenstate its "collapse" during the measurement. If we imagine the wave function drawn as a distribution of all possible measurement outcomes while they are still uncertain, the curve after the measurement becomes much more narrow, with only one option realized. The distribution determined by the wave function collapses to a narrow peak. If we measure the position, for instance, the wave that used to determine different likelihoods for different places of a particle collapses into a single spike just at the place where we have measured the particle. The process and the language of collapse may seem rather drastic, but they just reflect our change of knowledge accompanying any successful measurement.

If we do multiple measurements of different quantities one after another, such as position and momentum or energy, the wave function keeps collapsing on different eigenstates. If we measure the position first, the wave function becomes a narrow peak at some place. If we then measure the momentum, the wave function becomes a momentum eigenstate in which the wave length has a sharp value: an extended harmonic wave $\exp(ipx/\hbar)$, the opposite of a sharp peak in x. A second position measurement again collapses the wave to a narrow peak, but the intermediate momentum measurement has changed the wave function compared to what it had been. Position and momentum are so-called incompatible observables, whose measurements make the wave function collapse to almost opposite forms: one narrow, the other spread out.

One can think of measurements in quantum mechanics as idiosyncratic artists sculpting the wave function to their ideal. One artist forms the wave function, like clay, in a characteristic and peculiar way. The piece of art is not unique, but recognizable as the artist's work: whatever the outcome, an eigenstate is formed. A second artist going to work on the same material, after the first one is done, has a different idea of what a perfect piece of art should look like, and remodels the wave function to a new type of eigenstate. The wave function is perfectly malleable; no part is ever lost. All artists can realize their own imagination in the same material, no matter how messy the initial shape of the wave function or the intermittent meddling by other artists may have been.

Commutator To control meddling, we need order. It is not possible to measure position and momentum both to arbitrary precision, for that would require the wave function to collapse into an impossible eigenstate of both observables, narrow and spread out at the same time. This is the essence of the uncertainty principle: the variances of position and momentum measurements cannot vanish at the same time.

A good mathematical way to quantify the uncertainty in measuring two observables A and B is the commutator $\hat{A}\hat{B} - \hat{B}\hat{A}$ of the corresponding operators $\hat{A}$ and $\hat{B}$, abbreviated as $[\hat{A}, \hat{B}]$. The commutator indicates that we compare the outcomes of two different orderings of the measurements: first A and then B compared with first B and then A. Each observable makes the wave function collapse to one of its eigenstates. If the commutator does not vanish, the eigenstates cannot be the same; we obtain different results depending on whether we measure A or B first.

For position $\hat{x} = x$ and momentum $\hat{p} = -i\hbar\Delta/\Delta x$, the commutator, computed by the way it acts on a wave function Ψ, is $[\hat{x}, \hat{p}]\Psi = -i\hbar[x\Delta\Psi/\Delta x - \Delta(x\Psi)/\Delta x] = i\hbar\Psi$. In the second term, we obtain two contributions from the variations of x and Ψ, respectively, and the second one cancels the first term while the first one produces a constant. We read off $[\hat{x}, \hat{p}] = i\hbar$, a nonzero result showing the incompatibility of the two operators. The commutator never vanishes, but it is proportional to $\hbar$. It can be ignored in classical physics which knows nothing of $\hbar$

because of its relative smallness. However, it is important in quantum mechanics when we get close to the uncertainty limit.

The commutator is the key quantity that determines limitations by uncertainty relations. For position and momentum, the constant commutator implies a constant lower limit for the uncertainty product in $\mathbf{Var}_x\mathbf{Var}_p \geq \hbar/2$. More generally, for two observables, A and B, the commutator $[\hat{A}, \hat{B}]$ determines an uncertainty relation by

$$\mathbf{Var}_A\mathbf{Var}_B \geq \frac{1}{2}|\overline{[\hat{A}, \hat{B}]}| . \quad (4.18)$$

For position and momentum, the usual uncertainty relation follows. For other operators, one often obtains useful relationships that are much easier to see from formal calculations of commutators than from considering properties of varying wave functions.

The uncertainty relation derives from a general inequality valid for geometry in all spaces in which we can add vectors and compute their lengths (or norms). This Schwarz inequality reads $|v|^2|w|^2 \geq |v \cdot w|^2$. In three-dimensional space, one can confirm its validity by realizing that $v \cdot w = |v||w|\cos(\sphericalangle)$, with the angle $\sphericalangle$ between the vectors v and w, has an upper bound depending on the lengths $|v|$ and $|w|$: the cosine function never takes values larger than one.

Uncertainty relation Given two observables A and B and a state Ψ in which we determine variances and averages, we define the vectors $v = (\hat{A} - \overline{A})\Psi$ and $w = (\hat{B} - \overline{B})\Psi$. The variances can be written as $\mathbf{Var}_A^2 = |v|^2$ and $\mathbf{Var}_B^2 = |w|^2$, and their product is no smaller than $|v \cdot w|^2 = \mathrm{Re}(v \cdot w)^2 + \mathrm{Im}(v \cdot w)^2$. In these two terms, we compute $\mathrm{Re}(v \cdot w) = \frac{1}{2}[v \cdot w + (\overline{v \cdot w})^*] = \frac{1}{2}\overline{(\hat{A}\hat{B} + \hat{B}\hat{A})} - \overline{AB} = \mathbf{CoVar}_{A,B}$ and $\mathrm{Im}(v \cdot w) = \frac{1}{2}[v \cdot w - (v \cdot w)^*] = \frac{1}{2}[\hat{A}, \hat{B}]$:

$$\mathbf{Var}_A^2\mathbf{Var}_B^2 - \mathbf{CoVar}_{A,B}^2 \geq \frac{1}{4}|\overline{[\hat{A}, \hat{B}]}|^2 . \quad (4.19)$$

It follows that $\mathbf{Var}_A\mathbf{Var}_B \geq \frac{1}{2}|\overline{[\hat{A}, \hat{B}]}|$: subtracting the covariance squared can only make the product of variances smaller.

Schwarz inequality To prove the Schwarz inequality, we introduce a vector $u = w - v(v \cdot w)/|v|^2$, so that $v \cdot u = v \cdot w - |v|^2(v \cdot w)/|v|^2 = 0$. Its norm then equals $|u|^2 = w \cdot u - (v \cdot w/|v|^2)^* v \cdot u = w \cdot u = |w|^2 - |v \cdot w|^2/|v|^2$. With $|u|^2 \geq 0$, the inequality follows.

Conserved quantities Even the uncertain, fluctuating quantum world benefits from the constants of life. Uncertainty relations share a feature with conserved quantities: they are both controlled by commutators. Observables with commuting operators, $[\hat{A}, \hat{B}] = 0$, can be measured both with arbitrary precision. One example is momentum p and kinetic energy $p^2/2m$. Both can be measured at the same time because knowing p already determines the kinetic energy. There are less obvious examples for commuting operators, playing important roles in different systems. A general class of commuting operators is the energy together with any conserved quantity.

In classical physics, a conserved quantity has a vanishing time derivative according to the equations of motion; it always stays at its initial value. In quantum physics, we have a wave equation for Ψ instead of equations of motion for observables like position and momentum, and we cannot easily compute time derivatives of observables. However, we can compute time derivatives of averages, which make use of observables we measure and of wave functions to provide probabilities. The latter are subject to an equation of motion: Using the Schrödinger equation, we compute $\Delta\Psi/\Delta t = (i\hbar/2m)\Delta^2\Psi/\Delta x^2 - (i/\hbar)V(x)\Psi = -(i/\hbar)\hat{E}\Psi$. We also need the time derivative of the complex conjugate Ψ^* because it determines probabilities $|\Psi|^2 = \Psi^*\Psi$ together with Ψ. The complex conjugate flips the sign of every imaginary unit i in an equation, and therefore $\Delta\Psi^*/\Delta t = -(i\hbar/2m)\Delta^2\Psi^*/\Delta x^2 + (i/\hbar)V(x)\Psi^* = (i/\hbar)\hat{E}\Psi^*$.

The average of an observable O in a state Ψ is $\overline{O}(t) = \mathbf{Sum}_x \Psi(x,t)^* \hat{O}\Psi(x,t)$ $\Delta L(x)$. It changes in time according to the two time-dependent factors Ψ^* and Ψ: $\Delta\overline{O}/\Delta t = \mathbf{Sum}_x[(\Delta\Psi^*/\Delta t)\hat{O}\Psi + \Psi^*\hat{O}\Delta\Psi/\Delta t]\Delta L(x)$. If we insert time derivatives according to the Schrödinger equation, we obtain

$$\frac{\Delta\overline{O}}{\Delta t} = \frac{i}{\hbar}\mathbf{Sum}_x\left(\Psi^*\hat{E}\hat{O}\Psi - \Psi^*\hat{O}\hat{E}\Psi\right)\Delta L(x) = \frac{\overline{[\hat{O},\hat{E}]}}{i\hbar}\,. \tag{4.20}$$

An observable O is a conserved quantity, with constant averages $\overline{O}$ in all states, if its operator $\hat{O}$ commutes with the energy: $[\hat{O}, \hat{E}] = 0$.

Examples are the well-known conserved quantities in classical mechanics. If there is no force or a constant potential, the momentum is conserved: We confirm our preceding example because the energy is then just kinetic (perhaps up to a constant added to it), and we have $[\hat{p}, \hat{E}] = 0$ if $V(x)$ is constant. If the potential is spherically symmetric, all angular-momentum components commute with the energy and are conserved. However, as seen before, the angular-momentum components do not commute among themselves. These properties play an important role for the physics of atoms and the periodic table.

Energy-time uncertainty Energy can be borrowed, provided it is soon repaid. For some observable O and energy E, we have $\mathbf{Var}_O\mathbf{Var}_E \geq \frac{1}{2}|\overline{[\hat{O},\hat{E}]}| = \frac{1}{2}\hbar|\Delta\overline{O}/\Delta t|$. If we view $\mathbf{Var}_t^{(O)} = \mathbf{Var}_O/(|\Delta\overline{O}/\Delta t|)$, the time required for the average of O to change by its variance, as a version of time uncertainty, we obtain an uncertainty relation

$$\mathbf{Var}_t^{(O)}\mathbf{Var}_E \geq \frac{\hbar}{2}\,, \tag{4.21}$$

not unlike the position-momentum uncertainty relation.

However, the time variance as defined depends on the observable O used, and time does not have the same statistical properties as position or momentum or energy. We have considered time as a parameter in terms of which the wave function changes, not subject to individual measurements. Statistical properties of time are not intrinsically defined, without reference to other observables.

Still, the energy-time uncertainty relation has important applications. It implies further broadening of lines in emission and absorption spectra, for excited states going back to lower energy have a finite lifetime, an estimate for $\mathbf{Var}_t$. The energy variance then cannot be zero, smearing out frequency lines. The more unstable a state, the broader the lines for emission out of it (by measuring the broadness of lines, we can determine how stable states are). Also, unstable particles can be studied in this way: the more unstable ones can be produced with wider ranges of energies of particle accelerators; they appear as wider bumps in energy-dependent production rates.

Many woes There are several disconcerting bits related to the collapse of the wave function, often eliciting desperate-looking attempts of explanations. Measurements in quantum mechanics are thorough artists, shaping the wave function to their will. The collapse happens if we just measure an observable; we do not need to design an experiment with the purpose of shaping the wave function as an eigenstate. Also, while most measurements take some time, the collapse of the wave function happens in the instant the measurement result gets known. After all, the wave function and its collapse reflect the knowledge gained about the system, not the process of how the information is found. If we set up an experiment but do not register the result, or ignore it, the wave function does not collapse; our information about it does not change.

Wave functions can be of broad extension. For a pair of two particles created in a single event, such as the decay of another particle, we have one wave function to describe the whole system. If the particles move away in different directions, the wave function keeps spreading out. A measurement at some later time may act on just one part of the wave function; measurement devices are rather small compared to the distances the two particles can span between them. Even so, a measurement on one of the two particles affects the combined system. The whole wave function collapses, affecting even the distant particle whose properties we did not measure.

Examples are decays of spinless particles into spinning ones, such as two photons. To preserve angular momentum, the decay particles must spin in opposite ways. Just after the decay, each particle has a 50%-chance to spin one way, and a 50%-chance to spin the other way. But after one spin has been measured, the other must take the opposite value, with absolute certainty. Even if the particles are far apart, the wave function collapses instantly on the measured values of spin. There is no contradiction with relativity because we cannot transmit information in this way. Doing the measurement on one particle, we know what result a measurement on the other one would yield, but we cannot tell anyone there what would happen unless we use a conventional signal, traveling at most at the speed of light. Causality and the speed limit of relativity are not violated, but the apparent instantaneous action over large distances seems disconcerting (or spooky, as sometimes said).

The collapse process is of a different kind than other processes considered in physics. We describe changes in time by evolution equations of differential type, for instance, the wave equation. These equations are deterministic: Knowing the wave at one time allows us to predict what it will do later, by solving the wave equation. The collapse process, on the other hand, does not have time to evolve; it happens in an instant. And it is not deterministic: We know that the wave function will collapse into an eigenstate of the measured observable, but we do not know for sure which one (unless the wave function was an eigenstate already before the measurement). We have only probabilities for each eigenstate, bringing in the statistical nature of quantum mechanics.

Several attempts have been made to eliminate nondeterministic and other features. Eugene Wigner postulated that an observer's conscience, perhaps the most complex phenomenon in the universe, could trigger the collapse; after all, it is the observer who has to accept the blame for deciding to make a measurement. Relativity theory and quantum physics have indeed completed the transition from passive classical observers to self-conscious ones, aware of their motions and uncertainties. But trying to find a role for conscience in a strong, literal sense most likely goes too far.

Hidden-variable theories, for instance David Bohm's, search for underlying quantities more fundamental than the wave function, so that nondeterministic statistics would be seen as an imperfect device of an incomplete theory. The wave function would appear much like a fluid continuum, which we know not to be fundamental because the fluid consists of atoms and molecules, not of a continuous medium such as a density profile. However, no consistent formulation of quantum mechanics using hidden variables has been found, and one can even show by experiments that statistical aspects are necessary.

To disarm what appears as the brutality of the collapse of the wave function, cutting off large swaths of the wave and eliminating unrealized chances, Hugh Everett introduced the many-worlds interpretation. In this picture, the wave function does not collapse; it rather splits into all possible options according to the eigenstates it contains. Every one of the options is realized in its own universe, as part of a "multiverse" that branches out into many different parts at every turn of measurements. The suggested picture is not what one would call economical, and it does not lead to concrete new effects of quantum mechanics that could be tested.

Momentum space Space is nothing but the set of possible positions. From the perspective of measurements in quantum mechanics, we could equally well consider all possible momenta of a particle to characterize observations.

The wave function $\Psi(x,t)$ tells us, via $|\Psi(x,t)|^2$, how likely it is for collapse to occur at x when we measure the position at some time t. However, it collapses with some probability whenever we do a measurement, no matter what observable we are looking at. If we measure the momentum, the wave function collapses into a momentum eigenstate with value p, with different probabilities that depend on the state. For a complete description of momentum measurements and the likelihoods of their values, we do not use wave functions $\Psi(x,t)$, but rather functions $\Phi(p,t)$ that depend on momenta. (The cactus' arms have been closed, for the Greek Phi, but it loses nothing of its thorniness.)

In order to determine $\Phi(x,t)$, called the momentum-space wave function, we must write $\Psi(x,t)$ in such a way that the momentum eigenstates it contains become visible. We must call forth momentum as an artist to sculpt the wave function according to the momentum ideal, writing it as a combination of momentum eigenstates $\Psi_p(x,t) = \exp(ipx/\hbar)$. With coefficients $C_p(t)$, we must find a way to write $\Psi(x,t) = \mathbf{Sum}_p\, C_p(t)\exp(ipx/\hbar)\Delta L(p)$. Since $C_p(t)$ tells us "how much p" there is in the state, we identify the momentum-space wave function as $\Phi(p,t) = \sqrt{2\pi}\,C_p(t)$, and the probability to find a value p for momentum measurements at time t, or for the wave function to collapse into $\Psi_p(x) = \exp(ipx/\hbar)$, is $|\Phi(p,t)|^2$.

There is a well-established mathematical procedure to determine coefficients C_p for a given $\Psi(x,t)$, such that $\Psi(x) = \mathbf{Sum}_p\, C_p \exp(ipx/\hbar)\Delta L(p)$. The method was introduced by Joseph Fourier well before its application in quantum mechanics, or even before quantum mechanics itself became apparent. In fact, there are numerous applications in fields outside of quantum physics. An important use was perturbation theory of planetary orbits, by which one can describe the influence of other planets on a given one by contributions to the trajectory of shorter periods than the unperturbed Kepler ellipse. Astronomers wrote the coordinates of an orbit as $\mathbf{Sum}_n\, x_n \cos(\bigcirc nt/T)$ with the unperturbed period T, and summing over integers n. For terms with $n = 1$ and $n = 2$, we describe an elliptical orbit, the ellipse written as a perturbation of the radius of a circle: $x(t) = r(t)\cos(\bigcirc t/T)$ and $y(t) = r(t)\sin(\bigcirc t/T)$ with $r(t) = r_0 + s\cos(\bigcirc t/T)$. Using trigonometric identities, these functions amount to perturbed periodic ones with terms of $n = 1$ and $n = 2$. With attraction between all planets, a complicated system of differential equations results for all coordinates, solved by approximation writing all planetary orbits as perturbed Kepler ellipses with terms of $n > 2$. Approximate solutions for x_n then show how the orbits change.

A powerful feature of the Fourier expansion $\mathbf{Sum}_n\, x_n \cos(\bigcirc nt/T)$ is its invertability. If we know the x_n, we can sum up all terms and obtain the function expanded. However, we can also start with the full function and compute the coefficients x_n. Once we know the x_n, for instance, from meticulous observations of planetary orbits deviating from Kepler ellipses and computing the Fourier coefficients, we can draw conclusions about the forces that caused those terms to have the values found. With this procedure, Urbain Le Verrier analyzed the orbit of the planet Neptune, concluding that there should be another planet, unknown at that time. Le Verrier was able to compute a region in the sky in which the new planet should be seen at a given time, soon confirmed by direct observations by the astronomer Johann Galle. Uranus had been found.

Fourier transformation provides means to compute the momentum-space wave function $\Phi(p,t) = \mathbf{Sum}_x\, \Psi(x,t)\exp(-ipx/\hbar)\Delta L(p)/\sqrt{2\pi}$. Its inverse gives back the position-space wave function $\Psi(x,t) = \mathbf{Sum}_p\, \Phi(p,t)\exp(ipx/\hbar)\Delta L(p)/\sqrt{2\pi}$.

The momentum-space wave function shows how likely any possible measurement result for momentum is, computed as $|\Phi(p,t)|^2$.

Infinite square well Particles in an infinite square well just bounce back and forth between the walls. The walls are perfect reflectors, even in quantum mechanics because their infinite barrier prevents any tunneling leakage out of the well. From the classical viewpoint, at a given energy $E = p^2/2m$, purely kinetic, we can have only two possible momenta, $p = \pm\sqrt{2mE}$. In quantum mechanics, the possible momenta and the probabilities with which they are realized must be computed from $|\Phi(p,t)|^2$.

At fixed energy E_n of a stationary state in quantum mechanics, we have a wave function $\Psi_n(x) = C\sin(\frac{1}{2} \bigcirc nx/a)$. The function is harmonic, much like a momentum eigenstate $\exp(ipx/\hbar)$ or rather a superposition $\sin(\frac{1}{2} \bigcirc nx/a) = -\frac{1}{2}i[\exp(i\pi nx/a) - \exp(-i\pi nx/a)]$ with momenta $p_n = \pm\frac{1}{2}nh/a$ of the two signs, one positive and one negative. These two values look much like the classical values of momenta at energy $E_n = n^2h^2/(8ma^2)$.

However, the wave function is harmonic only inside the well; it vanishes outside. Unlike for a free particle not subject to any walls, the momentum has a direct relationship with kinetic energy only in the well. The wave function, in contrast to a point particle, does notice the difference, and possible momenta do not just take the values $\pm\sqrt{2mE_n}$.

> To see what momenta are realized, we compute $\Phi_n(p,t) = (C/\sqrt{2\pi})\mathbf{Sum}_x \sin(\frac{1}{2} \bigcirc nx/a)\exp(-ipx/\hbar)\Delta L(x)$, summing over all values of $0 < x < a$ in the well separated by $\Delta L(x)$. We can compute the sum by splitting $\exp(-ipx/\hbar)$ into its cos and sin contributions and using trigonometric identities. The result: $|\Psi_n(p)|^2 = 4\pi n^2(a/\hbar)[1-(-1)^n\cos(pa/\hbar)]/(n^2\pi^2 - p^2a^2/\hbar^2)^2$.

As shown by $|\Phi_n(p,t)|^2$ in Figure 4.9, the "classical" values $p_n = \pm\sqrt{2mE_n}$ are most likely, but others have a nonzero probability as well. Comparing the dependence of $\Psi_n(x,t)$ and $\Phi_n(p,t)$ on x/a and $pa/\hbar$, respectively, we see the uncertainty relation realized: For smaller a, $\Psi_n(x,t)$ narrows down and the position uncertainty decreases, while $\Phi_n(p,t)$ gets wider and increases the momentum uncertainty to respect the uncertainty principle.

Angular momentum Momentum comes with motion, linear for linear motion and angular for angular motion. Angular momentum depends on variations under rotation, just as linear momentum refers to variations under spatial shifts. Both types of momentum are crucial concepts to describe change. The larger the momentum, linear or angular, the faster an object moves along a curve, straight or circular. In quantum mechanics, the larger the linear momentum, the more the wave function oscillates along a Cartesian coordinate. The larger the angular momentum, the more the wave varies along an angle.

We can measure all components of linear momentum at the same time, even in quantum mechanics. Spatial oscillations along three orthogonal directions are independent of one another. To see how a wave function oscillates along the x-axis, we do not need to know or restrict how it varies along the y- and z-axes. Any uncertainty we might have for one oscillation does not affect uncertainties for the others. There is no uncertainty relation that would prevent us from measuring p_x and p_y precisely at the same time. The operators $\hat{p}_x = -i\hbar\Delta/\Delta x$ and $\hat{p}_y = -i\hbar\Delta/\Delta y$ commute and can be applied in any order, with the same result.

The three components of angular momentum quantify variations along circular curves, which get into one another's way. How a wave function varies if we watch it swirl around the x-axis depends on the way it varies around the y-axis. The trajectories described by both rotations not only intersect at the starting point (as they do for translations when we determine oscillations relevant for linear momentum). When the two rotating trajectories intersect the second time, at the latest, the variations along two different spherical angles influence each other. Angular-momentum operators do not commute, and measurements of the components are subject to uncertainty relations.

One can see the noncommuting behavior already for rotations by small angles. If we take a coordinate system, rotate it around the x-axis by some angle and then around the new y-axis by some other angle, the result differs from the opposite ordering by a rotation around the final z-axis. Imagining this behavior requires some mental acrobatics and spatial thinking. One can confirm it if one constructs a model coordinate system, for instance, with three pens. The two different orderings of rotations around two orthogonal axes differ by a rotation around the third axis, by an amount the sum of the two previous angles, assumed small.

When we read off angular oscillations from a wave function to determine components of angular momentum, first measuring L_x and then L_y differs from first measuring L_y and then L_x by the commutator $[\hat{L}_x, \hat{L}_y] = i\hbar\hat{L}_z$, the factor of $i\hbar$ coming from the usual term relating momenta to spatial variations. There are anal-

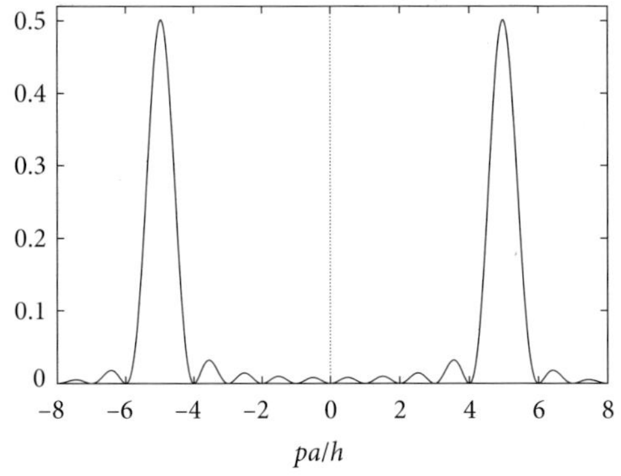

Figure 4.9 The likelihoods of measuring different momenta in a state with wave function $\Phi_{10}(p,0)$ in an infinite square well. The likelihood peaks at the classically expected values $p_n = \pm\frac{1}{2}na/h = \pm\sqrt{2mE_n}$.

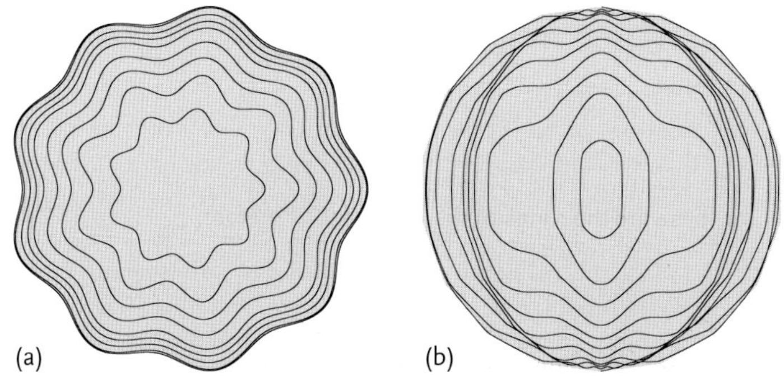

Figure 4.10 Cross-sections of a spherical wave function with sharp value of L_z. (a) Cross-section transversal to the z-axis all show the same number of oscillations in one circumference, and a sharp value for $-i\hbar\Delta/\Delta(2\pi \mathbf{Lo}/\bigcirc)$ can be extracted. (b) For cross-sections transversal to the x-axis, the number of oscillations depends on the x-value of the cross-section. No sharp value for L_x can be found.

ogous relations for the commutators of any other pair of the three components. Knowing the commutator, we obtain an uncertainty relation that limits the product $\mathbf{Var}_{L_x}\mathbf{Var}_{L_y} \geq \frac{1}{2}\hbar\overline{L}_z$ by the average value of L_z. Also this feature can be illustrated with variations of a wave function along different spherical directions: If L_z is sharp, an eigenstate with **Lo**-dependence $\exp(2\pi i\ell\mathbf{Lo}/\bigcirc)$ shows clear oscillations along the parallels, and none along meridians. The other components L_x and L_y are sensitive to variations seen when we rotate along a nonpolar axis, cross-sections of the full wave obtained by intersecting the sphere with planes at constant x or y. For planes through the sphere's center, we obtain great circles, or meridians along which only **La** varies but the wave function is constant. Meridians would indicate that L_x and L_y vanish. Along the other intersection circles, however, both **La** and **Lo** vary, and oscillations of the wave are seen: L_x and L_y cannot be sharp, but show variance; see Figure 4.10.

Angular-momentum algebra The commutator turn outs to be a powerful way to compute properties of angular momentum, such as eigenstates and eigenvalues. If we define $\hat{L}_\pm = \hat{L}_x \pm i\hat{L}_y$, the algebraic relationship $[\hat{L}_z, \hat{L}_\pm] = \pm\hbar\hat{L}_\pm$ that follows tells us that applying $\hat{L}_+$ raises the value of L_z, while $\hat{L}_-$ lowers it: For an eigenstate Ψ_ℓ of $\hat{L}_z$ with some eigenvalue $\hbar\ell$, such that $\hat{L}_z\Psi_\ell = \ell\hbar\Psi_\ell$, the state $\hat{L}_\pm\Psi_\ell$ is an eigenstate with eigenvalue $\hbar(\ell \pm 1)$. We see this by applying $\hat{L}_z$ to the new states and using the commutator: $\hat{L}_z\hat{L}_\pm\Psi_\ell = (\hat{L}_\pm\hat{L}_z + [\hat{L}_z, \hat{L}_\pm])\Psi_\ell = \hat{L}_\pm(\ell\hbar\Psi_\ell) \pm \hbar\hat{L}_\pm\Psi_\ell = \hbar(\ell \pm 1)\Psi_\ell$.
Starting with one eigenstate, we produce a whole ladder of eigenstates with eigenvalues shifted by integer multiples of $\hbar$. All these states have the same total angular momentum $\hat{L}^2 = \hat{L}_x^2 + \hat{L}_y^2 + \hat{L}_z^2$ because $[\hat{L}_\pm, \hat{L}^2] = 0$. Moreover, since $L_z^2 \leq L^2$, the ladder of L_z-eigenvalues cannot extend to infinite rungs but must be

bounded, as it turns out in such a way that $|\ell| \leq j$ if $\hbar^2 j(j+1)$ is the eigenvalue of $\hat{L}^2$.
The $2j+1$ different ℓ-values allowed for a given j give us the states we already used when we filled the shells of atoms in the periodic table. We also obtain restrictions for the possible values of j. We must be able to go from $-j$ to j in integer steps, for these are the rungs of the L_z-ladder. Therefore, $j-(-j)=2j$ must be an integer, and j can only take the values 0, 1/2, 1, 3/2 and so on. For $j=1/2$, we have the two possible spin states of $\ell=\pm 1/2$ realized for an electron. For integer j, we have the states possible for the angular momentum of an orbiting electron. Orbital motion can only come with integer j and ℓ, because the wave function then depends on **Lo** by $\exp(2\pi i\ell \mathbf{Lo}/\bigcirc)$, which gives us unique values for all points in the sphere only if ℓ is integer so that **Lo** and **Lo** $+ \bigcirc$ give the same value.
The fact that there is a finite number of possible states if we fix j provides useful examples of wave functions depending only on a few parameters. The operators $\hat{L}_z$, which acts by multiplying its eigenstates with eigenvalues $\ell\hbar$, and $\hat{L}_x$ and $\hat{L}_y$, whose action can be found using the ladder operators $\hat{L}_\pm$ then take the form of matrices.

Spin 1/2 The simplest spin turns either left or right. Just these two options are realized if $j=1/2$; there is only black or white and no shades of gray. The spin of an electron realizes these options. One can do experiments to see how electrons are deflected in a magnetic field. If we measure one component of the spin, the state collapses on an eigenstate according to the result obtained. We can do repeated measurements for different components of spin to test how successive collapses behave, and whether the probabilities and uncertainties coincide with what quantum mechanics claims.

Calculations are needed for predictions. If we choose one angular-momentum component, say L_z, to define eigenstates (since no two noncommuting components can have the same eigenstates), we are dealing with two states $\Psi_\pm$, and $\hat{L}_z\Psi_\pm = \pm\frac{1}{2}\hbar\Psi_\pm$.

Spin matrices We write these two equations as one matrix equation:

$$\hat{L}_z = \begin{pmatrix} \frac{1}{2}\hbar & 0 \\ 0 & -\frac{1}{2}\hbar \end{pmatrix} . \tag{4.22}$$

The columns and rows are arranged such that $\hat{L}_z\Psi_\pm = (L_z)_{+\pm}\Psi_+ + (L_z)_{-\pm}\Psi_-$ determines the components with the $+$-values always above or to the left of $-$-values.
For the other components, we obtain

$$\hat{L}_x = \begin{pmatrix} 0 & \frac{1}{2}\hbar \\ \frac{1}{2}\hbar & 0 \end{pmatrix} \quad \text{and} \quad \hat{L}_x = \begin{pmatrix} 0 & -\frac{1}{2}i\hbar \\ \frac{1}{2}i\hbar & 0 \end{pmatrix} \tag{4.23}$$

starting with the ladder operators $\hat{L}_{\pm}$ with $\hat{L}_{\pm}\Psi_{\pm} = 0$, and adding and subtracting them for $\hat{L}_x = \frac{1}{2}(\hat{L}_+ + \hat{L}_-)$ and $\hat{L}_y = \frac{1}{2}i(\hat{L}_- - \hat{L}_+)$. Multiplying these matrices as commutators indeed shows the typical relations for angular momentum.

In state Ψ_+, L_z always has the value $\frac{1}{2}\hbar$, but it is not an eigenstate of the other two components; their values are not sharp. The average of L_x equals the ++-component of $\hat{L}_x$, and the average of L_y the ++-component of $\hat{L}_y$. They both vanish: since a single measurement of L_x and L_y yields either $\frac{1}{2}\hbar$ or $-\frac{1}{2}\hbar$, just like L_z, these two values must be realized with equal probabilities for them to cancel each other on average: $\hat{L}_x$ has eigenstates $(\Psi_+ \pm \Psi_-)/\sqrt{2}$ with eigenvalues $\pm\frac{1}{2}\hbar$, written in terms of the eigenstates of $\hat{L}_z$. The coefficients in this expansion are $1/\sqrt{2}$, so that the probabilities, their norms squared, are both 1/2: it is equally likely to obtain either eigenvalue.

We compute the variances by first squaring the matrices, that is, multiplying each of them with itself. The variance squared is $\mathbf{Var}_L^2 = \overline{\hat{L}^2} - \overline{L}^2$, for each of the components. In Ψ_+, we read off $\overline{\hat{L}^2}$ as the ++-component of the squared matrix. For L_z, we obtain $\mathbf{Var}_{L_z} = 0$, as always for an eigenstate. For L_x and L_y, $\mathbf{Var}_{L_x} = \frac{1}{2}\hbar = \mathbf{Var}_{L_y}$. The product $\mathbf{Var}_{L_x}\mathbf{Var}_{L_y} = \frac{1}{4}\hbar^2 = \frac{1}{2}\hbar\overline{L}_z$ shows that the uncertainty relation is saturated; the product of variances takes the minimally allowed value.

Spin measurements Successive measurements of different spin components show the sculpting of the wave function and its perfect malleability. If we measure L_z first and obtain a positive value, the wave function collapses to Ψ_+, no matter what the initial state was. When we measure L_x in a second step, we have equal probabilities for its two values $\pm\frac{1}{2}\hbar$. If the result is again positive, the state becomes $(\Psi_+ + \Psi_-)/\sqrt{2}$. The value of L_x is now sharp; it must because, as assumed, we just measured L_x and found the positive value. The L_z-value, however, has become uncertain even though it was sharp just before the L_x-measurement. After the measurement, the state is a superposition of Ψ_+ and Ψ_- with equal coefficients. In a third measurement of L_z, we obtain the two possible values $\pm\frac{1}{2}\hbar$ with equal likelihoods. The L_x-measurement has restored equality among the two L_z-choices, irrespective of the result of the first measurement.

An actual measurement would make use of magnetic fields aligned along different axes in disjoint regions. We can first split an electron beam into two parts, one with spin up along the z-axis and one with spin down. This part of the experiment is a version of the historical Stern–Gerlach experiment if we shoot the beams on a screen to see where they land. At the screen, we measure the spin provided the beams have been separated far enough (for large enough magnetic field), and the state collapses. We see two separated blots for a whole beam, but if we were to send through electron after electron, each one would alight as a single dot on the screen, its wave function collapsing, either on the part associated with spin up or on the part for spin down.

The condition that the magnetic field be strong ensures that the state collapses. If the two split beams overlap on the screen, we cannot tell with certainty whether a single electron dot belongs to spin up or spin down. We would not measure the spin at all, and the spin state does not collapse. (What does collapse is the position state $\Psi(x)$: we measure the position when an electron alights at the screen.) If we direct an electron beam at a screen, the spin states do not collapse even though we are performing some version of the Stern–Gerlach experiment: there is always the magnetic field of Earth. This field, however, is so weak that, under normal circumstances, it does not cause a sizable splitting of the beams within their widths. The beams, after all, are at least as wide as the position uncertainty in directions transversal to the beam axis allows.

If we do separate the beams, another way to make the spin state collapse, without a screen, is to use just one of the beams for further experiments. We could send the spin-up beam through another region of strong magnetic field, aligned with the x-axis. While the z-component of the spin is sharp, there is an equal mixture of the two L_x-values. The beam is split into subbeams of equal intensities, now collapsed into L_x-eigenstates if we measure which beam a single electron is deflected into.

Many experiments can be devised in this way to test the rules of quantum mechanics. One usually uses photons instead of electrons and light rays instead of electron beams. Photons have sharp values of angular momentum when light is circularly polarized: the direction of the electric and magnetic field vectors of the electromagnetic wave rotates as the photon moves along the ray. Photon spin can be created and measured with polaroid filters. Depending on what spin components are represented in the ray and how the filter is aligned, only some part of the ray will be let through. The brightness behind the filter is a measure for the spin contributions. Photons do not have spin $s = 1/2$, but spin $s = 1$, described by new spin matrices.

Photon spin Photons are the most well-known particles with spin one. However, they are also massless and for this reason, do not have the full number of states expected. For a system with spin $s = 1$, we should have $2s + 1 = 3$ different states, with L_z-values zero or $\pm\hbar$. For a massless particle, the spin must be aligned with the direction of motion, either in the same or opposite direction. Only the two values $\pm\hbar$ can then be realized. (For electromagnetic waves, there are only two circular polarizations, and any other form such as linear polarization can be considered as a combination of the two. There is no third way corresponding to zero spin component along the ray axis.)

The absence of vanishing spin projection on the direction in which the photon moves is meaningful only for a massless particle, always moving at the speed of light. For any massive particle, we can find a reference frame in which the particle is at rest. How would we then determine the projection of the spin on the direction of motion? A photon is always in motion, and the projection can always be computed, in any reference frame. It is an invariant under Lorentz transformations because spin and velocity point in the same (or opposite) direction in any frame if they do so in one frame.

Classical laws show why the projection of photon spin on the direction of motion is zero. One of Maxwell's equations requires the combination $\Delta_x E^x/\Delta x + \Delta_y E^y/\Delta y + \Delta_z E^z/\Delta z = 0$ of derivatives of electric-field components to vanish, an equation called Gauss' law. If we assume plane-wave form for the components of the field in a quantized photon, $E^j = E_0^j \exp[i(k_x x + k_y y + k_z z)]$ with components k_j of the wave number, we derive that $k_x E^x + k_y E^y + k_z E^z = 0$. The wave vector points in the direction of motion, and the electric field determines the polarization (and spin). Since the projection $\vec{k} \cdot \vec{E}$ vanishes, the spin can only take the two aligned options.

Spin one A photon with spin one realizes only some of the $2s+1 = 3$ states. There are massive particles with spin one, for instance, the W and Z bosons associated with the weak interactions, and some mesons made from quarks. Those particles can realize all three states of the L_z-component. Also excited states of atoms can have orbital angular momentum $\ell = 1$ and realize all three L_z-states, as seen when shells are filled in the periodic table. It is therefore of interest to look at the three-dimensional space of states for spin one, and the corresponding spin matrices.

Spin matrices For spin one, spin operators are 3×3-matrices. If L_z is again the one chosen diagonal, we write its matrix

$$L_z = \begin{pmatrix} \hbar & 0 & 0 \\ 0 & 0 & 0 \\ 0 & 0 & \hbar \end{pmatrix} . \tag{4.24}$$

For the ladder operators and the other two components of $\vec{L}$, we have to work harder. They turn out to be

$$L_x = \begin{pmatrix} 0 & \hbar/\sqrt{2} & 0 \\ \hbar/\sqrt{2} & 0 & \hbar/\sqrt{2} \\ 0 & \hbar/\sqrt{2} & 0 \end{pmatrix} \tag{4.25}$$

$$L_y = \begin{pmatrix} 0 & -i\hbar/\sqrt{2} & 0 \\ i\hbar/\sqrt{2} & 0 & -i\hbar/\sqrt{2} \\ 0 & i\hbar/\sqrt{2} & 0 \end{pmatrix} . \tag{4.26}$$

Indeed, the correct angular-momentum algebra can be confirmed for these matrices.

As with $s = 1/2$, it is instructive to compute averages and variances in the three states of sharp L_z. In all of them, $\mathbf{Var}_{L_z}$ vanishes, as do $\overline{L}_x$ and $\overline{L}_y$: for sharp L_z, both L_x and L_y are distributed uniformly. Other variances do not vanish; we have $\mathbf{Var}_{L_x} = \hbar/\sqrt{2} = \mathbf{Var}_{L_y}$ in the two states with $L_z = \pm\hbar$, and $\mathbf{Var}_{L_x} = \hbar = \mathbf{Var}_{L_y}$ in the state with $L_z = 0$. The uncertainty relation allows variances according to $\mathbf{Var}_{L_x}\mathbf{Var}_{L_y} \geq \frac{1}{2}\hbar\overline{L}_z$, which is saturated for the states with $L_z = \pm\hbar$. For the state with $L_z = 0$, we would be allowed to have vanishing variances for L_x and L_y, but the uncertainty relation does not take the minimal value; L_x and L_y remain uncertain.

To compare the L_z-averages with L_x- and L_y-variances, we may draw the states on a sphere. All three components of classical angular momentum are sharp, and for fixed total angular momentum, they determine a point on the sphere $L_x^2 + L_y^2 + L_z^2 = L^2$. In quantum mechanics, not all spin components can be sharp at the same time. If one is sharp, the others are as unsharp as possible: the two different signs for their values are distributed uniformly. We cannot associate a single point or a sharp vector with angular momentum, but at best a circle spread out by the variances, determined as the intersection of the plane of constant $\overline{L}_z$ with the sphere $L^2 = s(s+1)\hbar^2$ of given total spin. The size of the circle depends on the variances of L_x- and L_y-components, which, as measures for the components together with the sharp L_z, lie on the sphere of radius $\sqrt{2}\hbar = \sqrt{s(s+1)}\hbar$: $\mathbf{Var}_{L_x}^2 + \mathbf{Var}_{L_y}^2 + \overline{L}_z^2 = 2\hbar^2$.

Higher spin With quantum uncertainty, spin spins. Not all the spin components can be measured at the same time as sharp values. Only one, say L_z, can be sharp while the remaining two are uncertain with uniformly distributed signs for the spin values. In an eigenstate of $\hat{L}_z$, $\overline{L}_x = 0 = \overline{L}_y$. The unsharp spin content in the remaining two components can be noticed only by nonzero variances $\mathbf{Var}_{L_x}$ and $\mathbf{Var}_{L_y}$ which take values such that the uncertain, washed-out spin vector (L_x, L_y, L_z) lies on a sphere of radius $\hbar\sqrt{s(s+1)}$: The components satisfy the equation $\mathbf{Var}_{L_x}^2 + \mathbf{Var}_{L_y}^2 + \overline{L}_z^2 = s(s+1)\hbar^2$. One cannot think of spin as a vector in a sharp direction, but as something spinning around the z-axis so fast that its precise direction is undetermined. We can only see that the endpoint of the vector lies on a circle, the intersection of the plane of constant $\overline{L}_z$ with the sphere of radius $\hbar\sqrt{s(s+1)}$. Spin is not a vector, but a cone.

For spin $s = 1/2$ and $s = 1$, we have worked with explicit matrices for spin operators, showing the spinning properties of spin. If we want to show the same features for large spin s, it would be tedious to find suitable $(2s+1) \times (2s+1)$-matrices L_x and L_y. (The sharp component L_z is always straightforward because we can choose it diagonal with components $\ell\hbar$, $\ell = -s, -s+1, \ldots, s$.) For general s, we must take recourse to abstract arguments, using what we know about algebraic relationships of spin components irrespective of how we represent them as matrices.

Angular-momentum algebra A basic property of spin components in quantum mechanics is their commutator relations

$$[\hat{L}_x, \hat{L}_y] = i\hbar\hat{L}_z\,, \quad [\hat{L}_y, \hat{L}_z] = i\hbar\hat{L}_x\,, \quad [\hat{L}_z, \hat{L}_x] = i\hbar\hat{L}_y\,, \tag{4.27}$$

which allow us to derive conditions for eigenvalues, as already used. We can also calculate some general properties of averages. For instance, if we use an eigenstate Ψ of $\hat{L}_z$, $\hat{L}_z\Psi = \ell\hbar\Psi$, two of the commutators help us compute the averages $\overline{L}_x$ and $\overline{L}_y$. In the average of the commutator in $\overline{L}_x = -i\overline{[\hat{L}_y, \hat{L}_z]}/\hbar$, we write $\overline{[\hat{L}_y, \hat{L}_z]} = \mathbf{Sum}_x\,\Psi(x)^*(\hat{L}_y\hat{L}_z - \hat{L}_z\hat{L}_y)\Psi(x)\Delta L(x)$. In each term in the sum, the operator $\hat{L}_z$ acting on the eigenstate produces just a number, $\ell\hbar$. We are left with the same terms $\Psi(x)^*\hat{L}_y\Psi(x)$ in the difference, which cancel. Therefore, $\overline{L}_x = 0$ and, by analogous reasons, $\hat{L}_y = 0$, no matter what the spin value is.

Viewing the variances as values of L_x and L_y, the spin components define a point on a sphere with radius $\hbar\sqrt{s(s+1)}$. Now, we use the relation $\hat{L}_x^2 + \hat{L}_y^2 + \hat{L}_z^2 = \hat{L}^2$ that defines the total angular momentum $\hat{L}$. In an eigenstate Ψ of $\hat{L}_z$, the total spin has eigenvalue $\hat{L}^2\Psi = \hbar^2 s(s+1)\Psi$. The average of an operator in one of its eigenstates is equal to the eigenvalue. Moreover, $\overline{L_z^2} = (\overline{L}_z)^2$, that is, the variance of an operator in one of its eigenstates vanishes. From the operator identity, we derive $\overline{L_x^2} + \overline{L_y^2} + \overline{L_z^2} = \hbar^2 s(s+1)$. The averages of L_x^2 and L_y^2 are not squares of the averages of L_x and L_y because there are nonvanishing variances. However, the averages $\overline{L}_x$ and $\overline{L}_y$ vanish, and thus we have $\overline{L_x^2} = \mathbf{Var}_{L_x}^2 + \overline{L}_x^2 = \mathbf{Var}_{L_x}^2$ and $\overline{L_y^2} = \mathbf{Var}_{L_y}^2$. The sphere identity $\mathbf{Var}_{L_x}^2 + \mathbf{Var}_{L_y}^2 + \overline{L_z^2} = s(s+1)\hbar^2$ follows for all spins.

5
The Universe III

Space and matter, in the present universe, are vast and, on average, cold. The overall matter distribution affects the expansion behavior of the universe, and local variations of matter concentrated in galaxies, stars or planets propel the actions of the contributing masses. Individual actors tell the fate of local plays, while all plays together show the universal story line of the theater of the cosmos. In many cases, the movements can be described and predicted with excellent precision by general relativity, sometimes even with the simpler Newtonian law of gravity. However, on occasion, the characters become so heated that a good dose of quantum weirdness can be seen.

The universe is not uniform in either space or time. It develops and evolves, growing out of a dense and hot phase fourteen billion years ago, driven by an immense, explosion-like cosmic event, the big bang. Subsequent expansion makes matter in the universe cool down, on average at least; but even in an older, colder, wiser universe, gravity tends to induce frequent relapses into habits of the heated past. A region that, by chance, came out of the big bang endowed with just a bit more density than its surroundings will, given enough time, become more concentrated by gravitational attraction, and seize matter from its neighborhood. Even the receding motion in expanding space can be overcome if enough mass is concentrated in small regions. Space keeps expanding, but in some part of it, matter falls toward a common center, bringing back densities and temperatures last seen closer to the big bang. Nuclear reactions, quenched by expansive cooling, reignite, making a newborn star shine its first light.

Strong variations of matter in space and time, hot episodes and regions, succeeded and surrounded by much cooler ones, show another aspect of the similarities in spatial and temporal profiles. If we imagine a thread through the universe, crossing eons in time but covering only minuscule distances in space, we see lined up hot spots within cold regions. The rare, but intense, hot ones are the big bang (or perhaps multiple such events) and the occasional star we happen to impale with our imaginary thread during its lifetime. In between, we witness vast, extended coldness. If we turn our thread around in four-dimensional space-time, so that it spans through space and crosses over a humble period in time, we see a picture not so different, threading our way through hot stars separated by wide and cold and almost empty space. The stars are a space-time flip of the big bang: extended

The Universe: A View from Classical and Quantum Gravity, First Edition. Martin Bojowald.

over some stretch of time but spatially confined, compared to the big bang which happened like a flash in all of space at once. Stars and the big bang share another feature that is important for us on Earth. They produce the rich collection of elements we enjoy and make use of in our own bodies and technologies.

5.1 Stars

Stars form when matter in a bounded region in the universe gets dense by gravitational attraction. Atoms and their nuclei come so close to one another that they can merge and enter reactions to form new and different elements. Not all types of nuclei require the same energy for them to form as bound states. Like electrons in an atom, filling their individual niches of energy shells still open after exclusion from the spots already covered, the protons and neutrons that make a nucleus have different states to occupy. Depending on how many protons and neutrons come together, the number of protons determining the element and the number of neutrons its isotope, some amount of energy is released when the nucleus is formed. Or, if a nucleus is to be disintegrated, or split as in a reaction of nuclear fission, the numbers of protons and neutrons in the initial and final products affect the required energy.

For a star, a seed investment of energy is paid by gravity, starting up the compression of matter assets. If matter constituents approach one another or fall toward a common center, their velocities increase, driven by the loss of gravitational potential energy. Individual atoms collide hard and often; their electrons are torn off them until their nuclei are stripped bare. Upon further compression, the nuclei collide. However, the protons and neutrons in a nucleus are much closer to one another than the electrons are to the nucleus. (A picture often used for nuclear reactions is the drop model, depicting the nucleus as a fluid of particles held together by their combined adhesive forces.) Nuclear particles are not stripped off one by one when nuclei collide at high energy.

Fusion When nuclei collide, their whole bodies partake in a heavy reaction, a merger and exchange of nuclear fluids. As a result, a nucleus of a new type, a new element is born. If the formation of the new element releases more energy than dismembering the initial nuclei into separate protons and neutrons would require, there is an overall energy gain. The star heats up even more than the gravitational energy increase would imply. Strong internal pressure builds up in the core of collapsing matter, which counters further gravitational compression. The new-formed star is stable for as long a time as it can sustain the reactions providing it with energy and pressure.

The first stars, formed after the big bang, only contained the light elements, that is, hydrogen, some helium, and trace amounts of lithium and beryllium. Helium has one of the most stable nuclei. Just as its electrons fill a complete set of states to form a whole shell, making it loath to chemical reactions in its noble-gas aloofness, the two protons and two neutrons it contains in its most common isotope combine to a system so well-adjusted to the nuclear forces that they are bound tight. Much energy is released when a helium nucleus forms from its constituents, the main source of star light.

When two protons and two neutrons form a helium nucleus, energy is released. However, we start with hydrogen: many free protons and just a few extra neutrons present in the stable heavy hydrogen isotope of deuterium. Adding a proton, or a nucleus of the common variety of hydrogen, to a deuterium nucleus does produce helium, but the less stable isotope with only one neutron. To arrive at the most stable helium isotope with its equal numbers of protons and neutrons, mergers are not enough. We need additional processes, an industry that can transform some part of the proton surplus into neutrons.

These processes are made possible by weak interactions. Under normal conditions as we encounter them on Earth, they lead to decay reactions, for instance, of the neutron itself. When free neutrons are produced, as happens in nuclear reactors, they are not stable but decay into a proton and an electron, together with another neutral and light particle called the antineutrino. The masses of a proton, an electron and an antineutrino add up to less than the neutron mass, and so it is favorable for a neutron to decay. However, in stars, if the significant energy that can be released by forming helium is thrown in as bait, the opposite reaction happens. One of the many protons is transformed into a neutron, a positron, the antiparticle of an electron, soon to disappear in annihilation, and a neutrino. The neutron would decay back into a proton if left alone, but in the dense core of a star it is much sooner grabbed by a less-stable helium nucleus to find its fulfillment, forming the stable version of helium. At last, the full amount of energy is released; the first cycle of nuclear fusion is complete.

Metals With opportunity comes progress. Fusion of hydrogen to helium is the main source of energy for a young star. The helium content is increased, making it likely for helium nuclei to collide as well. They can merge, too, and form some heavier elements that did not exist just after the big bang. Different elements are produced in characteristic fractions, depending on the likelihood of nuclear encounters required and the amounts of energy released. Iron and Nickel, some of the lighter metals with 26 and 28 protons, respectively, are the final elements that can be produced by fusion with energy gain. They slowly build up in the star, a waste product of nuclear fusion.

The metal-to-hydrogen ratio can be used to date stars, and thereby uncover the history of the formation of galaxies. The oldest stars in the Milky Way, identified by their low metallicity, have metal fractions of less than a hundred-thousandth of the Sun's. Our Sun, and the whole solar system including us, is rather young, a latecomer which formed after other generations of stars, with an element mix different from pure primordial hydrogen.

Red giant Hydrogen fuel disappears from the star, trapped in the nuclei of heavier elements it helps build. The star burns off its fuel and shines less brightly as it ages.

However, hydrogen depletion is not the end of the star. It enters a second spring when heavier elements jump into the fray, merging too until the most stable nuclei of iron and nickel are reached. Fusion reactions of elements heavier than hydrogen and helium do not produce as much energy per event, and they require even higher temperatures to happen. These reactions can be sustained only deep in the core of an old star.

Compared to its previous lifetime, the hydrogen-retired star burns quickly through its last resources. Higher temperatures make it bloated; the star swells up about a hundred fold. Future inhabitants of Earth will see a waxing Sun, reaching its red-giant stage in about five billion years and growing large enough to torch Earth. The surface temperature of such an enlarged star, with surface far from the hot core, will be lower than it used to be. Light emitted outside will have its peak intensity closer to the red part of the spectrum, as shown by Planck's formula. The combination of redness and girth has earned these stars the name red giant.

White dwarf All fuel is limited; at some time, the last reserves will be gone. With almost all iron in its core, an old red giant can no longer produce energy and heat. Its pressure falls, and the volume of all matter in the star can no longer be sustained. The star collapses on itself, beginning with its center.

Matter in a collapsing star gets denser and denser. In ordinary matter, there is much room between the elementary particles it is built from. Matter, to a large degree, is empty space: the space between nuclei and electrons in single atoms, combined with interatomic space in material structures. This space is much reduced in a collapsing star, but before all nuclei would collide, there are forces that can lead to renewed balance with the gravitational drive to compression.

Most of the atomic and interatomic space is the realm of electrons. These particles are tiny, but also light and fleet. They, or rather their wave functions, occupy a large part of the space in and between atoms. During collapse, the wave functions are confined to smaller and smaller regions. There is a limit to how close they can come to one another: Pauli's exclusion principle keeps the electrons apart not by any direct force, but by preventing two electrons from occupying the same state. This prohibitive force is strong enough to stabilize a collapsing star, counteracting its gravitational pressure. There is no fusion anymore and no heat is produced. The star is cold and much smaller than it used to be. However, it can exist for eons in this form of a white dwarf, with sizes not much larger than Jupiter's but of much higher densities.

When a collapsing star is stabilized, beginning in its core where it is densest, matter still falling in will be pushed back. A stellar shock wave forms that moves back outward and ejects much of the exterior matter ingredients. This explosion, the last bright signal from the star, is one of the different types of supernovae, important tools in astronomy. Those of type Ia are clean, distinguished by the absence of hydrogen lines in their spectra. They come from old stars that used up almost all their hydrogen before they exploded, rare events that happen just about once per century in each galaxy. In astronomy, they play an important role as standard candles, for instance, in measuring the acceleration of the universe. Supernovae also are the reason why there are elements heavier than iron and nickel, such as the precious metals, on planets formed after the first generation of stars, including Earth. During such brief but intense and desperate events, fusion reactions happen at abandon. There are no worries about the energy produced since no long-term structure needs to be maintained.

Neutron star For a heavy star, collapse continues, even beyond the dwarf stage. Pauli's exclusion principle, counteracting the gravitational pressure in white dwarfs, cannot be overcome, but the pressure subsides when electrons disappear. Instead of pushing electrons into each other, a process prohibited by Pauli's principle based on symmetry properties of wave functions, gravity can push electrons into protons. For these different kinds of particles, the exclusion principle does not hold: we can distinguish a proton from an electron, and no exclusion-inducing identity crisis occurs. What is more, weak interactions allow an electron and a proton to merge and form a neutron, ejecting a neutrino too. The neutrino interacts so weakly with other particles that it can fly through the whole star and escape. The single neutron takes much less space than an electron and a proton bound in some atomic state. When such reactions are initiated, the star collapses further, down to even smaller sizes and higher densities.

Neutrons, just like protons and electrons, are subject to the exclusion principle. When most electrons and protons have reacted, the core of the collapsing star consists of neutrons, matter so dense and strange that we hardly know how it might behave. It is comparable at best to the interior of a single nucleus, but blown up to macroscopic sizes. What we do know is that the exclusion principle for neutrons again helps the star counteract its own weight, preventing it from collapsing on itself. A neutron star is formed, much denser even than a white dwarf, with a radius of just about 10 km containing more than a solar mass. Gravity on the surface of a neutron star is so strong that the shape must be an almost perfect sphere (or ellipsoid for a rotating star). "Mountains" on a neutron star are just a few millimeters high; any higher protrusion would be flattened by the dense star's immense gravity.

Black hole Any force, even one as strong as the exclusion pressure of neutrons, can be overcome by gravity. Sometimes, a heavy star collapses, or two lighter objects orbit around each other and come closer and merge, a star-crossed pair of suns that end in complicit collapse. Or, an already collapsed neutron star could attract yet more matter from a neighbor star. The density increases, but only a limited, if large, value is allowed by neutrons excluding one another; they cannot be pushed closer.

Unlike electrons in a white dwarf, the neutrons in a neutron star (or their constituent quarks) are the only particles left and cannot react to form something even more concentrated. The different requirements of gravitational collapse and Pauli's exclusion can be reconciled only if no stable star is left. Instead, we reach a massive object of continuous collapse. There is no surface anymore because anything we could call a surface would be static and in violation of the condition of continued collapse. What we obtain is a realization of the strange, seclusive solutions we found in general relativity, a compact object with a space-time region delimited not by a surface on which one could land but by a horizon: A collapsing neutron star forms a black hole.

Neutron stars are small and dim, and cannot be seen on their own. Black holes are even darker. They may be hot and bright in their violent, collapsing interior, but the horizon does not allow any light produced to leave the black hole. Black holes can be detected only by indirect means. If they are part of a binary star system with a visible companion, orbiting properties, inferred from varying redshifts, can be used to estimate the mass of the invisible object, just as we measure the solar mass from orbiting behaviors of the planets using Kepler's laws. In close binary star systems, matter still hot and bright is often drawn from the visible star toward the invisible one, surrounding the latter in an accretion disk. The disk's inner radius gives us an idea of the size of the central object. If it is close to the Schwarzschild radius inferred from the mass estimate, the dark object is likely to be a black hole.

There are more refined techniques making use of properties of the accretion disk and its inner boundary. The accretion disk does not extend all the way to the horizon at the Schwarzschild radius, even though its matter is going to fall into the black hole. In an accretion disk, matter is orbiting around the black hole in near-stable, spiral-like orbits toward the massive center. Since this matter comes from the star orbiting around the black hole, it carries angular momentum as it falls. Angular momentum is conserved, its value maintained throughout the infall. Matter keeps orbiting around the center for a long time, approaching smaller radii. These regions form the visible accretion disk. When matter comes too close to the horizon, however, no stable orbiting is possible. The innermost stable circular orbit has a radius larger than the Schwarzschild radius. When matter reaches radii below this limit, it falls toward the horizon much faster than before. The region between the horizon and the innermost stable circular orbit is much less dense and bright than the stable part of the accretion disk. We can observe the position of the innermost stable circular orbit, and we can compute its radius with general relativity, depending on the mass and angular momentum of the black hole. Another method to estimate the mass is obtained.

Such indirect inferences often suffer from degeneracies: different physical parameters, for instance, the mass and angular momentum, determine one observational quantity. We need different, independent measurements to disentangle the contributions of both values to what we observe. For black holes, the mass/angular momentum degeneracy is not fully resolved with observations available at present, but there are plans for new effects to be used. An interesting one relies on properties of twisted light: light whose photons do not only carry spin, but also orbital angular momentum. The light rays they form have wave fronts of the shape of a circular staircase, not planar. A spinning black hole makes space-time around it follow the rotating motion, as a consequence of strong gravity influencing space-time. Like walking on a rotating surface such as a merry-go-round makes one spin and tumble, light emitted by the accretion disk in space-time dragged along by the rotating black hole acquires angular momentum. With advanced optical techniques built into modern telescopes, one could detect the angular momentum of photons and infer the rotation rate of the black hole independently of its mass. The degeneracy in observations would be resolved.

Hawking radiation A black hole is a continually collapsing object. Matter fallen through the horizon keeps falling toward the center, to reach it after finite time. General relativity cannot tell us what happens then, for the center of the black hole appears as a singularity in relativity's solutions. The theory and its equations break down and cannot be used to determine the further fate or whether the end is being reached. We must await further information about quantum theory in these dense space-time regions.

If we bring in quantum theory, the horizon is not quite as impenetrable from inside as it seems in general relativity. Classical matter falling through the horizon always has positive energy; it increases the mass and the horizon size even further. No classical form of matter can shrink the horizon to reveal at least a little bit of what's inside. Quantum matter is a different story. By the energy-time uncertainty relation, energy can, for brief amounts of time, be negative. If subzero energy fluctuations happen close to the horizon, with negative-energy quantum matter allowed enough time to fall into the black hole, the mass of the black hole and its Schwarzschild radius decrease.

One can think of this process as a particle-antiparticle pair emerging from the near vacuum around the black hole. Under normal circumstances, both particles would soon annihilate and leave nothing behind, but the horizon is not normal. Starting close to the horizon, one companion could fall through and join the collapsing matter before annihilation time. If the second partner escapes, a real and stable particle has been freed at the expense of the black hole's mass. All particles and their antiparticles emerge with equal likelihood. (Stars and black holes do not have large electric charges, which would turn away more particles or antiparticles.) Particles and antiparticles left outside from different pairs soon annihilate, forming photons just outside the horizon, which can leave the region. Black holes emit a weak form of radiation, called Hawking radiation. It does not come from inside, but rather from a thin layer around the horizon. Still, by making the mass and Schwarzschild radius shrink, Hawking radiation can break the strict confinement imposed by general relativity, but at a slow rate. For black holes of the usual stellar masses, the radiation is so weak that even the cold cosmic microwaves shining dimly on the horizon contribute more energy than Hawking radiation extracts. (Without cosmic microwaves, evaporation would still take a long time. A proton-sized black hole would evaporate in ten years. For heavier ones, the evaporation time increases by the third power of the Schwarzschild radius. A solar-mass black hole would need far longer than the current age of the universe since the big bang.) Black holes in our present epoch do not shrink, until the expansion of the universe has cooled down cosmic microwaves to even lower temperatures.

Gravitons The horizon is not only the boundary between accessible and secluded regions of space-time, it also shows an important separation, perhaps even incompatibility, between different laws of nature we use in cosmology. Nothing can escape a black hole through the horizon, not even light or the photons it is made of in quantum physics. Photons not only form light as elementary constituents, they also transmit the electromagnetic force. The gravitational force, if it allows a similar quantized picture, should be transmitted by its own kind of particles, called gravitons.

A quantum theory of the gravitational force is still being constructed, and we do not know for sure how it could be formulated in consistent terms. Some of the immense difficulties encountered are highlighted by a quantized graviton picture applied to a black hole. If nothing can escape through the horizon, also gravitons should be prohibited from traveling outward. However, we do feel the gravitational force of the black hole outside of it; we use it to detect black holes by estimating the mass contained in certain space-time regions. In any case, how do we reconcile the strong gravitational force outside of the horizon with the strict seclusion that should apply also to gravitons? The problem cannot be solved by assuming that gravitons, much like Hawking particles, are emitted from a thin layer around the horizon, for there is not enough mass in this region to generate the strong gravitational forces we see. Perhaps, this problem can be resolved only if we abandon the particle-transmission picture of quantum forces. It works well if the forces are not strong, for instance, in the case of light where spectroscopy supplies ample evidence of its validity, but it may be an approximation that fails when the force, like gravity around a black hole, gets strong.

5.2 Elements

The elements we find on Earth have been produced in stars, then released in the immense explosions called supernovae. When a new generation of stars and planets formed later, including our solar system, a long list of ingredients was available for a more diverse set of objects than hydrogen, the simplest element, could have provided. The first generation of stars cooked almost the whole supply from hydrogen, but the brief and fierce heat of the big bang did assist with preparations, fusing some of the hydrogen to helium and minor fractions of light elements. The ratio of hydrogen to helium just after the big bang, about three to one, determines the decomposition of the first generation of stars. By spectroscopy, one can deduce this ratio, and by theoretical investigations, one can compute how much helium should have been produced during the big bang. Both values agree, as one piece of evidence for the validity of the big bang model.

Big bang During the big bang, the temperature was higher than any limits we have reached in our laboratories, higher even than the temperature in the core of stars. It was higher than the temperature reached in the tiny collision spots of particles smashed together in accelerators. Like heat melts a snowflake and turns it into a uniform drop of water, the high temperatures of the big bang would have melted any element, any atom down to its elementary constituents, and perhaps beyond.

At current temperatures, atoms are stable because protons and neutrons in the nucleus and electrons in the shell are bound to one another by the forces between elementary particles. However, any force can be overcome if there is a will, or energy, and there was plenty of that in the intense flare of the big bang (or whatever earlier event might have caused it). Elements, atoms, and particles as we know them could not have existed; they formed only when the universe had cooled down to temperatures low enough to make the average energy of particles drop below what is required to smash one another. The steady cooling during sustained expansion determines the history of the big bang, with the early stages, as usual, shrouded in mystery.

Telling the story backwards, we have the last and most lucid phase of the big bang when electrons are allowed to bind stably to protons and the few light nuclei that were present. The primordial plasma of charged electrons and nuclei transformed itself into a gas of neutral atoms. The types of elements realized by neutral atoms depend on what nuclei had formed before. The neutralization process did not affect much of the history of the universe, except the small part played by humankind's intellectual endeavors.

Cosmic microwaves Lack of polarization leads to transparency. Neutralization made matter in the early universe transparent to light and electromagnetic waves. A plasma of charged particles does not allow these waves to travel far because photons are scattered or absorbed by free charged particles. Photons, after all, transmit the electromagnetic force acting on charges. By moving particles around, photons lose energy or are redirected, making it impossible to shine through the plasma. Charged particles bound in neutral atoms, on the other hand, absorb or scatter photons only when the photon frequency matches the discrete energy differences seen in quantum mechanics of atoms.

Neutral matter is much more transparent than a plasma. At the end of the big bang, neutralization rather suddenly made it possible for all the photons present in the hot dense universe to travel freely, when a temperature of about 4000 °C was reached (close to the surface temperature of the Sun). The photons and their radiation persisted through the ages, but participated in the cooling-down and redshifting. By now, we detect this radiation as cosmic microwaves, cooled down to just 2.7 °C above absolute zero, as one of the most important sources of information about the big bang.

Big-bang nucleosynthesis Nuclei other than simple protons were produced when the particle energies fell below values required to break open bound states of protons and neutrons. The temperature had to fall below what is realized in the core of the Sun, where fusion happens. When such a temperature was reached during the big bang, the phase of nucleosynthesis began. Protons and nearby neutrons caught each other's attention, attracted by strong nuclear forces. Protons became deuterons by binding one neutron, and sometimes unstable tritons by binding two.

Deuterons could fuse to form helium nuclei, and in rare cases the next few elements lithium and beryllium.

As the universe kept expanding, matter in it was diluted, making it less likely for bachelor protons and neutrons to meet. With the temperature, the energy of nuclei available for further fusion decreased. Nucleosynthesis came to an end, leaving the ratio of hydrogen and helium that later reignited (and reionized) in the cores of the first stars.

Before they formed nuclei, protons and neutrons existed as isolated particles. Today, neutrons are unstable and decay in about 15 min into protons, electrons, and antineutrinos. The reverse reaction, an electron and a proton forming a neutron (and ejecting a neutrino) is much less likely because the electron would need an energy sufficient to supply the mass difference of a neutron and a proton. During the big bang, on the other hand, many electrons could surpass this energy threshold; there was equilibrium between protons and neutrons transforming into each other. A sizable fraction of neutrons was sustained, which participated in nucleosynthesis once the temperature allowed light nuclei to be stable. When the temperature fell further to prohibit electrons from reacting with protons to form neutrons, all free neutrons decayed and disappeared from the pool; the ratios of different elements present during the big bang was set. While neutralization into atoms happened about 380 000 years after the densest moment of the big bang, the reactions of nucleosynthesis happened just minutes after this moment.

Big-bang nucleosynthesis depends rather sensitively on the expansion and dilution behavior of the early universe. Different temperature thresholds matter, depending on properties of elementary particles and nuclei. By this interplay of different laws of nature, comparing calculations with spectroscopic observations of the first stars, our understanding of the early universe can be tested even if there is no direct image of this phase, before neutralization released electromagnetic waves.

Baryogenesis Protons and neutrons are bound states of quarks, even smaller and more elementary particles. Quarks play a role similar to electrons in the hierarchy of elementary particles. By accelerator experiments, none of these particles have yet been smashed into even smaller bits, and no structure or finite radius has been seen.

By all accounts, we can view quarks and electrons as pointlike and elementary. Protons and neutrons, on the other hand, have a specific radius of about a millionth of a millionth of a millimeter. As examples of baryons, they are bound states of more elementary particles, the quarks. Hit with high energy, they break up into constituent quarks, but only briefly because quarks, unlike electrons, are subject to a strong force that does not allow them to exist as free particles. The force that binds protons and neutrons to nuclei, still quite strong, is a spillover of the force between their quarks.

When teased out of a proton and a neutron, quarks form new bound states. But at high temperature, particles have energies large enough to provide equilibrium between baryons smashed into quarks and quarks binding to form baryons. When the cosmic temperature fell below this threshold, the amounts of protons and neutrons available for the later stage of nucleosynthesis were determined. The process of forming protons and neutrons from quarks is called baryogenesis.

An unresolved question about baryogenesis is related to the fact that particles always have partner antiparticles, with which they annihilate to leave nothing but energy behind. Baryogenesis should not only produce protons and neutrons, but also antiprotons and antineutrons. Not only quarks should be assumed to be present, but also antiquarks, which then form antibaryons. So far, however, there has been no evidence of antigalaxies, that is, galaxies or other vast regions filled with antimatter instead of matter. Antigalaxies should be detectable even if they are far from ordinary galaxies. By stellar explosions and other events, galaxies eject some of their matter which, given enough time, can reach neighboring galaxies. Antimatter ejected by an antigalaxy will annihilate with matter from a neighbor galaxy, but no annihilation radiation has been seen.

One can solve this problem by positing that all antimatter produced during baryogenesis annihilated with matter. However, then we must explain why there is such a mismatch in the balance of matter and antimatter, a mismatch large enough to produce millions of galaxies from the difference. All we see now is survivors of this first war in the universe. How large must the combined amounts of matter and antimatter have been before annihilation, if the surviving matter can form all there is in the whole universe!

Some particle interactions treat matter and antimatter unequally, a possible origin of the mismatch in their amounts (but not the reason for the first war as an uprising of suppressed antimatter). However, it remains difficult to explain the vast amount of matter left over, resulting from tiny violations of matter/antimatter symmetry produced in a fraction of a second during baryogenesis.

Quark–gluon plasma Phases of the big bang before baryogenesis are even more mysterious. Early on during the equilibrium state of baryons being smashed into quarks and quarks again forming baryons, these reactions happened so rapidly that they can be seen as taking place in a sea of free quarks enmeshed with the particles that transmit the strong force between them. As analogs of photons for the electromagnetic force, the particles transmitting the strong force are called gluons (their interactions described by quantum chromodynamics, or QCD). At high density, the boundaries of neighboring baryons dissolve, giving rise to a merged phase of quarks and gluons, a quark–gluon plasma analogous to a plasma made from free charged particles not bound to atoms. Accelerator experiments have just come close to producing states of quark–gluon plasma, if only for fractions of a second. With these results, the early history of the universe can further be elucidated, by combining properties of the quark–gluon plasma with general relativity and the expansion of the universe.

Before baryons or the quark–gluon plasma formed, quarks and electrons must have come from somewhere, perhaps from even more fundamental particles yet unknown to us. Accelerator experiments have not revealed such particles, nor have they given hints to their existence. At this stage, the history of matter in the universe, writ so far, dissolves in the fog of ignorance.

Quantum field theory Experimental limitations to our knowledge play a crucial role when we try to describe high-density states in the universe, but also difficulties in our theoretical descriptions. Most experimental results in elementary particle physics agree well with theory, providing some of the most precise measurements ever made. However, puzzles remain when we extrapolate to energies far beyond those reached by experiments.

The theoretical basis relies on quantum physics applied not to a single particle and its motion, as in quantum mechanics, but to a collection of different particles allowed to interact with and react into one another. The position of a single particle is replaced with a so-called field, or a wave just like the wave function in quantum mechanics. The field itself, rather than the position, is then treated in a statistical way: There are averages, variances, and moments that tell us what field values we should expect. With these constructions, quantum field theory is even more complex than quantum mechanics. However, several approximations and visualizations allow us to investigate the processes relevant for accelerator experiments.

The energy of the fields tells us what dynamical processes can happen, just as the energy operator in quantum mechanics provides equations of motion. New quantum forces arise when the motion of average values, related to the classical values of observables, is influenced by the spreading of the wave, or by variances and other moments. In quantum mechanics, variances depend on averages of products of position values. In quantum field theory, variances depend on averages of products of field values, or products of the values taken by two different particles.

Feynman diagrams Quantum forces result from the influence of two-particle averages on single-particle averages, a formal mathematical term visualized by the picture of pair creation: Two particles can influence each other's motion if one of them splits into two, for instance, a less-energetic version of itself and a photon when the particle is charged, and the other one collides with one of the split-off particles. Two electrons repel each other if one electron emits a photon, which then hits the other electron and is absorbed, as in Figure 5.1. With appropriate values of the splitting and absorption rates, the resulting motion agrees with two electrons repelling each other by the electric force. Quantum field theory thereby provides a mathematical description for the photon-exchange picture of the electromagnetic force, or quantum electrodynamics (QED).

With a single photon exchanged, the classical behavior of the force is realized. Less often, it can happen that one electron splits off two photons, one after the other, which are then absorbed by the other electron. Or, one of the photons can form a short-lived particle-antiparticle pair with energy borrowed by the uncertainty relation, which then react with the other particles involved. An infinity of different processes can be imagined, more and more involved as new particles appear, and they all contribute to the force. Since the more complicated ones rely on pair creation and quantum physics, they provide quantum corrections to the classical force. By now, the sensitivity of accelerator experiments is large enough to be sensitive to many of these contributions, in perfect agreement.

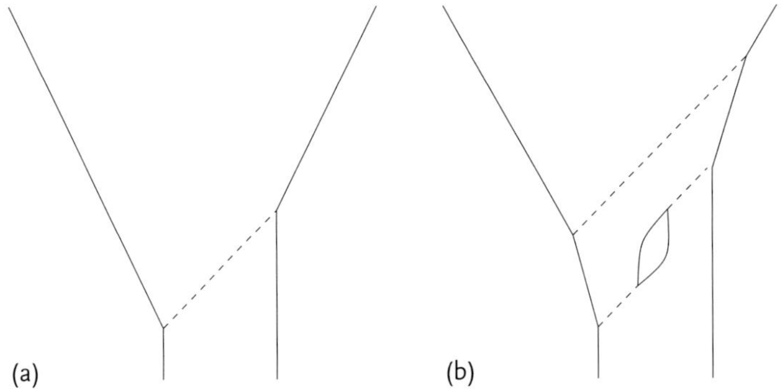

Figure 5.1 The electric force between two electrons (solid worldlines) acts in an elementary way by photon exchange (dashed worldlines). In the mathematical procedure illustrated by the Feynman diagrams, the main contribution comes from a single exchange (left), but multiple exchanges and particle-antiparticle pairs imply corrections whose presence can be measured.

With this combination of visualization and computational tools, developed mainly by Richard Feynman, particle physics has been made accessible to detailed investigations. Some puzzles remain. For instance, summing all quantum contributions to the force by complicated particle exchanges does not produce a finite number, and even most of the individual contributions do not produce direct finite results. Another procedure, called renormalization, hacking away infinities, is required to obtain finite values for the force. In mathematical terms, the approximations used in Feynman's expansion are not precise and well-defined, but they are well-developed for practical applications.[1]

A more crucial problem is encountered for strong forces. Quantum field theory works well when applied to the electromagnetic force, called quantum electrodynamics. But when we try to describe the strong force between quarks with these methods, calculations become much more complicated. With a stronger force, the contributions to interaction rates from particle exchange are much larger, even after renormalization, and we need many such terms to obtain the correct result. Present computational abilities are not sufficient for precise calculations, even with the help of supercomputers. One big question that remains open is to explain why quarks cannot exist freely but must bind to larger objects such as protons and neutrons, a property called confinement.

These computational problems are relevant when we try to describe the early universe with its quark–gluon plasma, when matter interactions were dominated by the strong force. Even gravity is much stronger at this ultra-dense state than we are used to. Under the circumstances of the big bang, we cannot rely on simple particle-splitting pictures to describe the interactions gravity mediates. A framework more complicated than ordinary quantum field theory is needed to describe

1) As Feynman advised in a different context of quantum physics, when conceptual and mathematical problems loom, it is often best to "shut up and calculate."

the complete history of the big bang, a theory called quantum gravity but available, as of now, only in incomplete terms.

5.3
Particles

Whatever the fundamental particles are, where did they come from? What formed them, and why are there particles and matter at all? Matter cannot exist without space and time to be, move and react in, but space-time can exist without matter. As general relativity shows, such empty stages may not even be boring. Empty space-time can contain black holes with their horizons, also if no matter is there to keep collapsing inside them. And gravity can produce its own energetic excitations, gravitational waves.

Current, empirically tested theories cannot provide answers to these questions; the emergence of particles and matter happens at densities too high to be accessible today. We know that there are different types of particles, reduced to much more elementary form than the diverse sets of different materials or even the elements of the periodic table would have suggested. However, what we currently describe as elementary particles are different excitations of independent fields, different potentialities of matter constituents carrying energy. These fields can be transformed into one another in particle reactions, but there are properties of particles that always remain untouched. For instance, in the nucleosynthesis reaction of a proton with an electron to form a neutron and a neutrino, no particle present initially stay till the end. However, a proton is replaced by another baryon, the neutron, and an electron by its relative the neutrino. Within these larger families of baryons and the electron-neutrino household, inheritance suffers no loss or gain.

In particle physics, we encounter new conservation laws, or conserved quantities such as the number of baryons which remain constant during all reactions. No known reaction can create a baryon if there was no baryon to begin with, and no known reaction can create an electron or a neutrino if there was no electron or neutrino to begin with. (We can, however, create a baryon and an antibaryon bound to annihilate, leaving the overall balance unaffected.) These laws are extracted from the particle intera ctions observed, all events analyzed at particle accelerator experiments obey them, but they cannot yet be decisively derived from an underlying theory.

Unification Different particle families have independent theories. There are the electromagnetic and weak forces, combined to the electroweak interactions, with their particle family of electrons and neutrinos and heavier cousins, such as muons. They are distinct from the family of baryons, bound by the strong force affecting only quarks and leaving the quark families intact. (There is also gravity which drags on all particles but does not change them.) To describe all particle interactions observed, we must assume that there are electron-neutrino families and their heavier analogs as well as the quark families; we must build them into the foundations of our theories. We cannot derive their presence from a more fundamental theory, and therefore we cannot explain how different particle families come into existence.

A powerful idea in physics is unification, which can be admired on the trajectory from separate electric and magnetic forces to the electromagnetic one, then to the combination with the weak force to electroweak interactions. There is hope that this path will continue to include the strong force and even gravity. Several candidate theories exist, but it remains unclear if one of them can be correct. As a common problem, unified theories combining the electroweak interaction with the strong force allow quarks to react and form electrons and myons, making the two quintessential particle families intermingle. However, if quarks can disappear in these reactions, the baryon number is no longer conserved and protons can decay. The decay time would be long, but in many cases not large enough to prevent all nuclei from decaying to electrons and neutrinos during the current lifetime of the universe since the big bang.

If one wants to explain how quarks, say, might have been formed from other particles, unification would be important. However, if quarks can be formed, they can also disappear, leading to the proton-decay problem. If we try to push unification to the final step, including gravity, we glimpse the fascinating potential of explaining how all matter particles could be formed out of space-time, the realm of the gravitational force. However, as with proton decay, the outlook is double-edged: If all matter particles can be formed from space-time at high energy densities, why do they not decay back into mere ripples on space-time, gravitational waves?

In a unified world, the weakest force has the longest breath because its fundamental excitations require the least energy to form. When there is an ample supply of energy, as during the big bang, more energetic states, including all elementary particles and their bound objects such as nuclei, form as the universe expands and cools down. When the temperature is too low to sustain production reactions, the bound states remain stable for some time. However, the reverse of the production reactions will happen, even if they are unlikely. As the universe keeps expanding, matter unified with gravity and space-time slowly disappears. In the end, there will be only space and time, and even they can come to an end in a final singularity. Only quantum gravity can tell us whether ultimate unification can be realized, and what the final fate of space and time is.

Particle production If the potentiality of one type of particles, or its field, exists in a theoretical formulation of quantum field theory, particles can be produced. These processes do not explain why a certain type of particles exists at all, for we must assume that there is a field, built into our theories. We cannot explain in this way why there are photons because we must assume that our theories contain a description of the electromagnetic field. However, with the interactions we know, photons will be produced in some way, given reactions of other particles or fields, for instance, the transitions of an electron between different energy levels in an atom.

In quantum field theory, the potentiality of particles is realized by the vacuum, the quantum state in which the field has the lowest possible value. Different particles or fields do not live in isolation but are coupled to each other by particle interactions, and they couple to space-time by gravity. Energy can be transferred from one field to another one, exciting the quantum state. In measurements, we notice the excitation as the presence of particles accounting for the energy difference in the two states.

By describing particles as waves, quantum physics does not view a single particle as an independent object, distinct from empty space. Rather, a particle is an excitation of a specific wave form, out of the vacuum state of the field. Particle creation by interactions is not a strict creation process from nothing; it is a transformation between two states of an existing field. Even if the initial state is empty vacuum, nothing is created. There is a transfer of energy, whose result we perceive as the presence of a new particle. A true creation process would explain how the field itself and its vacuum state can come into being, perhaps out of space-time or even out of nothingness. Such creation processes are not (yet) part of physics. Only complete unification of all fundamental physical laws could change this, but even with unification creation may never be seen.

Inflation Gravity is the ultimate fulfillment. Given the presence of a field and its vacuum state, gravity can excite it. If space-time behaves in a certain way, much like the shaking ground of an earthquake exciting tsunami waves, it can transfer energy to fill the matter vacuum with particles. In cosmology, accelerated expansion ensures productive conditions. Acceleration early on in the universe, even before the stages of the quark–gluon plasma, baryogenesis and nucleosynthesis, is an interesting assumption to explain why we have particles and matter. In the early universe, tsunami waves are productive rather than destructive: They form the seeds of all matter in the universe. The literal meaning of "tsunami," harbor wave, seems appropriate, for particle production comes from waves the big bang sent out, harboring entire galaxies in its hot dense womb.

Accelerated expansion early on in the universe, or inflation, is useful also for other questions. It is one option to solve the horizon problem of Friedmann cosmology. These multiple applications of the same idea make cosmologists somewhat more ready to accept the fact that general relativity does not easily allow conditions for accelerated expansion. Several models with special matter forms and interactions, tailored to provide the right, long-lasting amount of inflation, do exist; but they still appear rather contrived even after several decades of twisting and tuning. Most cosmologists hope that there is a better, more general reason for accelerated expansion, found perhaps in exotic particle properties at high densities or in the quantum nature of space-time itself, but no clear mechanism has yet emerged. Nevertheless, the prospect of explaining particle production out of the matter vacuum is tantalizing enough to keep working on the inflationary scenario. Most importantly, inflation leads to the right kind of tiny irregularities in the matter distributions just after the big bang, in good agreement with observations of temperature variations in the cosmic microwave background.

Klein–Gordon equation Structure forming in the early universe may be described by a single function $u(T, x)$ in space-time, satisfying a Lorentz covariant wave equation

$$-\frac{1}{c^2}\frac{\Delta_T^2 u}{\Delta T^2}+\frac{\Delta_x^2 u}{(a\Delta x)^2}+\frac{\Delta_y^2 u}{(a\Delta y)^2}+\frac{\Delta_z^2 u}{(a\Delta z)^2}-3\frac{H}{c}\frac{\Delta_T u}{c\Delta T}+\cdots=0\,.$$

The Hubble parameter $H = (\Delta a/\Delta T)/a$ contributes a term, showing how the field reacts to the expansion of space-time. Its consequence is friction that slows down oscillations of the field as a consequence of expansionary dilution. The dots refer to forces acting on the field by self-interaction, assumed to be small during inflation.

It is useful to rewrite the equation in several steps. First, we decompose the spatial dependence of u in plane waves, just as we do in quantum mechanics to bring out the momentum dependence. We write $u(T, x) = \mathbf{Sum}_k u_k(T) \exp(i\boldsymbol{k}\cdot\boldsymbol{x})\Delta V(k)$ and obtain the wave equation $\Delta_T^2 u_k/(c\Delta T)^2 + (k/a)^2 u_k + 3(H/c)\Delta u_k/(c\Delta T) + \cdots = 0$. By rewriting the equation for $v_k = a u_k$ and using (so-called conformal) time t such that $\Delta t = \Delta T/a(T)$, we eliminate the single time variation of u_k:

$$\frac{\Delta^2 v_k}{\Delta t^2}+\left(c^2k^2-\frac{\Delta^2 a}{a\Delta t^2}\right)v_k+\cdots=0\,.$$

An exponentially expanding universe, $a(T) = a_0 \exp(HT)$ as the main example of inflationary acceleration with constant $H^2 = \Lambda$, has $t = -1/(Ha)$ and $a^{-1}\Delta^2 a/\Delta t^2 = 2/t^2$. The wave equation then has solutions

$$v_k(t) = A_k \exp(-ikct)\left(1-\frac{i}{kct}\right) + B_k \exp(ikct)\left(1+\frac{i}{kct}\right) \qquad (5.1)$$

with two functions A_k and B_k depending only on k, not on t. Solving the Klein–Gordon equation shows how the field changes, but we must still know what values it had at some initial time, or what we should choose for A_k and B_k.

Inflationary cosmology combines aspects of general relativity and quantum physics. Inhomogeneous structure in the universe is assumed to evolve according to relativity, without quantum corrections even at rather high density. Initial values of inhomogeneity at the beginning of inflation, however, are determined by quantum physics. Structure and all matter is posited to begin in the vacuum state, by classical laws a state of nothingness which would always stay nothingness. The quantum vacuum, by contrast, is not nothingness; it is empty of particles, but it does not relinquish the potentiality of particles guaranteed by quantum field theory. While average particle numbers do vanish in the vacuum, the whole wave function and its variances do not. Accelerated expansion turns these values into real particles, leading to inhomogeneity and matter.

Vacuum A good choice for initial values of a field just produced is the vacuum. The vacuum or ground state is the state of lowest energy, as determined by the energy operator. The Klein–Gordon equation belongs to fields with energy

$$E = \frac{1}{2}\mathbf{Sum}_k \left[p_k^2 - \left(c^2 k^2 - \frac{1}{a}\frac{\Delta^2 a}{\Delta t^2} \right) v_k \right] \Delta V(k)$$

with the momentum $p_k = \Delta v_k / \Delta t$ of v_k. At early times, t near negative infinity, $(1/a)\Delta^2 a/\Delta t^2 = 2/t^2$ is almost zero; every wave number k contributes a term to the energy just of the form of the harmonic oscillator. We only need to choose the "mass" parameter as $m = 1$ and the "frequency" as $f = ck/(2\pi)$. We do not obtain the right units of mass, but this does not matter for mathematical solutions by analogy. We use our results for the harmonic oscillator to solve the new equations for the ground state of all modes v_k. For fixed k, with the position variance of the harmonic oscillator, we have $\mathbf{Var}_{v_k} = \sqrt{\hbar/2k}$. Moreover, the time dependence is $\exp(-2\pi i f t) = \exp(-ikct)$. In our classical solution (5.1), we should therefore choose initial values with $B_k = 0$ to eliminate contributions with $\exp(ikct)$, and $A_k = \sqrt{\hbar/2k}$.

Assuming fields to be in their vacuum before matter is produced seems a safe assumption. Quantum mechanics then tells us what initial values we should use for evolution. At this stage, general relativity's harbor wave comes into play during inflation and enlarges the tiny vacuum values to cosmic proportions. After billions of years, galaxies will follow the pattern of small irregularities resulting at random by particle production out of the initial vacuum. While individual proportions are random, statistical properties of the galaxy distribution follow a certain scheme that still refers to the vacuum numbers, and the cosmic microwave background has density and temperature variations showing the same pattern as it was about 380 000 years after inflation. With detailed CMB observations and galaxy maps existing now, the match is perfect. What is required before all these gravitational magnifications of structure by attraction started, is a nearly scale-free distribution of inhomogeneity, with modes v_k likely to occur all with the same value no matter what wave number k we look at.

Power spectrum Evolution out of the vacuum leads to an inhomogeneity function $v_k(t)$. Cosmologists express its form by the power spectrum $P(k) = \frac{1}{2}\pi^{-2}k^3|v_k/a|^2$, where v_k is evaluated at the time when it is no longer subject to inflation, a time reached when the friction term in the Klein–Gordon equation begins to dominate the wave-number term, or when $ck = aH$. With our previous solutions for A_k and $B_k = 0$, $k^3|v_k/a|^2 = \frac{1}{2}\hbar a^{-2}k^2[1 + a^2H^2/(ck)^2]$. Evaluated at $ck = aH$, the power spectrum $P(k)$ is proportional to H^2 and does not depend on k. Inflation produces the correct form of a scale-free spectrum required to match with CMB and galaxy observations.

Inflaton In inflationary cosmology, the field u_k plays a dual role, and even the conceptual underpinning of these two roles is quite dualistic. Assuming that the universe expands in an accelerated way, exponential as used in our solutions, a uniform, scale-free distribution is generated, ready to seed large structures in agreement with observations. We ignored self-interactions of u_k in our calculations, corresponding to a potential energy $V(u)$ which may be large but does not much depend on u. The self-interaction force $-\Delta V/\Delta u$ and its own dependence on u are small and do not contribute much to the wave equation. These conditions, which characterize u as an inflaton field, also help to induce accelerated expansion in the first place.

To see the influence of an inflaton field on cosmic expansion, we need to consider averages of u taken over all space at any given time. The energy provided by the average enters the Friedmann equation and determines how the scale factor a changes in time. If $V(u)$ is large but does not depend much on u, the inflaton's energy consists almost entirely of potential. The rate of change of the spatial average of u obeys an equation similar to the Klein–Gordon equation, just without any spatial changes, or with wave number $k = 0$. If u_0 starts at some value and is subject to small self-forces, its rate of change remains small and the kinetic energy can be ignored. In the Friedmann equation $[\Delta a/(a\Delta T)]^2 = 8\pi G d/(3c^2)$, the energy density $d = V(u_0)$ is purely potential. With a small rate of change of u_0, $V(u_0)$ and therefore d is close to constant. Solving the Friedmann equation under these conditions results in $a(T) = a(0)\exp[\sqrt{8\pi G V(u_0)/3}\,T/c]$, an exponential solution with constant acceleration. The inflaton not only develops scale-free inhomogeneity u_k, it also induces accelerated expansion of just the right kind by the energy contribution of u_0.

Inflation, or negative pressure, is possible in spite of gravity's usual attractive behavior. The condition of a large potential with weak self-force, however, is rather contrived. Many models of particle physics can be constructed in which a single field u with a self-interaction potential $V(u)$ appears, but the dependence on u and therefore the self-force is usually strong. For a power spectrum as scale-free as required to match with observations, one needs a long phase of inflation, the universe expanding by a factor of about exp(60), or 20 orders of magnitude, in a fraction of a second. The potential $V(u)$ for such long inflation must be so constant that its slope is one by a trillion, but still not quite zero. Moreover, although we assume the inhomogeneous modes u_k for $k \neq 0$ to start out in their vacuum state with vanishing quantum average, the inflaton mode with $k = 0$, driving inflation, must start with an immense value: The initial potential energy $V(u_0)$ is close to the Planck density, even though the potential increases so slowly from its minimum. The inflaton must start far away from the minimum of its potential, where the quantum mechanical ground or vacuum state would position it.

In spite of its great success in explaining the emergence of structure and all matter out of the vacuum, cosmologists do not show unconditional love for inflation. An unambiguous derivation of an inflaton with its flat potential from particle physics, or some other, perhaps gravity-induced mechanism, is still missing.

Matter The inflaton is too weird to be matter. It functions as the driving force of inflation, by generating negative pressure, as well as the recipient of structure and inhomogeneity out of the initial vacuum of the modes. This structure, evolved forward in time, is compared with our observations of the cosmic microwave background and the galaxy distribution. The latter refer to ordinary matter, and so a transfer of energy from the inflaton to matter must happen. There must be interactions between the inflaton and matter strong enough to facilitate the transfer, yet so weak that they do not interfere with the generation mechanism in which force terms involving u are small.

During inflation, the inflaton slowly approaches the minimum of its flat potential in an almost nondynamical, but not quite static state. Since the expansion rate of the universe features as a friction term in the Klein–Gordon equation, the minimum is not reached right away and inflation can go on for a long time. Once it does reach the minimum, the inflaton starts oscillating, with shrinking amplitude again due to friction. Oscillations are more dynamical than slow decline, and by interactions and resonances ordinary matter forms can be excited. Thus one imagines that the primordial structure is transferred from the inflaton to visible matter, but this process of reheating, as it is called, remains to be developed.

Different matter forms evolve in the expanding universe, depending on their rates of attraction compared with repulsion by internal pressure. Long after the inflaton has decayed by settling in its minimum, invisible matter again plays a role. The current cosmological scenario assumes the presence of cold dark matter, matter almost devoid of internal pressure and not subject to reactions that would allow it to form stars or other bright objects. Astronomers have found only indirect hints for such matter, first by rotation curves of galaxies, then by noticing that structure formation seems to require such matter. Cold dark matter follows the seeds of structure imprinted in the inflaton, and enlarges them by gravitational attraction. Matter we know and see by the light it generates follows these density centers and forms galaxies and stars where it aggregates. Ignition to stars happens at sufficient density, after a rather long epoch in which the universe was dark and dilute, after the last neutralization processes of the big bang.

In the dark age of the universe, there were two matter players, cold dark matter and the usual one we know, and gravity as a mediator. Under gravitational attraction, cold dark matter, with almost no internal pressure, collapses to dense centers. Visible matter is drawn to these density centers once they have formed, generates high pressure, and thereby counteracts gravitational collapse. A pattern ensues, called acoustic oscillations, in which the two-player system of collapsing cold dark matter and pressurizing matter keeps shifting energy around. A characteristic distribution of dense regions is formed, denser than the average matter distribution but not too dense thanks to the pressure, with a typical size of regions of space almost emptied by attracting matter to the denser parts. The precise distribution depends on what was there initially, but statistical properties of the distances between dense regions can be computed without much knowledge of the initial state. They can also be extracted from observations of the cosmic microwave background and the galaxy distribution, with excellent agreement.

More and more details of the distributions come in reach of satellite observations. Polarization of the microwave radiation gives hints of what intensity gravitational waves had in the early epoch. Deviations from the vacuum state either by subsequent evolution or by excitations present initially, so-called non-Gaussianity, are being measured. Different models for the inflaton and its self-interactions can be tested and compared, or perhaps ruled out if the match is no longer precise. The earliest stages of the universe, even the initial state where all matter and perhaps space and time came from, has been made accessible to observations.

Beginning Can physics derive why the universe began and why it began the way it did? The inflationary scenario seems to come close to a positive answer, but it does rely on assumptions. Even if we accept the rather special properties of the inflaton and its potential, there are additional choices involved in selecting the initial state.

Why should the inflaton begin its evolution with an initial vacuum state? The empty vacuum may be a natural choice for something emerging from the nothingness of a universe just thrown into existence. Then again, general relativity tells us that the early universe was not only young but also dense and hot. The vacuum state of lowest energy is not what we would associate with high temperatures in dense, overpopulated space.

And, what beginning are we talking about? At what time should we place it? To arrive at the beginning, we should push the state of matter and space as far back in time as we can, to values of time as small as our evolution equations allow. If we do so using general relativity for the evolution of space and quantum physics for the state of matter, we end up at the big-bang singularity. With these solutions, there is always a time earlier than any state of a given density and temperature, a time with even higher density and temperature. If we push back the initial time as far back as we can, we do not arrive at negative infinity, but rather at some finite time, about 14 billion years ago, with density and temperature infinite; we arrive at a singularity. Owing to infinities, our evolution equations break down; they lose all their mathematical meaning. They are meaningful at later times, but no such time could be initial because there would be earlier regular times between the singularity and the one we choose.

Singularity In general relativity, the universe could have existed only for a finite time before now; any attempted extrapolation further back stops at the singularity. The universe existed for a finite time, and yet it did not have a regular beginning. In the presence of a singularity, there is no good choice of what to consider as initial time, let alone the question what state, vacuum or otherwise, should have been realized. A finite time without beginning is not meaningless, but it prevents us from formulating theories that could explain the initial state, and why the universe is as it is. Observations suggest that the universe did start out in a simple state amenable to theoretical exploration. Inflation goes a long way toward an explanation of the initial state, but as long as it remains footed on the singular solutions of general relativity, its predictive power is muted by the need for additional assumptions.

In order to understand the beginning, we must find a regular description of the singularity. We must complete the theoretical foundation by using quantum theory not only for matter in the universe, but also for space and time of the universe. We must know what kind of quantum state space-time had at a possible beginning. We need a complete theory of space-time and matter, both quantum, that does not fail us by developing singularities. With such a theory, the big bang might have been a true, nonsingular beginning, or there may have been a universe with space, time and matter before the big bang, or a beginning might always remain singular, unthinkable.[2] Only a detailed combination of general relativity with quantum physics, forming a new theory of quantum gravity, can provide the answers.

Mathlas stood for ages, carrying the heavenly sphere. Many a titan walked by asking for advice, which he generously gave. One was Physikles, ordered to bring a quantum of fruits from the fields tended by Mathlas' daughters. Physikles, not allowed to enter the fields, asked Mathlas to bring some fruit while Physikles would carry the heavenly sphere for him.

Mathlas was glad to be relieved of the sphere, glad that someone else was willing to support his heavy task. He went to the garden, picked some fruits and walked back to Physikles. But he had already made up his mind that he would leave his companion standing there, doomed to keep carrying the sphere for him. At last, Mathlas would be free to follow his dreams – he had always wanted to generalize the 3-sphere to higher dimensions.

When Mathlas returned to Physikles, all his dreams were shattered, like the pieces of the sphere he saw lying around. Unable to grasp infinity, Physikles had hacked a finite chunk out of the heavens, which he now proudly carried on his shoulder. Mathlas was furious, sent Physikles away, with his fruits, put the heavens back together, and went back to his task of holding them.

2) "For the unceasingly renewed question 'what led to this change?' allows the mind no final rest, however exhausted it may get: for this reason, a first cause is as unthinkable as a beginning of time or a boundary of space." Arthur Schopenhauer: *On the Freedom of the Will.*

6
Quantum Gravity

Modern physics is founded on two fundamental theories, two titans with ambition to rule under all circumstances. Everything we observe happens in space and time; our theory of space-time, general relativity, must always remain active, even when curvature and the gravitational force are so weak that most of the intricacies of relativity can be ignored. And everything we observe is built from something, from elementary constituents whose structure physics is trying to uncover; our general theory of constituents, quantum physics, describes everything by waves, even when our perception is not fine enough to resolve the wave's extension. With general relativity, space-time has been made dynamical, no longer a mere stage but a physical object in its own right, subject to laws and change. Quantum physics makes all physical objects, elementary constituents and large compounds alike, live with wavelike fluctuations and uncertainty.

If both theories always apply, we must be able to combine them. We rarely encounter situations in which violent dynamics of space-time and the delicate dissection down to the tiniest building blocks are relevant at the same time, but examples do exist: At the big bang or in black holes, as per general relativity, space-time and what it contains are crushed into regions no larger than an elementary particle could possibly occupy. And, even if we leave out these questions, in physics we cannot tolerate two absolute and feuding rulers. There is only one sovereign, Nature, and the laws she whispers in our minds, ever so delicately, must be consistent with one another. As long as we don't know how general relativity and quantum physics can both apply at the same time, we cannot be sure that these theories are complete and correct; they might have to be extended, just as they extended the laws of classical mechanics in different ways. Indeed, the singularities of general relativity and questions clouding measurements in quantum mechanics' collapse of the wave function indicate that gaps remain in these theories.

How do we combine relativistic space-time physics with quantum wave functions? What does it mean to have fluctuations and uncertainty of time, maybe even of the direction of time? Can the wave function of space-time, the wave function of the universe, collapse on different, possibly widely separate times when a measurement is made? Our concepts of space and time are so elementary, so primeval, that uncertainty in them is hard to fathom. (Bending them by curvature was already a tough pill to swallow.) If there is a wave function that collapses on different

The Universe: A View from Classical and Quantum Gravity, First Edition. Martin Bojowald.

times, in an uncontrollable manner, the world we experience would not seem real; it would appear like a dream, a raw stream of consciousness, switching at random between different ideas and realities, jumping ahead and back in time.[1] The stories told by relativity and quantum physics, when combined to quantum gravity, might resemble a Proustian novel. However, is this apogee of writing a good model of nature?

As decades of research on quantum gravity have shown with abandon, efforts of combining general relativity with quantum physics are beset not only by conceptual problems but also, at a technical level, mathematical difficulties. Quantum physics is counter-intuitive and conceptually problematic, and it is difficult to visualize what the wave function means. However, the theory works; we can use it to derive predictions that are in excellent agreement with measurements. We are at liberty to ignore the difficulties of interpretation, and to choose just to calculate. In quantum gravity, this route is blocked. We do not even know for sure how to calculate.

Technical problems that make quantum gravity unlike any other quantum system we have explored are manifold. Difficulties result from the self-interacting nature of gravity, especially at high density or curvature where quantum gravity should be important. We have already seen the problems of singularities and of reconciling the strict capture of everything by black holes even though the gravitational force and its carriers must somehow escape. Gravity acts on anything that has energy, even on itself. In mathematical terms, the equations for space-time are nonlinear and difficult to solve: the potential R_3 in (2.36) is a complicated function of the metric.

Energy plays an important role in all of physics; in quantum physics, it guides our way to the dynamics of the wave function. The notion of energy is more difficult in gravity. The expression (2.36) compensates the matter energy. It therefore contributes negatively to an equation of energy balance, for instance, written as $-3c^2a^3H^2/(8\pi G) + E_{\text{matter}} = 0$ for the Friedmann equation of cosmology. While the matter energy E_{matter} is always positive, its gravitational counterpart is negative. Without a lower bound for the gravitational energy, there can be no ground state, no basic state such as the vacuum, or Ψ_0 for the harmonic oscillator, from which all other states are built by excitation. Without a ground state, there is no rest for quantum space-time to settle down to, not even the calm and steady Minkowski space-time of special relativity.

We refer to space-time when we do quantum physics. We construct excited states by acting with ladder operators on the ground state, and the ground state is determined as the state annihilated by a lowering operator. We have seen the construction for the harmonic oscillator; elementary particles are described by the same methods, with a set of ladder operators for each type of particle. To derive the ladder operators and the way they act, we expanded the work done on a particle as a sum of exponentials $\exp(-2\pi i f t)$ of sharp energy $E = hf$. We refer to time because

1) Think of a much less organized version of the dense and highly fluctuating city of Calcutta, which operates one-way streets that change directions at predetermined times.

energy in quantum mechanics is computed with the operator $\hat{E} = i\hbar\Delta/\Delta t$, motivated by observations of the photoelectric effect. If we try to obey general relativity too, we must allow arbitrary transformations of time t of nonlinear form, mixing it with spatial coordinates. After such a transformation, the old exponentials become unrecognizable, and we could not tell what energy we have.

The problem is even deeper. In quantum physics, we realize elementary constituents as minute excitations of states, implemented by the mathematics of ladder operators. In quantum gravity, a quantum theory of space-time, we should use ladder operators of time to derive what the elementary constituents of time are. However, if we must refer to time to construct ladder operators, our arguments become circular. Additionally, out of what state would we excite time? What could the ground state, the vacuum of quantum gravity be, a state characterized by the absence of any excitations, even of space-time?

Solving these problems in a consistent quantum theory of gravity remains a challenge, but the endeavor should be promising. With such a theory, we could hope to understand the most elementary concepts of nature, the microscopic properties of space and time, with a quantized structure that may be atomic and discrete. Space-time would appear on the same footing as matter, both made from constituents. We could perhaps shed light on the beginning and the end of time, modeled imperfectly by general relativity's singularities. The expanding universe would be a sustained interaction process, an interplay of space-time atoms with matter particles, unifying our view on matter, space and time. There is a stronger, more powerful sense of unification, as a theory that determines how gravity and matter interact, down to the minutest detail of their interaction strengths. Such a theory would be a theory of just about everything; everything would be predictable. We could calculate Newton's constant G and the speed of light c, rather than extracting them from observations before we use their values in Einstein's and Maxwell's equations. Such promises have kept research going for decades, and while it remains unclear if they can be realized, much has been learned.

Several approaches are being followed to formulate a quantum theory of gravity, foremost string theory and loop quantum gravity. We do not yet know what a successful theory will look like in the end. It could be string theory, some version of loop quantum gravity, none of the above, or all of the above. String theory and loop quantum gravity have almost complementary advantages and disadvantages; the scenarios they provide, even the mathematics they use, are so different that it is not easy to compare them. String theory's strength in unification and loop quantum gravity's success in describing some elementary properties of quantum space-time suggest that both theories may have to be combined in some way, to some kind of "stroop" theory, in spite of their differences.[2)]

2) According to the Stroop effect, a sunject's answers are delayed when questions are posed with conflicting information, for instance, when the name of a color shown is printed with ink of a different color.

6.1 Quantum Cosmology

At least at a formal level, it is straightforward to combine the equations of general relativity restricted to isotropic space with quantum physics. Instead of the position x of a particle we have the scale factor a of space-time, the mechanical energy replaced by the sum of terms that appear in the Friedmann equation. Treating a like x in quantum mechanics, introducing its average $\overline{a}$, variance $\mathbf{Var}_a$, moments, and even an operator $\hat{a}$ to act on a wave function of the universe, $\Psi(a)$, defines isotropic quantum cosmology, a simpler setting in which one can explore the intricacies of quantum gravity.

Wheeler–DeWitt equation Energy informs evolution. In general relativity, the role of energy is played by the expression (2.36), or, if space is isotropic, the terms $-3c^2a(\Delta a/\Delta T)^2/(8\pi G) + E_{\text{matter}} = 0$ of the Friedmann equation. Even if the expression is required to vanish, owing to the absence of an absolute time, the equation controls evolution.

There is one variable parameter in the spatial metric of FLRW type, the scale factor a of $\Delta s^2 = -\Delta T^2 + a(T)^2[(\Delta x)^2 + (\Delta y)^2 + (\Delta z)^2]$. When we apply quantum physics to space-time and its geometry, we do not have wave functions of positions x, y and z that appear in coordinate displacements, but rather wave functions of a. Quantum mechanics also requires us to know what the momentum is, related to the rate of change of a. In general relativity, the rate of change of the spatial metric h_{ij} is K^{ij}, appearing in the kinetic term of (2.36). As the rate of change of the metric, in which a appears squared, it takes the value $p_a = -3c^2a(\Delta a/\Delta T)/(4\pi G)$ in isotropic cosmology, with a factor so as to have the right units of momentum (if a has units of length). The energy expression in terms of the momentum reads

$$E = -\frac{2\pi G}{3c^2 a} p_a^2 + E_{\text{matter}} = 0 . \tag{6.1}$$

Guided by quantum mechanics, we introduce wave functions $\Psi(a)$, which tell us the values of the scale factor and its momentum by the action of operators $\hat{a}$ and $\hat{p}_a = -i\hbar\Delta/\Delta a$. Also, E will be an operator, using $\hat{a}$ and $\hat{p}_a$. Unlike in quantum mechanics, the exact form of the operator is not unique because it makes a difference whether we choose to act first with $\hat{p}_a^2$ or first with $1/\hat{a}$ in $\hat{E}$, or some other ordering. For now, we make the choice of $\hat{p}_a^2$ first to define a version of the energy operator. Again, unlike in quantum mechanics, the space-time expression for E always vanishes; we cannot equate it to an independent energy parameter extracted by $\hat{E} = i\hbar\Delta/\Delta T$ from wave functions. Instead, if the classical E is zero, the quantum wave function is required to be independent of T. We do not obtain a wave equation for changes in time, but rather a constraint equation, the Wheeler–DeWitt equation

$$\hat{E}\Psi(a) = \frac{2\pi G\hbar^2}{3c^2 a}\frac{\Delta^2\Psi}{\Delta a^2} + \hat{E}_{\text{matter}}\Psi(a) = 0 . \tag{6.2}$$

Scale factor Quantum physics replaces energies by distance scales, oscillation lengths of the wave function in space and time. If we divide by h, the energy becomes a frequency as per $E = hf$. In the first term of (6.2), a factor $G\hbar/c^2 = c\ell_{\text{Planck}}^2$ with the Planck length $\ell_{\text{Planck}} = \sqrt{G\hbar/c^3}$ is then left, multiplying $(1/a)\Delta^2\Psi/\Delta a^2$. We split the terms as c/a times $\ell_{\text{Planck}}^2\Delta^2\Psi/\Delta a^2$. The first factor equals one divided by the time required for light to move a distance a, amounting to a tiny frequency for a of cosmic scale. The second factor, the squared ratio of the Planck length by the oscillation length of the wave function in a, must then be huge for the product to equal a typical matter frequency $f_{\text{matter}} = E_{\text{matter}}/h$. The Wheeler–DeWitt equation requires the wave function to vary on scales much smaller than the Planck length, and the Planck length, if we insert the numerical values of the fundamental constants used, is tiny: $\ell_{\text{Planck}} = 1.6 \times 10^{-35}$ m (sixteen millionth of a quadrillionth of a quadrillionth of a meter). How can it be possible for a wave function to vary on such tiny scales when we don't know what space-time looks like at such an elementary level?

There are other problems with the initial form (6.2) of the Wheeler–DeWitt equation, or even with the assumption of a wave function $\Psi(a)$. The scale factor a does not give us an absolute distance or a size, but rather the relative scale. We cannot measure the value of a; only its changes as they appear in the Hubble parameter $H = \Delta a/(a\Delta T)$ are observable. If we are interested in a distance, we must combine a with the coordinate displacement Δx of the endpoints of a line, with invariant distance $\Delta s = a\Delta x$. We can change our coordinates by a constant rescaling, and the scale factor at any given time can be made tiny or huge. As a consequence, the oscillation length of the wave function in the Wheeler–DeWitt equation can be modified just by changing coordinates, but then there is not much physical information contained in $\Psi(a)$.

Geometry requires us to derive an alternative equation. The momentum is scale dependent just like the scale factor; it refers to the product $a\Delta a/\Delta T$ instead of the Hubble parameter. We can replace it with the Hubble parameter if we use $p_V = -c^2\Delta a/(4\pi G a\Delta T)$ instead, the momentum of $V = a^3$. The momentum is now scaling invariant, and the variable V can be interpreted as the volume of a region of size one in coordinates, $\Delta x\Delta y\Delta z = 1$. The value of V still changes if we rescale coordinates or vary the region, but we can make the energy equation insensitive to the choice if we restrict matter to be contained in the same region. The region can be large enough to encompass all we see in the universe; it does not restrict cosmology. We write $E = -6\pi G V p_V^2/c^2 + E_{\text{matter}} = 0$ and obtain a constraint equation

$$\hat{E}\Psi(V) = 6\pi\hbar c\ell_{\text{Planck}}^2 V\frac{\Delta^2\Psi}{\Delta V^2} + \hat{E}_{\text{matter}}\Psi(V) = 0 \tag{6.3}$$

for wave functions $\Psi(V)$.

We distribute the factors as c/ℓ_{Planck} times $\ell_{\text{Planck}}^3 V\Delta^2\Psi/\Delta V^2$. The first is one by the time required for light to traverse the tiny Planck length, amounting to a huge frequency. The second factor is the squared ratio of the geometric mean $\sqrt{\ell_{\text{Planck}}^3 V}$ by the oscillation length of $\Psi(V)$. The geometric mean is small, but larger than the

Planck volume ℓ^3_{Planck}. The ratio must be small to suppress the large c/ℓ_{Planck} down to the typical size of matter frequencies $E_{\text{matter}}/\hbar$. The wave function now oscillates on scales much larger than the Planck length, where our usual space-time picture should be trustworthy. Only for large matter energies in a small volume, or large energy density, will variations of $\Psi(V)$ be more violent, regimes in which quantum gravity should indeed be significant.

Planck size The Planck length is the doorkeeper of quantum gravity. Its value is determined by known physics: Newton's constant from gravity, the speed of light from electrodynamics or relativity, and Planck's constant from quantum mechanics. These theories alone do not tell us how quantum gravity should look, but the Planck length turns out to be the only parameter with units of length that can be formed from the constants of nature. Its value should play some role in the equations of quantum gravity, and indeed it appears in the constraint equation of quantum cosmology.

A wave function $\Psi(V)$ varies over distances the size of the Planck length, with $\Delta^2\Psi/\Delta V^2 \sim 1/\ell^6_{\text{Planck}}$, if the energy contained in a region of size ℓ^3_{Planck} is about as large as $E_{\text{Planck}} = c\hbar/\ell_{\text{Planck}} = \sqrt{c^5\hbar/G}$, the Planck energy. It amounts to a mass of $M_{\text{Planck}} = E_{\text{Planck}}/c^2 = \sqrt{c\hbar/G} \sim 0.02$ mg, a rather ordinary value. However, this dash of mass contained in a Planck volume amounts to an enormous density, the Planck density $M_{\text{Planck}}/\ell^3_{\text{Planck}}$ of more than one trillion solar masses compressed to the size of a proton. At such extremes, we should expect quantum gravity to show its cards.

Wave function of the universe The whole universe is a gambling game, at least in quantum cosmology. Everything there is, matter and space, is determined by a wave function, subject to statistics. The volume is specified only on average; it has variance and uncertainty. The rules of quantum mechanics apply to the wave function of the universe.

However, what does statistics mean if there is, by definition, just one universe? In quantum mechanics, we interpret the average $\overline{x}$ as the mean value of a large number of measurement results, all performed for systems in the same state. It may be delicate, but we are able to prepare a collection of atoms all in identical states, for instance using modern atom traps. However, we cannot prepare a universe in a given state, let alone two or even a large number of universes. In quantum cosmology, and by extension quantum gravity, we cannot trust all the laws we have grown accustomed to in quantum mechanics, nor can we go back to the certainty of classical physics.

A related problem refers to the role of observers. In quantum mechanics, we think of a collection of systems prepared in the same state, experimented on by us as external observers. In quantum cosmology, the system experimented on is the whole universe; by necessity it contains us observers as well. We have a wave function for the whole system, in contrast to quantum mechanics where the wave function of an observed system tells us what measurement results we should expect for a separate observer, and what statistical properties we have.

These problems have not been resolved. To go further, we avoid them by using the wave function not for statistical information it might contain, but rather as a means to compute quantum forces implied for gravity. The average $\overline{x}$ in quantum mechanics satisfies an equation of motion that amends the classical one for x by terms that depend on $\hbar$, thereby showing their quantum origin. We can use these quantum corrections to classical physics even if we do not observe the wave function. In quantum cosmology, analogous terms tell us what the gravitational force looks like at high density, and how the expansion of the universe and the structure of space-time might change.

Time Evolution needs time. Quantum cosmology only provides a constraint equation for the wave function, without a change in time because the total E, containing gravity and matter, always vanishes. The derivative $i\hbar\Delta/\Delta T$ of the Schrödinger equation is replaced by a constraining zero in the Wheeler–DeWitt equation. How can we derive quantum corrections to evolution equations if there is no evolution, no time?

In classical cosmology, the constraint equation is just a rewrite of the Friedmann equation, an evolution equation used for cosmological solutions for the expanding universe. Time appears lost after going to quantum cosmology because it is not absolute. Time is not an external parameter, but part of the system, with properties determined by space-time geometry and encoded by gravity's contribution to the total E. If we would equate the quantum $\hat{E}$ to $i\hbar\Delta/\Delta T$, we would take a step backwards to the prerelativity thinking of a time separate from physical entities.

In relativity, change is realized in relations, detecting change of one quantity by reference to the values taken by another one. Time as an external parameter was never realistic, for we do measure time by relations, by reference to the revolving Earth or the turning hands of a clock. An external parameter is just simpler than a separate physical variable. Quantum cosmology forces us to take the realistic view and identify change by relating different physical quantities to one another.

Relational time For evolution, we need at least two players. One takes the role of time, the other evolves. So far, we only have one quantity in our equation, the volume V, but we have not yet specified what our matter contribution E_{matter} should be. Matter has constituents, described in physics by fields or other dynamical parameters. Such a parameter, once specified, can evolve relative to the volume, or vice versa.

To be specific, let us assume that matter is of a form with $E_{\text{matter}} = \frac{1}{2}p^2/V$ with the momentum p of some matter quantity. In this way, we assume matter energy to be purely kinetic, of a form that occurs in physics for a scalar field as used for inflationary cosmology, albeit one without mass. Since the total energy $E = -6\pi G V p_V^2/c^2 + \frac{1}{2}p^2/V$ does not depend on the matter variable, but only its momentum p, p remains constant: there is no potential or a force that would change its value. Solving $E = 0$ for the constant p, we obtain $p = \sqrt{12\pi G} V p_V/c$

(up to a positive or negative sign that remains undetermined when taking a square root).

The momentum p, in quantum cosmology, bevomes an operator $-i\hbar\Delta/\Delta t$, where t is initially the matter parameter. By calling it t, we indicate that we let matter play the role of time, with reference to which the volume evolves. Our quantum equation for p appears as an evolution equation for the wave function $\Psi(V,t)$ of the universe, obeying $\Delta\Psi/\Delta t = \sqrt{16\pi G}V\Delta\Psi/(c\Delta V)$.

We decided to stay away from the statistics of the wave function and instead use it to derive quantum corrections to classical equations. As in quantum mechanics, we replace p by the average $\overline{p} = \sqrt{12\pi G}(\overline{V}\overline{p}_V + \mathbf{CoVar})/c$. With a classical expression quadratic in V and p_V, the quantum average is corrected by adding the covariance. There are no coupling terms, products of averages and moments, just as it was realized for the free particle and the harmonic oscillator. As in those two examples of quantum mechanics, quantum evolution of $\overline{V}$ is identical to classical evolution of $V = a^3$ as per the Friedmann equation. Solutions of the Friedmann equation with our matter choice are all singular: they reach the value $V = 0$ at some finite time, where the energy is infinite. Quantum cosmology of the form used here does not resolve the singularity problem, for at least one model remains singular.

Other forms of matter or new ingredients such as inhomogeneous geometry remove the quadratic nature of p and lead to interesting quantum corrections. New terms depending on the variance of V or the covariance could imply quantum forces to counteract gravitational collapse to a singularity. However, such forces depend on the state, and it is difficult to tell what state we might have near a singularity. If we want to resolve the classical singularity problem, we must bring forth more advanced properties of quantum space-time, referring not only to its variances but also to the structure of its constituents. Additionally, we should take a closer look at matter, for matter at high densities could behave in unexpected ways.

6.2 Unification

Guessing what matter could be like when densities are extreme is a gamble that one would not want to take when the whole cosmos is at stake. We have experienced surprise after surprise, discovery after discovery, as physicists and engineers improved the technology to accelerate and control particle beams at high energies. Ernest Rutherford, in 1911, used electron beams to probe the structure of atoms, discovering evidence for the nucleus. In the course of the twentieth century, when higher energies and momenta were achieved to probe the forces over ever smaller distances (the momentum $p = h/l$, after all, determines the wavelength l of a particle ray used to resolve small structures), measurements not only became more precise. They also, unexpectedly, showed signs that new particles were produced, unstable particles of masses so large that they had to await high-energy experiments for their resurrection. So many different types of particles were found that intricate classification schemes had to be invented to control the flood of new data, much like the collection of lines of atomic emission and absorption spectra in the nineteenth century.

No one would have guessed the existence of all these particles, let alone their properties. What do we have to expect when we imagine a cosmos collapsing to ever higher densities and temperatures, like our own universe watched in temporal reverse? With growing temperature, the average energy of all particles increases; they move with ever more rapid speeds and hit each other with more violence. Energies achieved so far in particle accelerators would soon be surpassed, leaving us ignorant about what matter there might be. Matter influences the dynamics of space-time; without knowing its properties, we could not see in detail how space-time behaves.

As more and more particles were found and measured, patterns emerged, patterns that, viewed with the imagination of physicists and described by the powerful mathematical tool of group theory, turned out to be ordered to a high degree. No longer were different particles just classified in random terms; they followed the inevitable code of mathematics. As quantum mechanics had brought order to the irregular forest of emission and absorption lines, an extended version, quantum field theory, had begun to arrange particles in more manageable form.

Not only book-keeping improved, organizing properties such as masses, charges and spins of the particles measured. Combined with the powerful idea of unification, quantum field theory was also able to predict new particles that were indeed measured when sufficient energies were reached. With some assumptions about our theories, we can avoid guessing, or rather shift the rules of the game to our advantage. With the strict control of mathematics, our imagination is held in stronger bounds when we devise our theories, compared to what we could think of just deciding on particle properties. Additionally, the more phenomena and properties of the different particles and forces at ever higher energies we incorporate in a unified quantum field theory, the less wiggle room there is.

Maxwell had unified the electric and magnetic forces, to this day the exemplar of unification. All of a sudden, the speed of light was taken from our choices; it became a number predicted by theory, its value as inevitable as the motion of a planet. With new information about particles, unification was taken one degree higher by the inclusion of the weak force, for instance, making electrons interact with neutrinos. The new, electroweak theory predicted some new particles in the correct way. There is another force, the strong force that keeps protons and neutrons bound in nuclei. Including it in a unified framework is more difficult, but proposals do exist. With more phenomena included, our choices have become more constrained.

When we come to the final force, gravity, unification implies quantum gravity; we would, after all, combine a quantum field theory with general relativity. Such a theory would predict the behavior of gravitons, the carriers of the gravitational force, and show what space-time is at high density. We are even less sure about such a theory, compared to unification with the strong force. Including space-time could be too tough a requirement, with all the puzzles we see when we just think about what quantum gravity could be. Even if quantum gravity exists as a consistent theory, it need not imply unification (while complete unification would imply quantum gravity). These questions notwithstanding, tantalizing prospects have been found by research on unification.

Field theory Fields sprout everywhere. Localized particles placed at single points or occupying small regions are idealizations of narrow versions of extended waves or fields. In quantum mechanics, particles are realized as instances of the wave function, defined on all of space. Classical physics also knows of phenomena which cannot be reduced to something happening just at a point. Foremost, there is the electromagnetic field that not only tells us what forces charged particles and currents encounter; it has its own life, with excitations traveling through space.

Fields are position-dependent phenomena with their own dynamics and interactions. A single-component field is a function $s(x, t)$ on space-time, called a scalar as distinct from vector or tensor fields. An example for the latter is the electromagnetic field, described in relativity theory by the field-strength tensor, a multicomponent matrix. A propagating scalar field is subject to a wave equation resembling the differential equations of the electric and magnetic fields, for instance, the Klein–Gordon equation

$$-\frac{1}{c^2}\frac{\Delta_t^2 s}{\Delta t^2} + \frac{\Delta_x^2 s}{\Delta x^2} - \frac{M^2 c^2}{\hbar^2} s = 0 \tag{6.4}$$

with a number M, the mass of particles corresponding to excitations of the field.

Instead of zero on the right-hand side of the equation, there could be more complicated terms, functions of s to indicate self-interactions of the field, or products with other fields to include coupling terms. Irrespective of these terms, the differential part of the equation is general; it is determined by symmetry, by the condition that the equation remain true if space and time are subjected to a Lorentz transformation. An equation for a field on space-time must be invariant under all transformations that do not affect the geometry of space-time but only change our description of it by coordinates. The form of the invariant Minkowski line element, recognizable in the combination of time and space derivatives in (6.4), does not give us any choice other than (6.4).

If we look for solutions of the Klein–Gordon equation, we may first assume that there is no spatial variation, or $\Delta s/\Delta x = 0$. The equation $\Delta^2 s/\Delta t^2 = -(M^2 c^4/\hbar^2)s$ then resembles some of our equations encountered in quantum mechanics, with similar solutions $s(x, t) = A_+ \exp(iMc^2 t/\hbar) + A_- \exp(-iMc^2 t/\hbar)$ for two constants $A_\pm$, looking like wave functions in quantum mechanics with rest energy $E = Mc^2$. However, the analogy is only formal: the scalar field $s(x, t)$ is not a wave function.

Klein–Gordon equation If we allow for spatial variation of $s(x, t)$, the Klein–Gordon equation is a partial differential equation and more difficult to solve. We can simplify the procedure by trying different forms of spatial variations, all with uniform wavelength l: $s(x, t)$ is then proportional to $\exp(ikx)$ with $k = 2\pi/l$. Inserting this position dependence lets us solve for the time dependence as before, but with a different constant:

$$s_k^{\pm}(x,t) = \exp(ikx)\exp\left(i\sqrt{k^2+\frac{M^2c^2}{\hbar^2}}ct\right).$$

Any other solution can be written as a combination of different ones with uniform wavelength: $s(x,t) = \mathbf{Sum}_k[A_+(k)s_k^+(x,t) + A_-(k)s_k^-(x,t)]\Delta L(k)$.

A scalar field has energy given by

$$E(s,p) = \mathbf{Sum}_x\left(\frac{p^2}{2M} + \frac{\hbar^2}{2M}\left(\frac{\Delta s}{\Delta x}\right)^2 + \frac{1}{2}Mc^2s^2\right)V(x) \tag{6.5}$$

with the size $V(x)$ of small patches of space in which we have field value $s(x,t)$ and field momentum $p(x,t) = \hbar\Delta s(x,t)/(c\Delta t)$. The electromagnetic field has a similar form of its energy with sums of all its components, where pc would run through the components of the electric field and $\Delta s/\Delta x$ through the components of the magnetic field. Since photons are massless, the s^2-term is absent in this case.

Quantum field theory A wave of a wave is a quantized field, a wave function for a phenomenon of waves. As quantum mechanics describes a particle at position $x(t)$ by a wave function $\Psi(x,t)$, quantum field theory describes a field $s(x,t)$ by a wave function $\Psi[s(x,t)]$, a function of a function. The increased complexity of functions of functions can be avoided by referring to fields of uniform wavelength. Any field can be written as a sum of those simpler ones, and uniformity reduces the general functional freedom to just one parameter l. We may look for wave functions for fields of fixed uniform wavelength l, called modes, and later see how they are to be combined for the whole field.

Quantum theory tells dynamics by energy. For a single mode, the scalar-field energy reduces to $E_l = \frac{1}{2}(p_l^2 + (h^2/l^2 + M^2c^2)s_l^2)/M$, with $p_l = \hbar\Delta s_l/(c\Delta t)$. Except for different constants, this energy looks just like the energy of a harmonic oscillator, one that would have frequency $f = (h/ml)\sqrt{1 + l^2M^2c^2/h^2}$ if its mass were m. We know how to quantize a harmonic oscillator: we have a ground state $\Psi_0^{(l)}$ and ladder operators $\hat{a}_{\pm}(l)$ that bring us to excited states. In quantum field theory, there are ladder operators for all modes, that is, all uniform wavelengths, and for all fields or particles. Quantum field theory is a maze, an arcade game with an infinity of ladders. However, in elementary reactions of particle physics, just a few ladders are climbed up or down, and only by a few steps.

Interactions result from product terms $s_1(x,t)s_2(x,t)$ with two fields of different types, just as a single harmonic oscillator interacts with an electric field $e(t)$ by the work term $q\overline{x}(t)e(t)$ added to its energy. The oscillator position is quantized to $\hat{x} = \hat{a}_+ + \hat{a}_-$ in terms of ladder operators; a field mode is quantized to $\hat{s}_l = \hat{a}_+(l) + \hat{a}_-(l)$. In the energy, products of several fields and their modes imply products of ladder operators, which transfer excitations by lowering one field's rung and raising another. A numerical factor of the product of fields, called the coupling constant, determines the interaction strength, or how likely it is for such a transfer of excitation to occur.

Interaction terms, products of fields in the energy, are visualized by drawing a vertex, an intersection of directed straight lines. For each lowering operator $\hat{a}_{-}^{I}(l)$ of wavelength l and particle type I, we draw a line toward the vertex; for every raising operator, a departing line. The vertex swallows the lowered excitations, and spews out those raised, as one elementary reaction. By combining different vertices, more complicated reaction can be formed, realized as successions of the elementary ones. Complicated graphs result, illustrating the detailed histories of all interaction processes (Figure 5.1).

Richard Feynman quantified the graphs which bear his name, by associating a number to each vertex and mathematical rules to their graphical combinations. Reaction rates or scattering probabilities can be computed if all the coupling constants are known, multiplying products of ladder operators or the vertices of Feynman graphs. If the coupling constants are small numbers, the dominant contribution to an interaction is obtained for just one or a few vertices; any additional vertex will contribute another small factor of the coupling constant. In quantum electrodynamics, the quantum theory of electrons and photons interacting on their ladders, the few-vertex approximation is so good that it provides some of the most precise agreements of theory and experiment.

Unification Symmetry unifies. If there are transformations that do not change observations, different-looking objects are identified as one. Space and time have been unified to space-time, and the electric and magnetic fields to the electromagnetic one. If fields are identified, they no longer interact in arbitrary ways. Their coupling terms and constants can be derived from symmetry transformations, and then put into Feynman graphs to compute reaction rates.

We may view any collection of scalar fields $s_1(x, t)$, $s_2(x, t)$ and so on as a single, multicomponent field $(s_1, s_2, \ldots)$. These are components of one field rather than just many independent fields if we know transformations between them. For instance, the electric field is a three-component field, with transformations between all components under a spatial rotation. The electromagnetic field is a six-component field, an antisymmetric matrix, with transformations by Lorentz boosts. If we try to unify even more fields as different components of a new one, we need new types of symmetry, for all space-time symmetries have already been used to unify electromagnetism.

Independently of space-time, mathematics allows the appearance of abstract symmetry transformations, mediated by a group, a set of transformations that can be multiplied, with certain rules. If there is a group action that mixes field components, such as (s_1, s_2) mapped to $S \cdot (s_1, s_2) = (S_{11}s_1 + S_{12}s_2, S_{21}s_1 + S_{22}s_2)$ by multiplication with a matrix

$$S = \begin{pmatrix} S_{11} & S_{12} \\ S_{21} & S_{22} \end{pmatrix},$$

invariance of reaction rates under the symmetry requires a specific form of coupling terms. Lorentz invariance gave us a strict relation between time variations and space variations of a field subject to the Klein–Gordon equation. Invariance under an abstract group would relate the possible products in coupling terms.

The aim of unified field theory is to find one universal symmetry relating the fields of every single particle, force, or wave we have observed. The aim is to find a unique theory that leaves no room for choices of its coupling constants or mass parameters, a pan-nomon that rules everything by its equations, much as the speed of light follows from Maxwell's unified electromagnetism. The aim is to discover the only sentence by which nature can be written, the unique verse of the universe.

Extra dimensions More space-time means more symmetry. The wave equation and the combined field-strength tensor of electromagnetism have the form they have because they are required to respect Lorentz transformations, extending rotations to space-time. Space and time, the electric and magnetic fields, energy and momentum, and all the quantities that appear together in four-vectors or four-tensors are unified, or part of a unified theory, because they can be transformed into each other if only an observer's viewpoint changes. Instead of looking for new symmetries, we may look for new space-time, new dimensions beyond those we experience.

In 1921, the mathematician Theodor Kaluza, followed by Oskar Klein, made the first attempt to unify the newly found gravitational theory of Einstein's general relativity with Maxwell's electromagnetism. Electromagnetism can be formulated in terms of four potentials instead of six field components, just as we express a force by a potential. The magnetic field does not allow the usual potential as a single function, called a scalar potential, because it can deflect a charge without changing the velocity's magnitude; a scalar potential, by contrast, works by playing velocity-dependent kinetic energy against a position-dependent term while keeping the total energy constant. In a mountainscape or potential landscape, we cannot always avoid changing our speed as we walk around.

To accomplish changes of direction at constant velocity, we need a vector potential: a vector field $\vec{A}$ from which the components of the magnetic field are derived as $B^x = \Delta_y A_z/\Delta y - \Delta_z A_y/\Delta z$, $B^y = \Delta_z A_x/\Delta z - \Delta_x A_z/\Delta x$, $B^z = \Delta_x A_y/\Delta x - \Delta_y A_x/\Delta y$. There seems to be no gain, with three components of the field expressed by three components of the potential. However, by writing $\vec{B}$ as derivatives of another vector, we produce the electric field derived from its scalar potential P in a way respectful to space-time: $E^i = \Delta_{x^i} P/\Delta x^i - \Delta_t A^i/\Delta t$. If we view P/c as the time component of a four-vector with spatial components $\vec{A}$, $\vec{E}$ and $\vec{B}$ are computed by the same pattern from the components of $\boldsymbol{P} = (P/c, \vec{A})$, taking derivatives by components of $\boldsymbol{x} = (ct, \vec{x})$.

The electromagnetic field is computed from the four-potential by derivatives, and the gravitational force follows from curvature, or derivatives of the metric tensor g_{ab}. The potential is akin to the metric, but they are separate objects in space-time, that is, in four-dimensional space-time. If we had one more dimension, we could fit all the components into one tensor, still reflection symmetric as required for the metric:

$$G = \begin{pmatrix} 0 & P/c & A_x & A_y & A_z \\ P/c & g_{tt} & g_{tx} & g_{ty} & g_{tz} \\ A_x & g_{xt} & g_{xx} & g_{xy} & g_{xz} \\ A_y & g_{yt} & g_{yx} & g_{yy} & g_{yz} \\ A_z & g_{zt} & g_{zx} & g_{zy} & g_{zz} \end{pmatrix} . \tag{6.6}$$

If we also extend our Lorentz transformations to five-dimensional space-time, they will mix the metric with the four-potential, general relativity with electromagnetism. Invariant equations must then combine gravity and the electromagnetic force in a specific way, allowing only one universal coupling constant instead of two free ones, G and c. Such a theory would allow us to derive Newton's constant from the speed of light.

A new, unseen dimension may not be problematic. It could be short, so that we never move much along it and do not notice it, or something could prevent us from moving. In five dimensions, fields and forces depend on five coordinates instead of four, but we could just happen to live in a region in which fields are almost constant along the fifth dimension. These are assumptions that seem rather contrived in order to explain away observations of a new dimension that we pulled out of a hat. However, unification, the relation of coupling constants of different forces, would be powerful enough to justify assumptions about space-time. Such a theory would also be testable even if we could not see the new dimensions by direct means: unified force laws predict new interactions of the known particles, which we can compare with measurements. For instance, in (6.6) the top left component may be nonzero, as a new field component.

If higher-dimensional theories can be consistent with the known behavior of electromagnetism and gravity, they could accomplish unification. However, the "if" is big. The original idea of Kaluza and Klein did not work out since electromagnetic forces and gravity are just too different to be unified in a simple way. The specific form of higher-dimensional theory is delicate, and new ingredients as well as even more dimensions are required. These extra structures and tunings, though not the idea itself of higher dimensions, give unification a high price of contrivance.

Extended objects A point is our best idea of something very small. When we think of elementary particles as pointlike or spherical with small radius, we implement a negative feature: the absence of recognizable structure on such tiny scales. We describe the position of an object by a single set of coordinate values because it is the most relevant aspect of motion, the latter formulated as a one-dimensional curve, that is, the worldline in space-time. When we zoom in and consider the size of the object, its shape and orientation in space, we must include additional degrees of freedom, new parameters that can vary if the position does not change. We have to know much more about the object to build a good theory; we must look closer and be able to resolve its structure.

In quantum physics, all objects are described by wave functions. Defined as functions all over space, waves are not pointlike. Nevertheless, we stay true to the point picture because we say that the value of the wave function at some point determines how likely it is to measure a particle at that point. In some cases, we may choose to ignore known substructures as an approximation, for instance, when we view the nucleus of a hydrogen atom as a pointlike proton, constituting the central source of the Coulomb potential, instead of a complicated bound state of quarks. In other cases, with quarks or electrons, muons and neutrinos, we have not yet been able to detect any substructure. Accelerator experiments have shown that they cannot have a radius larger than 10^{-18} m (about a thousandth of the measured proton radius). They may well be pointlike, or they may by extended with a size smaller than the observed upper limit. If they are extended objects, the question is what their shape may be.

While the final answer about substructures of elementary particles can only come from experiments of sufficient resolution, some theoretical indications speak a clear verdict against points. The vanishing dimensions of a point imply divergences, infinities as we can already see them when we look at the Coulomb potential at $r = 0$. Quantum field theory can resolve some of these problems. The closer we are to an infinity in the potential at the central charge, the more likely it is for brief flashes of particle-antiparticle pairs to be produced, with energy borrowed by appealing to the time-energy uncertainty principle. One of the partners, the one with charge of sign opposite to the central charge, will be attracted to the center, the other one repelled. Electric charges in quantum field theory have a tendency to shield or screen themselves by this process, called vacuum polarization, drawing neutralizing charge out of the vacuum.

Quantum field theory gives a substructure even to point particles, for we should view the whole collection of the central charge together with polarization charges as one extended object, a rather complicated-looking one to be sure. A mathematical procedure called renormalization allows us to account for the polarization charges by adapting the numerical values of coupling constants. After this step, interaction processes of charged particles have a good mathematical description with finite results.

Gravity rejects the polarizing tendency of quantum field theory. Newton's potential diverges like Coulomb's, but we already know that it is not correct when its value is large compared to c^2. General relativity takes over with a more complicated mathematical construct, the multicomponent metric instead of a single potential function. The divergence, however, remains. Some of the metric components in Schwarzschild's solution diverge at $r = 0$, and although this singularity is shielded by the horizon, calamity remains for observers trying to do physics within the black hole. Here, near the singularity, vacuum polarization should happen in a quantum field theory of gravity. Particle-antiparticle pairs should be produced, the more likely the closer we are to the singularity. Gravity's democratic nature then stands in the way of polarization: the gravitational force treats all masses and energy forms as equals; they are all attracted. There is no negative mass, and the two partners of a particle-antiparticle pair are both attracted to the center. The central mass is not shielded or neutralized; it is reinforced by vacuum polarization, making the divergence even worse than it appears in its classical incarnation. When it comes to gravity, we must be careful about the substructure of particles, or all masses, as viewed in quantum field theory.

String worldsheet A point is simple, but singular. Pluralistic objects are more complicated because they require more choices, more parameters for a complete specification; then again, the additional freedom and inclusiveness might be a boon for bringing together a large class of particle interactions. One can strike a balance of divergence and diversity by looking at one-dimensional lines, or strings instead of points.

A string can move, but also vibrate; there are different ways of realizing energy. An extended string as elementary object is more versatile. It can appear as a particle when its size is not resolved, and its vibrational modes constitute an independent form of energy. A single string could encompass what to our pointy eyes would appear as different types of particles. A theory based on strings is of interest not only to avoid divergences, but also as a new option and new hope for unification.

A point particle moves along a worldline in space-time; a string sweeps out a worldsheet. We replace timelike curves in space-time with two-dimensional surfaces, or deformed cylinders with a timelike axis. Equations of motion for a particle are obtained by maximizing the proper-time duration $\mathbf{Sum}_T \Delta T = \mathbf{Sum}_T \sqrt{1 - |\vec{V}(T)|^2/c^2}\,\Delta t$ along the worldline $\boldsymbol{x}(T)$. A good analog for a string worldsheet is the proper area $\mathbf{Sum}_X \Delta A(\boldsymbol{X})$ summed over all events $\boldsymbol{X}$ on the surface, or the proper girth of the tube swept out in space-time by the string. As a two-dimensional surface in space-time, the worldsheet is parameterized by two coordinates instead of just one proper time, as functions $X^a(T, S)$ for the space-time components of the four-vector $\boldsymbol{X}$. At this time, we also include the possibility of extra dimensions, so that $\boldsymbol{X}$ is not a four-vector, but a D-vector, with $D \geq 4$ the space-time dimension. We keep our option of Kaluza–Klein unification in addition to the new modes of the string.

String action The area $\Delta A(\boldsymbol{X})$ is computed as $(\Delta A)^2 = -h_{00}h_{11} + h_{01}h_{10}$ with

$$h_{IJ} = \mathbf{Sum}_{a,b=0}^{D-1} g_{ab} \frac{\Delta X^a}{\Delta y^I} \frac{\Delta X^b}{\Delta y^J} \tag{6.7}$$

if we use coordinates $y^I = (T, S)$ on the surface. The metric h_{IJ} is induced by the space-time metric g_{ab} on the worldsheet as a surface in space-time.

Like proper time in terms of coordinate time, the worldsheet area requires the computation of a square root, a rather complicated function when it comes to quantum theory, at least compared with the quadratic polynomial of the harmonic oscillator's energy. As the physicist Alexander Polyakov realized, it is more convenient to use an expression for the area which produces the same classical equations but is, at least in simple and still interesting cases, quadratic in the functions $X^a(T,S)$.

Polyakov action Using an auxiliary metric m_{IJ} on the worksheet (with inverse m^{IJ}), we define

$$\Delta a = \mathbf{Sum}_{a,b=0}^{D-1}\mathbf{Sum}_{I,J=0}^{1} g_{ab}\frac{\Delta X^a}{\Delta y^I}\frac{\Delta X^b}{\Delta y^J} m^{IJ}\sqrt{m_{01}m_{10} - m_{00}m_{11}}$$

$$= c\mathbf{Sum}_{a,b=0}^{D} g_{ab}\left(\frac{\Delta X^a}{\Delta S}\frac{\Delta X^b}{\Delta S} - \frac{1}{c^2}\frac{\Delta X^a}{\Delta T}\frac{\Delta X^b}{\Delta T}\right) \qquad (6.8)$$

with m^{IJ} assumed in the second step to be of Minkowski form. Summed over all events on the worldsheet, $a = \mathbf{Sum}_{T,S}\Delta a(T,S)$ defines the Polyakov action. It takes its maximal values if m_{IJ} is such that $\Delta a = \Delta A$ agrees with the area element of the worldsheet, and if $X^a(S,T)$ satisfies the wave equation $\mathbf{Sum}_{I,J=0}^{1} m_{IJ}(\Delta^2 X^a)/(\Delta y^I \Delta y^J) = 0$.

If the space-time metric g_{ab} is constant, as it is for Minkowski geometry, the Polyakov action is a manageable quadratic polynomial in the rates of change of X^a. With curvature, the metric $g_{ab}(X^c)$ cannot be constant and an additional and more complicated, perhaps nonpolynomial dependence on X^c results. However, if we first quantize the Minkowski version, small curvature at least can be accounted for by suitable approximation schemes to take into account the extra X^c-dependent terms, just as we use semiclassical approximations, effective equations, and other methods to extend results for the quadratic harmonic oscillator to more complicated potentials.

Conformal invariance Physicists crave symmetry, and string worldsheets have plenty. Although we wrote our expressions with reference to two coordinates T and S, we are free to change them without modifying the area or the action. The string is invariant under coordinate transformations on its worldsheet, just as general relativity is invariant under arbitrary changes of space-time coordinates. Under such transformations, the two fields $X^a(T,S)$ and $m_{IJ}(T,S)$ change.

If we change space-time coordinates for the string, area and action again remain invariant. The area, after all, refers to a geometrical quantity that cannot depend on how we label points. The formulas realize this expectation by referring to space-time scalars with all indices a and b duly paired and summed over. However, under such transformations, the form of g_{ab} and its dependence on X^c change; if we start with Minkowski space in some of the usual coordinates, the metric may look much more complicated after a transformation. Only so-called isometries, that is, transformations that leave the metric unchanged, such as Lorentz transformations of Minkowski space-time, are realized in simple terms.

A more surprising and most important symmetry is conformal invariance, a specialty of the two-dimensional nature of the worldsheet. We are allowed to change the worldsheet metric m_{IJ} not just by responding to coordinate changes, but by arbitrary rescalings, mapping $m_{IJ}(T, S)$ to $L(T, S)m_{IJ}(T, S)$ with some positive function $L(T, S)$ on the worldsheet. At every point, we can change our scale at will. In the action (6.8), the square root is multiplied with $L(S, T)$, and the inverse metric m^{IJ} with $1/L(S, T)$. Both factors cancel, leaving the action unchanged. Going from worldlines of point particles to two-dimensional worldsheets of strings has a welcome consequence, a new symmetry thanks to a fortunate cancellation. This conformal invariance allows elegant mathematical constructions and derivations in quantum field theory; it is one of the reasons for the power and appeal of string theory.

String modes Strings vibrate. Like a musical string, a fundamental string has different modes, or harmonics that appear as clear and distinguishable notes, elementary dispatches of the melody. For physical strings, the modes are elementary in a physical sense, elementary particles in their mathematical realization.

> It turns out that conformal invariance allows us to map any worldsheet metric m_{IJ} to the two-dimensional version of Minkowski space-time. The wave equation satisfied by the string functions $X^a(T, S)$ then reduces to the Klein–Gordon equation $-\Delta^2 X^a/(c\Delta T)^2 + \Delta^2 X^a/(\Delta S)^2 = 0$, solved in general form by $X^a(T, S) = \mathbf{Sum}_n\{A_n^+ \exp[in(S - cT)] + A_n^- \exp[in(-S - cT)]\}$ with complex numbers A_n. For a closed string of the shape of a deformed circle, X^a must take the same value at two endpoints of an interval for S, $X^a(T, S) = X^a(T, S + 2\pi)$. The mode parameter n then takes integer values. Inserting the mode expansion in the action, we obtain a quadratic expression for infinitely many A_n instead of functions $X^a(T, S)$.

Quantum strings Quantum strings vibrate and fluctuate. A glimpse can be obtained by treating the quadratic system, arrived at after assuming Minkowskian g_{ab} and expressing strings by modes, like a collection of harmonic oscillators. We should also account for the symmetries, making sure that they are respected when strings are quantum.

It is not always obvious to extend symmetries to quantum theory. A point particle, for instance, always looks rotationally invariant, but its quantum mechanical wave function may have anisotropic or direction-dependent features. Sometimes, breaking symmetries makes things more interesting, but often it is fatal. The shape of objects need not be symmetric; if we assume it so, it is only for convenience. Geometrical symmetries are another story, for they are crucial in determining what is to be considered a geometrical property or one contingent on choices of observers or coordinates. Lorentz symmetries, and their extended form as arbitrary coordinate transformations of general relativity, have played an irreplaceable role in this book. If these symmetries were violated, we would lose our geometry, our understanding of the fundamental notions of space and time; we could no longer do geometry or physics. If quantum theory would not respect these symmetries, we might as well give up.

It is not known what a quantum theory of strings respecting all space-time transformations could look like, nor whether it exists. However, one can ensure that the special Lorentz transformations if g_{ab} is Minkowskian are realized, as well as conformal invariance of the two-dimensional worldsheet. Even this restricted symmetry is not at all an easy win: it is possible only if space-time has more than four dimensions (six additional space directions), and requires a new type of particles related to the known ones by yet another symmetry, namely, supersymmetry. It can be realized only in a handful of different versions, with no other choice for the usual multitude of particle parameters such as coupling constants and masses. We are as close to unification as seems possible in present-day physics, and this unification includes the gravitational force, for one of the string modes behaves just like a graviton, a quantum gravitational wave.

If our world had ten dimensions and would show us elementary particles with supersymmetry, for each elementary particle a partner of equal mass and (unlike its antiparticle) equal charge but different spin, string theory would be an obvious choice for a unified theory. However, our world is more complicated, by the standards of the string perspective. To match string predictions with observations, we must assume that those extra dimensions are so short that they have not been resolved yet. Supersymmetry can be realized only at high energies not yet reached by particle accelerators, just as the symmetry relating electromagnetic and weak interactions was confirmed only when an energy threshold of about 90 proton masses had been surpassed.

In designing shapes of rolled-up dimensions and break-down scenarios of supersymmetry, additional ingredients are used much less unique than the theories of high-dimensional and supersymmetric strings. It has been estimated that there are more than one broken-down string theory for every single proton in the universe, a vast overindulgence. Perhaps, there are additional consistency requirements that will come from full invariance under all coordinate transformations, not just Lorentzian ones, or from a democratic implementation of all background metrics g_{ab}, not just the Minkowski one. After all, Lorentz symmetry requires the dimension of space-time in string theory to have one given value, so the much larger set of unrestricted coordinate transformations should have even more severe implications, that is, if it can be realized at all. Small curvature can be implemented, as demonstrated by studying the additional interactions implied by a position-dependent $g_{ab}(X^c)$. The situation appears hopeful, although much work remains to be done for a consistent coverage of all space-times.

In the meantime, some research explores what physics could be with vast nonuniqueness. It could be that the actual universe is much larger than what we see, and elsewhere the laws of physics might be different. For every possible breakdown scenario of string theory, there could be one region of the universe that matches it. What we see around us would just happen to be so because other physical laws would not allow life of our form to exist. Such anthropic arguments dispense with prediction and, if strong enough, rather try to explain why we observe what we observe. However, the predictive power of a unified theory is relinquished. To save a beloved theory, we invent alternate realities. Not even a universe is enough; the different parts that realize all string breakdowns form a multiverse, with disconnected regions out of mutual contact.

It might well be that what we've probed so far (not much more, after all, than our past light cone) is only a small piece of a larger cosmic puzzle. There may be surprises, even new physical laws beyond our current horizon. With the speed of light as the maximum, we can only wait, continue to observe the universe, record what we see, and compare with theories, just as we always do in science. At some point, it could happen that our telescopes catch a light ray coming from a region of a different string framework; the idea of the multiverse is, in principle, testable. However, with many different versions to be accounted for, more than one per proton, the multiverse turns into an omnivorous omniverse, encompassing everything in its girth. If nearly every physical law can be realized, science is sentenced to death by taking its predictivity.

There is no way around the realization of general space-time transformations as symmetries. Going from Lorentz transformations to arbitrary coordinate changes, in full generality, should require strong consistency conditions, restricting string scenarios to manageable size. However, it is a dreadful problem. The strategy of string theory, as powerful as it turns out to be for unification, seems difficult to follow in order to understand quantum space-time. Starting with the quadratic model of strings in Minkowski space-time and working slowly toward curvature by including weak interactions is tedious. It has provided information about quantum properties of space-time, but not in direct ways. For some questions, it might be better to address quantum space-time without a specific simplifying choice. It would be more difficult to find detailed properties of different kinds of matter, but questions of space-time may be in closer reach.

6.3
Space-Time Atoms

Quantum mechanics explains the atomic structure of matter, as well as its long-term stability. From quantum gravity, we may expect a similar deed for space-time.

In the scientific viewpoint of matter on the smallest scales, the two opponents of continuous fluids and discrete atoms battled each other for centuries, even while the wave versus the corpuscle nature of light was being debated. The idea of smallest, indivisible objects appeals to our sense of finiteness; but then, if we can cut matter down to some size, why not go further?

Chemists found the idea of atoms useful to describe, organize and memorize the rules of reactions, with atoms as handy units combined in different ways. Unity does not imply indivisibility under all circumstances, but chemistry gave rise to important questions whose answers, found by researchers such as John Dalton and Amedeo Avogadro, had a side effect of strong hints for the existence of atoms as discrete building blocks. Physicists, at that time, were elaborating the laws of heat transfer and energy, deemed substances that flowed between material bodies. Important physical quantities seemed fluidlike, and a continuum view of the world persisted. However, also in these circles, atomic ideas were appealing, for instance, to explain heat as a degree of motion and vibration in matter constituents, as proposed early on by Mikhail Lomonosov.

Without much hope to resolve the scales expected for the size of atoms, many physicists remained doubtful about their reality, as useful as they might be as theoretical constructs. Even Ludwig Boltzmann, the Austrian physicist who, toward the end of the nineteenth century, completed and perfected Lomonosov's old idea of explaining heat phenomena by the motion of atoms, remained skeptical, until, it is said, early in the twentieth century, just at the time when quantum mechanics was being developed. In any case, it was not quantum mechanics, another theoretical framework just as important as his favorite, microscopic thermodynamics, that turned the tides. The most convincing piece of evidence for Boltzmann and most of his colleagues was experiments.

Microscopes, still based on the refraction of light, remained far from being able to resolve atoms. However, another one of Einstein's early insights, in 1905, had shown that one can test properties of atoms by indirect observations on much larger scales. Einstein analyzed Brownian motion, a vibrating movement of microscopic objects such as pollen grains suspended in a liquid. Molecules in the liquid move according to heat, hit the pollen grains in rather random fashion, and make them move. Every single hit is minuscule, but their sheer number magnifies the effect to something visible in a microscope. Statistical methods show how likely it is for the grains to move a certain distance in a given time, depending on the temperature. Observations soon found agreement, establishing atoms as real. Another fifty years later, in 1955, Erwin Müller had developed field ion microscopy, a new and more powerful form of magnification, the first to produce images of atomic resolution. These latter developments were important for materials science, but they were no longer needed to establish atoms as real.

Planck length How big is an atom? This question may seem difficult to answer, requiring a good deal of quantum mechanics, not to mention ideas of what an atom is made of, and how so. However, with a few ingredients, a simple answer, one based just on the consideration of units and dimensions, comes rather close to the correct value.

Let us look at a hydrogen atom. We know that it contains an electron, even if we have not seen much about how it moves and what else there might be in the atom. An electron has two characteristic parameters, its charge e and mass m_e. We also know that hydrogen could not be stable without quantum mechanics: a classical picture would have the orbiting electron radiate electromagnetic waves, lose energy, and quickly fall into the singular center of the Coulomb potential. Quantum mechanics' most popular parameter is Planck's constant $\hbar$. There is only one possibility to combine these three values to a quantity with units of length, the Bohr radius $a_0 = \hbar^2/(m_e e^2)$. It evaluates to about 53 billionth of a millimeter, and agrees with the average radius of an electron orbit computed in quantum mechanics and with actual measurements.

Physics is not as simple as dimensional considerations. For hydrogen, we can guess the radius quite well (although most would remain unconvinced without a solid theoretical foundation). For other, heavier atoms, the question is more difficult. If we look at uranium, we must account for $Z = 92$ electrons, a rather large number. Our result depends on how we include Z. Do we multiply a_0 with Z to provide more room for the extra electrons? Or divide by Z because the additional charge should be neutralized by a more highly charged nucleus, which pulls all electrons closer? We see two factors of the electron charge e in the denominator of a_0, so why not divide by Z^2? Depending on our choice, the results differ by almost six orders of magnitude. Much more must be known about the structure of atoms to derive, rather than guess, their radii.

Dimensional arguments provide rough ideas, but must be supported and improved by detailed theories. In quantum gravity, a variant of the Bohr radius is often used to estimate scales of its microscopic structure. We no longer refer to electrons, but still to quantum physics. We should use $\hbar$, and in addition, the numbers that characterize space-time: the speed of light c of relativity theory, and Newton's constant G of the gravitational force. Again, a unique length parameter can be formed, the Planck length $\ell_P = \sqrt{G\hbar/c^3}$. It evaluates to 10^{-35} m, about a billionth of a billionth of the distances resolved with our current best microscopes, particle accelerators. There is no hope to probe such small space-time structures by direct images; we cannot expect observational guidance in our quest to find the form of space-time atoms. We must rely on theory, without the safety net of experiments for missteps of our imagination.

Empty space Emptiness contains much. Matter in its most contemplative state, the vacuum, has in it the seeds of much of its more excited manners.

From the ground state of the harmonic oscillator, we can generate all other states by ladder operators, and their properties are predetermined. From ground state fluctuations, we can compute the variances in excited states, and their energies follow because the ladder operators raise the ground state energy by fixed amounts. In particle physics and its theoretical framework, quantum field theory, we have ladder operators for all types of particles, and for all of them with different wave lengths or momenta. Starting with the field-theoretic ground state, the vacuum, we generate all kinds of particles in different states of motion, see how they interact by product terms of ladder operators in the energy, and how they disappear in annihilation. Empty space, as the vacuum is often called, is not devoid of information; on the contrary, it holds the key to all the other states.

In string theory, we encounter the vacuum when all string modes are unexcited, with expansion coefficients $A_n^{\pm}$ playing the role of ladder operators. The string vacuum is similar to the general one of particle physics, except that unification with gravity implies also the absence of gravitational waves, or gravitons, not just of matter particles and photons. The vacuum of string theory is emptier than the common empty space.

However, empty space, or emptier space, is not empty yet. It still contains something, namely, space, and space with time is recognized in general relativity as a physical and dynamical object. The absolute vacuum should be devoid even of excitations of space, devoid of geometry, distances and extensions. The vacuum of quantum gravity, a quantum theory of space-time geometry, must be emptier than emptier. This emptiest space, deserving the name "space" by its barest mathematical properties of a manifold, a set of points, is almost unimaginable, but it is the key to quantum gravity.

Loops How do we climb up from emptiest space? As often, the harmonic oscillator provides guidance. We can use its quantum theory in string theory on Minkowski space-time because the string action is quadratic, just like the harmonic-oscillator energy. Such a simplification does not exist for gravity in general, where energy is rather given by the complicated expression of the Hamiltonian constraint (2.36). However, gravity is a consequence of geometry. It seems more suitable to measure excitation levels by extensions rather than energies, and there are geometrical expressions much simpler than the gravitational energy.

A convenient choice of quantities to build ladder operators has given rise to loop quantum gravity, one of the theories that try to unravel the elementary structure of quantum space-time. Toward the end of the 1980s, Carlo Rovelli and Lee Smolin introduced such operators, making use of a formulation of general relativity with variables used earlier by Abhay Ashtekar. They did not refer to the spatial metric and its rate of change, but rather to fields more akin to those encountered in electrodynamics. More familiar mathematical techniques can then be used, although several creative adaptations to the requirements of relativity and space-time physics had to be made. The 1990s experienced a flurry of new results, made possible in particular with key mathematical input from Jerzy Lewandowski; a detailed framework of quantum geometry was established. Although the gravitational energy and dynamics remains complicated to the present day, some properties of the full quantum behavior of gravity have been found with constructions proposed by Thomas Thiemann in 1996.

Geometry is described by a metric h_{ij}, but there are alternative ways. If one provides at each point of space three vectors which are to be considered as mutually orthogonal and of unit length, we can derive all geometrical relationships. We may interpret the three vectors, called a triad, as the axes of a local Cartesian system with units marked on them. The vectors are allowed to change as we move around, and so the axes may rotate; we do not have a global coordinate system for all of space, but rather what amounts to a local inertial frame of space. The axes allow us to do geometry, and the triad serves as a substitute of the metric. We denote it as $\vec{E}_k$, a set of vectors labeled by an index k taking the values 1,2,3. Each vector has a rate of change, or a momentum which we call $\underset{\rightarrow}{A}_k$. These fields enjoy several mathematical similarities with the electric field $\vec{E}$ and vector potential $\vec{A}$ of electromagnetism, just in triplicate.

Covectors We have moved the arrow position on $\underset{\rightarrow}{A}_k$ to distinguish between two types of vectors. We have vectors $\vec{v}$ to represent directions in space, the primary motivation to introduce these mathematical objects. Given a metric g_{ab}, we compute a number $|\vec{v}|^2 = \mathbf{Sum}_{a,b} g_{ab} v^a v^b$ for every vector, its length squared. While components v^a of the vector change if we use different coordinates, as do the metric components g_{ab}, the length remains invariant. Modifying factors from coordinate transformations cancel in the sum, indicated in formula by the opposite positions of upper and lower indices.
We may view $v_a = \mathbf{Sum}_b g_{ab} v^b$ as a new type of vector, distinguished from a direction by components written with subscripts. The combination $\mathbf{Sum}_a v_a v^a$ is invariant: it amounts to the length squared of $\vec{v}$. Coordinate changes of v_a are such that they cancel those of v^a in the sum. In general, we define our second type of vectors, called covectors, as $\underset{\rightarrow}{w}$ with components w_a such that any combination $\mathbf{Sum}_a w_a v^a = \underset{\rightarrow}{w} \cdot \vec{v}$ with a vector $\vec{v}$ is invariant. We can always turn a vector into a covector and vice versa, as shown by the transition from v^a to v_a, and the notion of covectors seems superfluous. However, we use the metric in the process, and in quantum gravity, the metric is physical and to be turned into an operator. It is then dangerous to hide metric components; instead, we work with explicit denotations of the vector type by arrow or index positions.

Holonomy-flux Space is built and measured in flux. A geometrical feature of space geometry computed from the triad is the analog of electric flux through a surface S, called F_S. The flux is related to the area of the surface, and like electric flux, it has a rather simple expression in terms of the fields; it is a manageable object even in quantum theory.

Flux With three independent fields in the triad, there are three components of flux, $F_{S,k} = \mathbf{Sum}_y \vec{E}_k(y) \cdot \underset{\rightarrow}{n}(y) \Delta A(y)$ summed over all points y on the surface. The orientation in space of each piece $\Delta A(y)$ of the surface is characterized by a covector $\underset{\rightarrow}{n}(y)$, the normal. If the surface piece obeys the equation $f(x^a) = 0$, the normal is computed as $n_a = \Delta_{x^a} f / \Delta x^a$. Dividing by Δx^a implies that we have a covector whose coordinate change cancels vector changes such as Δx^a itself. The triad is paired with a covector, and must be a vector to produce an invariant flux.

Paired with electric flux is magnetic flux through a surface, as a measure of the strength of the magnetic field. In electromagnetism, one often writes the magnetic flux in terms of the vector potential that provides the magnetic field. The gravitational analog makes use of the three vector potentials $\underset{\rightarrow}{A}_k$, in a form called holonomies along closed curves, the boundaries of surfaces of magnetic flux.

Holonomy Given a curve C, the magnetic flux is $B_{C,k} = \mathbf{Sum}_s \underset{\rightarrow}{A}_k(s) \cdot \vec{t}(s)\Delta s$ summed over all points s on the curve with tangent vector $\vec{t}(s)$. If the curve is parameterized as a function $\vec{p}(s)$, the tangent vector has components $t^a = \Delta p^a/\Delta s$. As with the flux, we see that transformation properties of $\underset{\rightarrow}{A}_k$ are determined by the subsets of space they are associated and paired with. The components of magnetic flux are exponentiated to define holonomies $H_C = \exp(i\mathbf{Sum}_k T_k B_{C,k})$, with some coefficients T_k to describe their mixture. Instead of working with all possible choices of numbers T_k, one often chooses three matrices from whose components one can extract all holonomies, no matter what numbers T_k are used. A convenient choice is the spin matrices (4.22), (4.23) encountered in quantum mechanics, $T_k = L_k/\hbar$, $k = x, y, z$.
A holonomy H_C depends on the vector potentials $\underset{\rightarrow}{A}_k$ and has components related to spin matrices. If there were just one vector potential $\underset{\rightarrow}{A}$, holonomies would be exponentials $\exp[i\mathbf{Sum}_s \underset{\rightarrow}{A}(s) \cdot \vec{t}(s)\Delta s]$, or a combination of sine and cosine using $\exp(2\pi i a/\bigcirc) = \cos(a) + i\sin(a)$. Given all three $\underset{\rightarrow}{A}_k$, we generalize the exponential to spin matrices. For a short curve of length Δs tangent to $\vec{t}$, dropping the summation, we define $\underset{\rightarrow}{A}_k \cdot \vec{t}\Delta s = Ar_k$ with $\mathbf{Sum}_k r_k r_k = 1$, some kind of polar-coordinate version of the components. Then, $\exp(iA\mathbf{Sum}_k T_k r_k) = \cos(\bigcirc A/4\pi) + 2i\mathbf{Sum}_k r_k T_k \sin(\bigcirc A/4\pi)$, a function that behaves like an exponential: $\exp[i(A + B)\mathbf{Sum}_k T_k r_k] = \exp(iA\mathbf{Sum}_k T_k r_k)\exp(iB\mathbf{Sum}_k T_k r_k)$ using matrix multiplication. More general versions $H_C^{(j)}$ are obtained with the choice $T_k = L_k^{(j)}/\hbar$, matrices for spin $j > \frac{1}{2}$.
Spin matrices are examples for noncommutative multiplication, and therefore the order of factors in holonomies along extended curves must be carefully defined when an explicit value is to be computed. (Gravity in Ashtekar's formulation is an example of a non-Abelian gauge theory.) For most considerations, just the general form of holonomies is relevant and the ordering is not so crucial.

From holonomies around closed curves, or loops, loop quantum gravity builds its ladder operators. A ladder operator of the harmonic oscillator has a commutator $\hat{E}\hat{a}_+ - \hat{a}_+\hat{E} = hf\hat{a}_+$; in this way, a state Ψ with energy E, such that $\hat{E}\Psi = E\Psi$, is raised to a state $\hat{a}_+\Psi$ with energy $E + hf$: $\hat{E}(\hat{a}_+\Psi) = \hat{a}_+(\hat{E}\Psi) + hf\hat{a}_+\Psi = (E + hf)\hat{a}_+\Psi$. If we substitute the quantum flux $\hat{F}_S$ for $\hat{E}$ and a quantum holonomy $\hat{h}_C$ for $\hat{a}_+$, we obtain a similar relation provided the surface and loop used intersect: $\hat{F}_S\hat{h}_C - \hat{h}_C\hat{F}_S = 8\pi I\ell_P^2\hat{h}_C$. The constant 8π is included by convention, and I, called the Barbero–Immirzi parameter, is introduced because the actual scale of quantum gravity may differ from the Planck length, computed by dimensional arguments. Quantum holonomies act as ladder operators, raising the excitation level of geometry as measured by fluxes.

Holonomy-flux algebra The vector potential is the momentum of the electric field, obeying per component and point in space a commutator relation just like $\hat{x}\hat{p} - \hat{p}\hat{x} = i\hbar$ in quantum mechanics. Instead of $\hbar$, units of area for $\vec{E}_k$ and dimensionless $\underset{\rightarrow}{A}_k$ require a factor of ℓ_{P}^2, also proportional to $\hbar$. Holonomies as exponentiated forms of magnetic flux are reproduced in the commutator, giving rise to the relationship used.

In a field theory such as gravity, commutator relations are realized pointwise: field values at different points are independent, and should therefore commute. We may view them as a large collection of independent copies of the quantum mechanical commutators, labeled by points in space. Operators for electric fields and vector potentials instead of fluxes and holonomies obey $\hat{\underset{\rightarrow j}{A}}(x_1)\hat{\vec{E}}_k(x_2) - \hat{\vec{E}}_k(x_2)\hat{\underset{\rightarrow j}{A}}(x_1) = 0$ if $j \neq k$ and x_1 and x_2 are two different points in space. Only if the labels and points are equal should the field components behave like position and momentum in quantum mechanics, for one is the rate of change of the other. In this case, the commutator should be the number $8\pi i I \ell_{\mathrm{P}}^2$; however, one of the difficulties of field theories implies that the actual value is infinite. A single point at which fields are evaluated is just too singular to produce a finite result. As with the extended objects used in string theory, albeit in a form quite different by mathematical and physical principles, holonomies and fluxes draw out singular points and give finite commutators and excitations.

Spin networks The rungs of a ladder show how one climbs up. Ladder operators in quantum mechanics and quantum field theory determine increments of energy. Holonomies as ladder operators of geometry relate building blocks of space, constructed in smallest, atomic steps.

A single holonomy is associated with a curve; a holonomy operator excites geometry along the curve it is based on. We visualize the structure of space in a state obtained by exciting geometry as a network, the collection of all curves used in any one of the holonomy operators that appear in the construction, Figure 6.1. Space appears as a mesh of connected and interlinked loops. To build space atom by atom, we apply holonomy operators for a suitable number of loops, and we can use the same holonomy several times to raise further the excitation just along one loop. We add more atoms of space, or we enlarge existing atoms by further exciting them, two elementary processes that together form the growth and expansion of space of the universe.

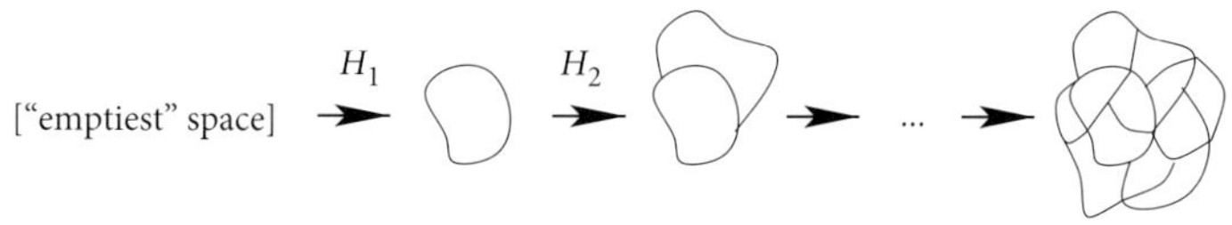

Figure 6.1 Starting with the emptiest, unextended space, loops create space described by spin networks.

The network keeps track of loops used in holonomies in order to characterize all our states. We need a second type of label to indicate how much a single loop has been excited, for instance, referring to the number of times the corresponding holonomy operator has acted. It is more convenient to refer to something analogous to spin numbers in quantum mechanics, even though our networks are not rotating. Holonomies have components collected by reference to the same spin matrices that appear in quantum mechanics, and the same kind of quantum numbers offers itself for a mathematical description. Combining spins and networks, the atomic states of quantum space are called spin networks.

Spin-network function As we add the classical spin vectors of different particles to one total angular momentum, we combine the quantum spins to joint spins of different sizes. A given number of n particles, all with spin $\frac{1}{2}$, can combine to different values of spin j with relative weights $D_j(n)$. We multiply components of the spin matrices, depending on what spin states the particles are in. Two spin-$\frac{1}{2}$ particles have L_z-values up to $\hbar$, the two $\frac{1}{2}\hbar$ adding up. Matrices for $\hat{L}_z$ produce all components in (4.24) from products of those in (4.22), the middle zero appearing twice: The first particle may have $L_z = \frac{1}{2}\hbar$ and the second $L_z = -\frac{1}{2}\hbar$, or vice versa. Two spin-$\frac{1}{2}$ particles produce all components for spin one, plus an additional spin zero. We write $D_0(2) = \frac{1}{2} = D_1(2)$.
In a spin network, n is the number of excitations of a single loop with holonomy H_C, which using $D_j(n)$ is reexpressed in terms of j: the product H_C^n of n holonomies equals $\mathbf{Sum}_j D_j(n) H_C^{(j)}$ with $H_C^{(j)}$ a holonomy whose T_k are chosen as $L_k^{(j)}/\hbar$, a matrix with $(2j+1)^2$ components instead of the four of Pauli matrices for spin $\frac{1}{2}$. A spin network, as a function of the vector potentials, then equals $\mathbf{Prod}_C P[H_C^{(j)}(\underset{\rightarrow}{A}_k)]$, where P picks out components of holonomies to result in a function. Spin-network functions are labeled by the curves C used together with the spins j on each of them and a parameterization of all possible choices P of components, combining the $D_j(n)$.

In a material atom, we have a number of elementary particles, namely, the electrons, protons and neutrons, all with spin $\frac{1}{2}$. These particles are at separate positions, although their wave functions may overlap, and so it is possible to assign a spin to each of them. However, many physical processes, for instance, the emission of photons as electromagnetic waves when an electron changes its state, happen on scales encompassing more than a single particle. For such questions, it is the combined spin of all particles that matters when the energy of the whole system in a magnetic field is computed, just as we add the angular momenta of all parts of Earth when we consider the whole rotating planet. We combine the spins of all particles in an atom, first of those in the nucleus because they are closer together, then the spins of the electrons. The energy of every single electron, a moving charge or a current, depends on the combined spins: a current produces and interacts with a magnetic field, and the energy of a spinning or orbiting charged particle depends on its orientation with respect to the magnetic field. Combined spins have much to say about atomic energies, interactions, and their binding states in molecules.

A spin network of loop quantum gravity combines the spins of its links to form a state, a function of the vector potentials $\underrightarrow{A_k}$. Instead of matter energies, geometrical properties of space depend on combined spin values, not related to actual rotation but good mathematical analogs. The area of a surface, just like the flux, is determined by summing up $8\pi I \ell_P^2 \sqrt{j_C(j_C+1)}$ for all curves C that intersect the surface. It is not the metric of general relativity, summed over all points, or integrated in a continuous fashion, but individual, discrete spin contributions that endow a surface, on its own just an unsized mathematical set, with area and extension. The volume of a region depends, in a rather complicated manner, on how the spins of links intersecting within the region combine (referring to the coefficients P of spin networks). Again, the geometrical measure provided by volume is discrete: As we enlarge the region, just including more and more points, it does not always grow in geometry, in terms of the size assigned to it by a spin network state. The volume increases only when it has grown enough to engulf another intersection point of network links; it inches up by a quantum jump of Planckian size, a tiny Planck volume ℓ_P^3 multiplied by the spins as excitation levels.

Geometry, size and extension all come from the spins of a spin network. Space appears as a giant molecule, with links and their intersections as atomic building blocks, their combined spins affecting not energies but areas and volumes. A single spin network is an eigenstate in which area and volume take sharp values. A general state is a sum of different spin network states, a superposition of different spatial geometries. As an atom can be in a superposition of different energy states, making energy measurements nonsharp and fluctuating due to variance, quantum geometry in general appears in superpositions and fluctuates. Surfaces and regions no longer have sharp areas and volumes; when measured, their values are determined by statistics of the state. There are probabilities for measurement outcomes. Some values might be more likely, but others are possible too. Geometry and space are subject to the laws of quantum physics; they fluctuate.

Quantum geometry Space seems a land of plenty. Even the tiniest region we can imagine is made of unnumbered excitations of geometry, links and spins of spin networks. We usually picture space as a continuum, with points everywhere, all places available. When there is a surplus of everything, we don't need to think about gaps and lack of access; the richness of the continuum is a good approximation. However, this view is not exact in loop quantum gravity, for if we look closely, we see space at the Planck scale as a patchwork, porous and deprived.

An atom seems empty to us, with not much space occupied by its particles, much emptiness between the electrons and the nucleus. Space is even more empty. There is nothing between the links of a spin network, just mathematical points, but no space because there is no extension; between the links, no surface would have area, and no region volume. It can be taxing to look at places that have almost nothing left to them, but it can teach us much. Emptiest space, like the ground state or the vacuum, is the humble origin from which everything else in physics, that is, all its fields, its powers and the material world grows.

Starting with a dense mesh of a spin network (or several ones in superposition), filling up the spacious world of abundance we are used to, we can imagine going down to the essentials by cutting off link after link, removing thread by thread, taking away spatial atom by spatial atom. Areas and volumes shrink; space becomes more sparse and discrete. After a long series of steps, we arrive at a state with just a few links and low spins, a network near the ground state of geometry, bare of almost all excitations. When we remove the last link, a space is left endowed only with what it needs to exist in the lofty mathematical world, namely, points, though with no physicality, nor geometry. Its points are not extended. However many points we gather, no volume grows.

State of Hell Emptiest space is far from our experience, and it may seem a distant curiosity, an exotic spectacle, a mathematical indulgence. Nonetheless, it plays a real role in cosmology, for most space-time solutions of general relativity start or end in a singularity of space collapsed to zero size. This state may not seem empty; after all, matter densities and temperatures diverge. However, if the volume vanishes, every finite amount of matter, even the smallest one, would imply infinite density. When there is not even volume, matter is secondary. At a singularity such as at the big bang, space is as empty as it gets, and everything comes from it.

Infinite density is a direct consequence of vanishing volume. Diverging temperature is not as obvious, but follows by a mathematical analogy. On the side of geometry, we have vanishing volume, realized in the ground state Ψ_0 of geometrical excitations. In quantum mechanics, we have characterized states by their moments, averages of powers and products of position and momentum expected in measurements. In quantum geometry, holonomies and fluxes play the role of basic operators, and their moments in a state can be used instead of a wave function.

> For moments of holonomies, that is, averages of all powers $\hat{H}_C^n$ along some curve C, we compute $\langle\Psi_0|\hat{H}_C^n|\Psi_0\rangle = 0$ if $n \neq 0$, and $\langle\Psi_0|\hat{H}_C^n|\Psi_0\rangle = \langle\Psi_0|\Psi_0\rangle = 1$ otherwise, a result valid for all ladder operators as used already for the harmonic oscillator. All averages of the flux and its powers vanish, for no geometry is excited.

In terms of geometry, the state Ψ_0 is the ground state, with the smallest possible volume, zero. In terms of energy, the situation looks different. Energy is difficult to compute if we don't have a clear idea of what matter we have in such a state, but a mathematical analogy helps us. Infinite temperature simplifies mathematics. At finite temperature, many parameters must be specified to know the state. The state could be one of sharp energy, for example, an eigenstate of the energy operator, or any superposition of different energy states. We need infinitely many parameters for an exact state, for instance, given by the amplitudes and probabilities to find the different energy states in superposition. If the temperature is infinite, we have as much energy as we want to excite any state; all energies contribute to our infinite-temperature state with equal likelihood. Having specified the probabilities, we can compute averages.

Density matrix The probabilities of a state are collected in the density matrix, an operator whose average $\langle \Psi | \hat{D} | \Psi \rangle$ in some test state Ψ equals the likelihood of finding the properties of state Ψ when we measure the original state. An equilibrium state at temperature T has density matrix $\hat{D} = \exp[-\hat{E}/(k_B T)]/Z$ with a constant Z normalizing to total probability one, the energy operator $\hat{E}$, and Boltzmann's constant k_B of thermodynamics, as used in Planck's formula. Low energies are found more often than high ones, and the larger the temperature, the more likely it is to measure high energies. The detailed behavior depends on the energy operator, usually in a complicated way. However, at infinite temperature, the exponential is energy-independent, and $\hat{D}$ equals a constant $1/Z$. All energy states are assigned the same likelihood.

An infinite-temperature state and emptiest space Ψ_0 have the same averages for all powers of holonomies and fluxes. Even without knowledge of matter in these extremes, mathematics tells us that vanishing volume goes along not just with diverging density but also with infinite temperature, earning this state another moniker: State of Hell. The analogy with the state of the big-bang singularity is complete.

Time Time is experienced by change. Even if matter is absent, geometry changes when space-time is dynamical. Time progresses even in empty space.

We have described a process of changing spin networks, removing links to approach the State of Hell. However, these changes were not dynamical; they were willful, performed by our imagination to compare different geometries. In the space-time fabric of our universe, such processes go on even without our doing. Links and excitations of geometry are removed when space collapses, for instance in a black hole, and they are created when space or the whole universe expands.

Geometry is dynamical, but there is no absolute time, no reference to describe change. Evolution is relative, one parameter varying as another one moves, just as we measure motion with respect to the Earth or a clock's hands. As the volume of space increases, provided by all intersection points in a spin network, other quantities, geometrical or material, change too. Relations of these variables tell the story of the universe.

As links and excitations change, the universe grows and evolves. We can imagine adding links in many different ways, but there is just one universe, which at any given instant can grow in only one way. Cosmic dynamics must be constrained, just as the metric cannot expand at will, but rather follows Einstein's equation. The spin networks of loop quantum gravity must be subject to equations showing how they evolve.

In quantum mechanics, the energy operator determines the Schrödinger equation and all dynamics of wave functions. In general relativity, we do not have an energy expression, but rather the Hamiltonian constraint which, as in quantum cosmology, is translated into a state equation with no apparent time parameter. Sometimes, we can treat one of the physical variables just like time, but only for limited regimes. We must work with different clocks, a mathematical inconvenience, but also a realistic feature: the clocks we use cannot be good under all circumstances. In cosmology, matter can be so extreme, with densities and temperatures so high that no common clock can function.

Hamiltonian constraint Dynamics in relativity is constrained. It cannot be arbitrary, for it must respect the space-time whole and its symmetries. General relativity tells us the classical form, but it is based on a spatial continuum that is not present in quantum gravity. The dynamics is to be adapted, in a discrete way that takes into account spatial atoms, but does not do violence to space-time symmetries. At this stage, we encounter the biggest challenge for loop quantum gravity and related approaches.

Dynamics is formed by elementary processes of generating links and excitations, mediated by single operations by holonomies. If we know the classical form of the constraint, playing the role of energy in mechanics, we see how it depends on the vector potential $\underrightarrow{A}_k$ and what combination of holonomies as ladder operators it gives rise to.

Energy In electrodynamics and other interactions of particle physics described on Minkowski space-time, we have energy expressions of the form $\mathbf{Sum}_{x,k}\Delta V(x)(|\vec{E}_k(x)|^2+|\vec{B}_k(x)|^2)$ with electric-type fields $\vec{E}_k$ and magnetic-type fields $\vec{B}_k=\underrightarrow{\nabla}\times\underrightarrow{A}_k+\mathbf{Sum}_{m,n}C_{kmn}\underrightarrow{A}_m\times\underrightarrow{A}_n$. The structure constants C_{kmn} vanish for the electromagnetic fields, and have specific forms for the weak and strong interactions. Gravity on any space-time has a Hamiltonian constraint

$$\frac{c^2}{16\pi G}\mathbf{Sum}_x\left(\Delta V(x)\frac{\mathbf{Sum}_{k,m,n}C_{kmn}(\vec{B}_k\times\vec{E}_m)\cdot\vec{E}_n}{\sqrt{\frac{1}{6}|\mathbf{Sum}_{k,m,n}C_{kmn}(\vec{E}_k\times\vec{E}_m)\cdot\vec{E}_n|}}+\cdots\right)$$

for the same kind of fields (not writing some more complicated terms). Here, C_{kmn} is antisymmetric, changing sign whenever two indices are interchanged, and $C_{123}=1$. If we write $\vec{B}_k$ in terms of $\underrightarrow{A}_k$, and use holonomies with their form as ladder operators in quantum theory, the atomic dynamics follows.

Quantum geometry puts strong impositions on the atomic dynamics of space-time. We cannot refer to continuous derivatives; at this elementary level, we are brought back to finite differences and their ratios as rates of change. Our usual differential equations, such as the wave equation, are approximations in a discrete world; if we look closely, we should see deviations, for example, in the dispersion of waves, but these effects are tiny enough to have escaped detection to date.

There are other modifications implied by more specific properties of loop quantum gravity. By using holonomies as ladder operators, we work with exponentials of $\underset{\rightarrow}{A}_k$ instead of polynomials, as the classical Hamiltonian does. These functions, and the corresponding dynamics, differ from each other by terms of higher powers of $\underset{\rightarrow}{A}_k$: if x is small, $\exp(x) \approx 1 + x$ (see Figure 1.2), but powers of x must be added when x grows. When $\underset{\rightarrow}{A}_k$ takes large values, at high curvature, these differences, called holonomy corrections, could be sizable. However, in common regimes, these effects are not yet detectable.

Finally, there is a more indirect consequence of discrete quantum geometry: The gravitational Hamiltonian requires us to divide by a function of $\vec{E}_k$, or flux operators in loop quantum gravity. A flux operator vanishes often, whenever its surface is not intersected by the links of a spin network. We cannot divide by such an operator without producing infinities. A closer look reveals that discreteness also helps here. Division by $\vec{E}_k$ can be written as a form of finite differences, in the limit when the step size vanishes. In discrete quantum geometry, we still take finite differences, but do not go all the way to zero step size. A finite value then remains, differing from the classical divergence by another effect, called inverse-triad correction.

> **Inverse triad** We interpret the function $\mathrm{sgn}(E)/\sqrt{|E|}$ of E as the limit of $(\sqrt{|E+h|} - \sqrt{|E-h|})/h$ for h very small: $\sqrt{|E+h|} = \sqrt{|E|}\sqrt{1+h/E} \approx \sqrt{|E|}(1+\frac{1}{2}h/E)$. At $E = 0$, the original function diverges, as does the limit for $h \to 0$. However, the difference for $h \neq 0$ remains finite, vanishing at $E = 0$.

Quantum space-time Space and time are fluctuating, uncertain, and discrete, just like matter. Not all these properties are easy to grasp. Discreteness is unproblematic, for it is rather its opposite, the continuum, that requires an additional step of abstraction. Fluctuations of space and time, on the other hand, are revolting. When Charles Dickens described fluctuations in *A Christmas Carol*, he had no problems nailing those of matter, by the example of limbs as a type of constituents: "For as its belt sparkled and glittered now in one part and now in another, and what was light one instant, at another time was dark, so the figure itself fluctuated in its distinctness: being now a thing with one arm, now with one leg, now with twenty legs, now a pair of legs without a head, now a head without a body: of which dissolving parts, no outline would be visible in the dense gloom wherein they melted away. And in the very wonder of this, it would be itself again; distinct and clear as ever." However, he was unable to imagine fluctuations of time, which remain sharp, with clear ticks marked by the regular beats of "now."

Physicists still struggle with fluctuating space-time. Most implications are therefore studied in semiclassical or other approximations, in which one need not consider the full wave function, but rather investigates quantum effects by corrections implied to classical equations. New forces may arise, or modifications of classical space-time symmetries may change the laws of relativity. By a detailed analysis of cosmological or other processes, one may find sizable implications making those new terms detectable.

Most corrections arise at high density, and are otherwise suppressed by tiny factors of d/d_{Planck}, the matter density relative to the Planck density. In cosmology, the density is the main parameter that rules, by the Friedmann equation, the expansion history of the universe. To circumvent the tight constraints of dimensional arguments, allowing only tiny d/d_{Planck}, we need some other parameter, or a large dimensionless number that can magnify small values if it appears in the right places. Discrete space-time offers a clear candidate for another scale, or a large dimensionless number: the average size of spatial atoms or their density as a new scale, or their number in any given region as a large dimensionless parameter. We just have to hope that this number appears such that it magnifies effects, at least in some predictions.

Inverse-triad corrections turn out to be of the right form. They do not refer to the density but rather to the atomic scale, via fluxes. They have a recognizable geometrical effect: a long calculation of differences between elementary changes in quantum geometry occurring in different orders shows that the hypersurface-deformation algebra, the most elementary law of general geometry, is modified to $T[N_1]T[N_2] - T[N_2]T[N_1] = S[f(\vec{E}_k)(N_2\Delta N_1/\Delta x - N_1\Delta N_2/\Delta x)]$, with a function $f(\vec{E}_k)$ of inverse-triad form $\sqrt{|E|}(\sqrt{|E + \ell^2_{\text{Planck}}|} - \sqrt{|E - \ell^2_{\text{Planck}}|})/\ell^2_{\text{Planck}}$. Discrete space, via inverse-triad corrections, slows down or speeds up propagation; for linear deformations, as in Figure 2.42, we no longer obtain the expected relation $\Delta x = v\Delta t$, but rather $\Delta x = fv\Delta t$. Speeds change, and even basic notions of uniform motion are modified. Implications in cosmology can be significant, and we return to a last glimpse at the universe, probing its most fundamental scales.

7
The Universe IV

As we look deeper and deeper into the workings of the world, we find hints about the structure of the smallest as well as largest things, the building blocks of matter and space-time and their compounds up to cosmic scales. We have come a long way, unraveling how atoms and some particles such as protons are built. We understand how space-time reacts to energy, and how its curves and shape make matter move around.

However, when we try to combine the smallest with the largest, when we try to describe how a dense piece of space-time or the whole universe behaves when it is small, our current theories, general relativity and quantum physics, are insufficient. We are still unable to combine them to a consistent whole. If we only bring together their most characteristic parameters, Newton's constant for gravity and Planck's constant of quantum mechanics, they give us one more hint about the tiniest scales: there should be a transition from structureless continuum to some kind of chiseled latticework of a mesh-size as small as the Planck length. However, this size is so tiny that observations are far from resolving it, and will remain so for a long, long while. We cannot use experiments to probe the smallest scales; we must rely on intricate theoretical considerations.

There seem to be different ways of combining general relativity with quantum physics. Since none of those known are complete and fully consistent, we cannot be sure if theory alone will be a good guide. Taken in separation, general relativity and quantum physics have their own weaknesses, such as singularities. Trying to resolve them to improve our theories is the main strategy in the absence of experiments. With such complicated theories, our options are limited. They may be too much constrained to allow any consistent combination – nature may, after all, turn out to be more complex than what we can describe with mathematics. Or, theoretical constraints may be too weak for unique predictions, in which case it would still be interesting to explore the possibilities. Additionally, the option of one and only one combined theory is still on the table, a theory that would tell us what must happen under given circumstances without having to look and experiment – provided we could solve its equations.

For now, we cannot be sure about full consistency, let alone the uniqueness of our ideas regarding quantum gravity. When we apply existing constructions to questions about the real universe, we are probing nature as much as we are test-

The Universe: A View from Classical and Quantum Gravity, First Edition. Martin Bojowald.

ing our own thoughts. As we try to use theory to uncover new phenomena, we might end up discovering that we are on the wrong track. It could be an interesting place nonetheless, much as the place Columbus arrived at by "mistake." We always learn from exploration and mathematical model-building, even if some ideas do not make their way into the theory aimed at. It is always illuminating and stimulating to apply theory to the real world, perhaps all of it in cosmology, even if some oft-made claims, for instance, when we use mathematics, almost as metaphors, to talk about solutions for the whole cosmos, seem to border on hubris. We should keep in mind that specific results may be overturned with new knowledge, and that scientific statements about some ultimate questions, such as the beginning of time or the end of the universe, may not always be the ultimate answers.

7.1
Big Bang

General relativity shows us the dynamics of space-time. Quantum physics of matter in expanding space-time tells us how atoms and particles emerge from the vacuum, how they begin. A combination of general relativity with quantum physics should help us understand how space and even time begin and grow. What does nature look like, with space apace and time still prime?

As per general relativity, rewinding the expansion of the universe leads to a state of complete collapse, all matter and all of space squeezed into a single point. The density and temperature are infinite, and our equations, no longer producing finite numbers, lose their meaning. Within this theory, we reach a limit of space-time, a singularity; the theory is incomplete and cannot give us a full view of the cosmos. Time in its solutions is limited, but it does not begin, for a beginning requires a meaningful initial state. The only meaning of a singularity is that our theories break down when it is reached.

General relativity shows how space-time evolves, but it cannot tell us how it all began. At high, singular-looking densities, quantum physics is essential, quantum physics even of space and time. Some ingredients of quantum gravity are needed to see what happens deep within the big bang.

Time Time has a finite beginning or has gone on forever. It is pure logic, with a dose of classical thinking about time, that gives rise to this statement, irrespective of general relativity or quantum gravity. Most cosmological models as well as general world-views therefore fall into two classes, linear models with a beginning and subsequent progress of time, and cyclic ones in which things happen time and again.

Both scenarios, the linear and cyclic one, have their problems. General relativity seems to indicate a linear view with a beginning of time, but what could at best be seen as a beginning is a singularity, the breakdown of theory. Bringing in some properties of quantum gravity might change this failure in two ways: a nonsingular, though still absolute beginning, might be obtained from equations that remain valid at the highest densities, or those equations might extend space-time to time before the big bang. These are the two most common ideas in models of quantum cosmology.

Models differ in the way they tame the singularity. Somehow, the ultimate meltdown of space-time, caused by ever-attractive gravity, must be prevented. There are strong quantum forces that can stabilize matter even at the densities of white dwarfs or neutron stars, but all known ones will fail when collapse progresses further. A theory of quantum gravity could avoid the singularity if it gives rise to new forces strong enough to counteract gravity, or if it changes the behavior of gravity itself, making it repulsive when curvy space-time has gulped too much matter.

String theory provides examples for new forms of matter as an implication of relativity and quantum physics combined. A string vibrates in modes appearing as different particles, which in higher dimensions can result in unification of the known forces and predicted new ones. Some versions of string matter, with names such as string gas, tachyon condensates, or fuzzballs, have been proposed and analyzed with an eye on their hold on the singularity. In some cases, strings can help the universe make a stitch in time that saves (more than) nine, a stitch in space-time that saves the whole universe before its full collapse. Matter in the universe would be stabilized at high density, that is, at a nonsingular state that can serve as the beginning of the cosmos, including even time.

Theories of quantum gravity that focus on space-time structure instead of unification do not provide candidates for new particles or forces, but they can show how gravity itself might change. In general relativity, the gravitational force is a consequence of space-time curvature, of deformations of the space-time continuum. If space-time has a fundamental structure, it becomes less rigid when the universe is small and dense; its parts have more freedom to move without the continuum's connectivity. They may move right through one another, in a regular transition instead of singular collapse. Or, the altered form of space-time might make matter move in different ways, not by gravitational attraction or the focusing of classical geodesics, but with a repulsive component. If space is discrete, it cannot support an arbitrary amount of energy and momentum: the wave function of matter cannot oscillate with wave lengths smaller than the discrete scale. The density is bounded, by about the Planck density when the discrete spacing, the size of spatial atoms, is the Planck length. Loops of loop quantum gravity provide material for the stitch in time.

Wave function Particles are not points but wave functions. If we can describe the whole universe by one consistent wave function of everything, it cannot collapse to a single point. Some extension, at least quantum fluctuations, will always remain.

Before string theory and loop quantum gravity were applied to cosmology, the Wheeler–DeWitt equation for wave functions of the universe was the main tool to analyze the quantum cosmos. It gives rise to new forces, as always in quantum physics when the motion of a whole wave function is considered, but they are not strong enough to stop the total collapse under all circumstances. Instead, the equation allows one to analyze how, or with what types of wave functions at the singularity, meaningful evolution could be achieved. The singularity was not resolved as a consequence of new or altered forces derived from quantum gravity, but rather by postulating what the wave function, what quantum fluctuations should imply.

Without knowing forces active at high density, one cannot derive how space-time evolves at or through the densest part of the big bang. More specific scenarios rather than general principles then play the central role. The tunneling phenomenon of quantum mechanics and mathematical aspects of curved surfaces have been most prominent.

With the tunneling proposal, Alexander Vilenkin suggested that the wave function of the newborn universe should look just like the wave function after tunneling. The wave function of the universe does not tunnel through a physical barrier as in quantum mechanics; after all, the singularity problem to be solved by quantum gravity is not the penetration, but the absence of a force strong enough to build a barrier against total collapse. The wave function of the universe tunnels, as it were, through the realm of our ignorance; it tunnels into existence from nothing, a brief (perhaps too brief) name for all that we don't know about the universe beyond the big-bang singularity. Words used in specific scenarios are not always to be taken as literal descriptions. By necessity, they are metaphorical: with relativity's mathematical equations, also our method of discourse regarding nature breaks down at a singularity. Only full-fledged quantum gravity could turn mere metaphors into faithful descriptions of the world. Without it, a key part of metaphors, their target or tenor, remains unknown and vague.

With mathematical instead of physical images, one can try to find a nonsingular formulation of the beginning of space-time by closing it off in itself, like a closed surface which does not rip off or pinch off, but is connected everywhere to its own smooth self. Such a surface has no boundary, a model for the no-boundary proposal of James Hartle and Stephen Hawking: The wave function of the universe should be such that space-time is closed in itself at the big bang. Quantum physics or wave functions do not seem necessary to close off a surface, but space-time in general relativity cannot be closed in itself; the theory has no means for a nonsingular beginning or end of time.

Quantum physics is more flexible. As Hartle and Hawking suggested, it might convert all of time into space, not just in part as Lorentz transformations do. At high density, space-time could be replaced by four-dimensional space, with four space dimensions and no time, a Euclidean region with geometrical laws resembling more those of Euclid's rather than Minkowski's. In general relativity, it is not possible to connect four-dimensional space to space-time without a singularity; edges of space and space-time, that is, borders and beginnings, are just too different to connect. However, with quantum fluctuations and superpositions, and accompanying fuzziness, the incompatible fringes could be washed out to make them match. No precise mechanism had been given by Hartle and Hawking, but some properties of wave functions could still be computed.

Bounce cosmology Cremator-creator events, processes that combine destruction with construction, are not uncommon on cosmic scales.[1] Stars cook up the elements and release them in their cataclysmic supernova deaths, processing the seeds, hydrogen and helium with trace amounts of other light elements, provided by the hellish heat of big-bang nucleosynthesis. Perhaps, even space and time can be born only if something worthy of their grandness perishes; perhaps they can be reborn only if another universe dies.

Ideas of physical theories for a cyclic cosmos are older than the big-bang model, but not much older than the concerns about it. In the 1930s, Richard Tolman suggested that exotic forms of matter at high density could circumvent the attractive nature of gravity, providing repulsion that can turn space-time collapse into expansion. Several concrete scenarios were analyzed only decades later, and by now there are promising indications that properties of the collapse phase could lead to cosmic structure similar to what inflation implies, without the need to postulate a new inflation particle or field.

However, even after decades of research, one central problem found by Tolman remains unresolved: Given an infinite amount of time, and perhaps an infinite number of cycles of collapse and expansion, disorder or entropy, the waste product of physical processes, should by now have piled up[2] to leave nothing but a wasteland, a random mix of radiation and particles flying around, minding their own business. Structures as we see them as galaxies, solar systems, planets and life, should be impossible. Gravity can explain how ordered structure arises from uniformity, as in the process that starts with an almost homogeneous distribution of cosmic background radiation and culminates in the aggregates we observe as galaxies. However, it remains difficult to factor gravity into the equation of entropy increase. Cosmologists remain uneasy about an infinity of cycles.

Irrespective of what such long-term evolution entails, one can explore the options for bounces allowed by physical laws. For matter to be repulsive under gravity, its energy must be negative. Such matter has never been observed, but in quantum theory, with energy-time uncertainty, it is possible for energy to change by any amount provided its old value is restored soon after, when a time $\Delta t = \frac{1}{2}\hbar/\Delta E$ has passed. Energy can fluctuate to negative values, to be repulsive. To stop the collapse of a whole universe, however, this borrowed resistance is insufficient. Energy would have to fluctuate to the negative side at just the right moment, and by an enormous amount.

If matter is too weak, space-time itself might have to bend. The atomic, discrete structures, for instance, of loop quantum gravity, are not easy to control by

1) The influence of the cosmos on humankind has often found impressions on cultural ideas in destruction combined with new beginnings. The Mayan culture, for one, had such a philosophy, with one transition "predicted" to happen on 21 December 2012.

2) Disorder or entropy seems a faithful companion of the progress of time in thermodynamics. The modern understanding is the opposite of what Homer described: Penelope promised her suitors to choose one among them when she had finished weaving Laertes' burying shroud. To postpone the decision, she undid her day-work at night, gaining more time, that is, turning back time by destroying order.

mathematics, but they have suggested a simple mechanism which has become rather popular in loop quantum cosmology. In this theory, one uses exponentiated quantities such as $\exp(i\ell H/c)$ with a length parameter ℓ, or the real combinations $\sin[\bigcirc \ell H/(2\pi c)]$ and $\cos[\bigcirc \ell H/(2\pi c)]$, instead of the Hubble parameter $H = \Delta a/(a\Delta T)$, for only these objects, derived from holonomies $H_C^{(j)}$, can act as ladder operators of geometry. The Friedmann equation $H^2 = 8\pi G d/(3c^2)$ is then seen as a small-H approximation to an equation that refers to a trigonometric function instead of a quadratic polynomial in H. If the Friedmann equation is replaced by $\sin(\ell H/c)^2 c^2/\ell^2 = 8\pi G d/(3c^2)$, reducing to the classical form for c/H much larger than ℓ, the matter density d is bounded by an amount of $3c^4/(8\pi G\ell^2)$. If $\ell \approx \ell_{\text{Planck}}$ is close to the Planck length, the maximum energy density is about the Planck density. Collapse cannot proceed unhindered; modified gravity corresponding to the altered form of the Friedmann equation must stop it or turn it to expansion.

Bounce solutions of a universe with loop effects for matter at high density can be found. However, modifying gravitational laws is dangerous business. We might be able to fight singularities, though at a high price. General relativity is covariant; it obeys the equivalence principle and is insensitive to what coordinates we choose to describe space and time. If we mess with its equations, even in ways motivated by some properties of quantum physics, these fragile symmetries, introduced for good reasons and responsible for the great successes of the theory, are destroyed. A rigorous analysis of quantum space-time is required, not just a modification of an equation for isotropic universe models; but it remains incomplete due to lack of control on full loop quantum gravity. Although cre(m)ator solutions may seem possible in simple models, it could turn out that the exorcist called upon beats the singularity by killing space-time.

Before the big bang Repulsive gravity can stop collapse. In loop quantum cosmology, the Friedmann equation is modified in a way suggested by spatial discreteness, or the atomic nature of space. At an elementary level, atomic time leads to a constraint

$$(V+\ell^3)\Psi(V+\ell^3) - 2V\Psi(V) + (V-\ell^3)\Psi(V-\ell^3) = -\frac{\ell^6 \hat{E}_{\text{matter}}\Psi(V)}{6\pi\hbar c\ell_{\text{Planck}}^2} \quad (7.1)$$

of discrete form, a difference equation. The step-size ℓ^3 is finite, that is, no limit of ℓ going to zero is taken as in derivatives, and to be determined by quantum gravity. In simple cases, $\ell \sim \ell_{\text{Planck}}$ is Planckian, but it could be different and even change with time, depending on the total volume $|V|$ of space.

We have just alluded to another difference with the original Wheeler–DeWitt equation, namely, that V determines the volume by its absolute value $|V|$, but could be negative. The value represents the "electric" field used in loop quantum gravity, or a geometrical triad, the flux which can change sign if space is turned inside out. It is difficult to imagine three-dimensional space turned inside out. However, as a mathematical operation, we just invert our frame axes, making coordinates run the opposite way along them. On all three axes, which would be defined as part of a local inertial frame if space is curved, we map the unit 1 to what used to be -1, or multiply all coordinate differences with -1. The product $\Delta x \Delta y \Delta z$ changes sign, corresponding to a change of sign of V.

With negative V available, there is a possibility for space-time and a universe "before" the big bang. We may view the wave function for positive V as the one of our expanding part of the universe. For negative V, loop quantum cosmology provides a new wave function not seen in the old quantum cosmology, a wave function for a mirror universe with space turned inside out. If we can connect the two independent parts of the wave function, we see whether and how the universe evolves through the big bang. Vanishing V would no longer be a singularity because evolution would not stop.

A difference equation determines evolution to smaller V by recurrence, by successively solving it for $\Psi(V-\ell^3)$, starting with initial values chosen at some $V_0+\ell^3$ and V_0. We first solve for

$$\Psi(V_0-\ell^3) = \frac{2V_0\Psi(V_0) - (V_0+\ell^3)\Psi(V_0+\ell^3) + (\ell^6/6\pi\hbar c\ell_{\text{Planck}}^2)\hat{E}_{\text{matter}}\Psi(V_0)}{V_0-\ell^3},$$

and by the same scheme, keep going to $\Psi(V_0-2\ell^3)$ and so on. The wave function is determined by the initial values, without any problems in solving the equation, until the number $V_0-\ell^3$ we must divide by becomes zero. For given ℓ, there is always one value of V when this happens, producing infinity just when we are about to compute the value of $\Psi(0)$, the value of the wave function at the singularity. Again, the singularity seems to hold its power even over quantum physics.

Upon close inspection, the situation is improved. We are unable to determine the value of $\Psi(0)$, but we may still try to find $\Psi(-\ell^3)$, the first step through the looking glass, into the unknown of flipped space. We find $\Psi(-\ell^3) = \Psi(\ell^3) - (\ell^6/6\pi\hbar c\ell_{\text{Planck}}^2)\hat{E}_{\text{matter}}\Psi(0)$. One of the unknown $\Psi(0)$ disappears because it is multiplied with $V=0$, vanishing at this recurrence step. The matter term still refers to $\Psi(0)$, but irrespective of this value, the matter energy vanishes for $V=0$: Inverse-triad corrections make $\hat{E}_{\text{matter}}(0)$ vanish for all forms of matter. We obtain $\Psi(-\ell^3) = \Psi(\ell^3)$, a simple relationship without ambiguity. The next step of the recurrence determines $\Psi(-2\ell^3)$, and again any reference to the unknown $\Psi(0)$ drops out. Starting with initial values, the wave function is determined for all positive and negative V, "before" and "after" the singularity. We are not able to find the value of $\Psi(0)$, but it does not matter; it is not needed to compute how the wave function evolves through the big bang.

With our difference equation, the two parts of the wave function for positive and negative V are related by $\Psi(-\ell^3) = \Psi(\ell^3)$, indicating an exact mirror image at negative V, a mere copy of our expanding universe. More realistic models change the relationship, making it more complicated, though also more interesting. There are additional parameters such as those for an anisotropic shape of space, and the wave function depends on all of them. Even if we keep a single V, the matter energy $\hat{E}_{\text{matter}}$ is known to depend on the orientation of space; it changes if space turns its inside out. Experiments with elementary particles have shown that the weak interaction is not invariant under mirror reflection; the matter energy changes under switching V with $-V$. After $\Psi(-\ell^3) = \Psi(\ell^3)$ is reached, the recurrence refers to $\hat{E}_{\text{matter}}$ evaluated for negative V, making the history different from what has been computed for positive V.

End of time Space-time can avoid its end by killing time. This suggestion may sound futile and suicidal, but it can be realized in sparing terms. Quantum-gravity modifications mess with space-time and its symmetries, but detailed calculations show that damage is avoided if equations are chosen in a specific, careful way. Covariance of space-time is then not destroyed, but it is altered. It takes a new form with quantum corrections, a form that makes space-time at Planckian densities behave like four-dimensional space. If time turns into space, collapse can be stopped and the singularity is avoided. In this way, loop quantum cosmology has led to a detailed form of cosmic history akin to the no-boundary wave function, but derived from quantum gravity.

Loop quantum cosmology, with its difference equation, has provided the first concrete indication that quantum gravity could replace the big-bang singularity with a bounce, a cre(m)ator event turning contraction to expansion, demise to growth. The key to this conclusion is an equation for the wave function, leaving open what space-time looks like in the high-density phase of the big bang and "before," in the new part of the universe. We have surrounded timelike references with quotation marks, for without knowledge of what space-time we have, we cannot be sure if what we think of as time is indeed time.

We are dealing with a wave function because we have quantized space-time, but in pictures such as collapse, expansion or a bounce, we imagine the classical notion of time, an ordered succession of sharp moments. If time is quantum, it should fluctuate, disturbing time's order until it is unrecognizable. If time loses its order, or its arrow, it no longer appears as time to us. Time without order is just like space.

The no-boundary wave function has suggested that all of time may turn into space when it is quantum. Loop quantum cosmology can analyze quantum space-time in detail. It changes space-time by modifications of the hypersurface-deformation algebra, taking the form $T[N_1]T[N_2] - T[N_2]T[N_1] = S[f(N_2 \Delta N_1/\Delta x - N_1 \Delta N_2/\Delta x)]$ with f not always of the classical value $f = 1$. The basic law of motion is modified to read $\Delta x = f v \Delta t$, slowing down or speeding up propagation. At Planckian density, near the bounce point, f differs much from 1. It even reaches the value zero, stopping all motion no matter how large a velocity v we try. If there is no motion, time has ended, albeit in a nonsingular way, without infinities. When $f = 0$ at a certain value of the density, the time direction disappears, only leaving space.

The density for $f = 0$ is smaller than the maximum density reached at the bounce. For values between those limits, f does not vanish, but is negative. In terms of motion, a velocity in one direction would imply motion in the opposite direction, indicating even stronger resistance to motion than $f = 0$ would imply, a vehement allergic reaction instead of mere apathy. Geometry provides a simpler interpretation: Negative f amounts to undoing the sign change required to go from rotations to Lorentz boosts. (Redraw Figure 2.42 with Euclidean right angles.) If we rotate by a positive angle, a point on the positive part of the x-axis moves to the left, toward negative values, and vice versa. For a rotation, we expect a negative sign in the angle-displacement relationship. If quantum space-time leads to a negative f, we turn these arguments around and conclude that it has all its time turned into space. "All which may sound preposterous; yet there are conditions under which nothing could keep us from losing account of the passage of time, losing account even of our own age; lacking, as we do, any trace of an inner time-organ, and being absolutely incapable of fixing it even with an approach to accuracy by ourselves, without any outward fixed points as guides."[3] Time is lost, even for the universe.

The bounce region of quantum space-time is not a temporal transition phase because there is no time. It connects collapse to expansion because its points are not separate from those found in the true space-time regions, but its geometry is not of space-time form. There is no evolution from collapse to expansion, even though some information, but not all, is transmitted. Loop quantum cosmology, rendering the singularity regular, strikes a balance between cyclic and linear models; it gives us creative license: It combines collapse with expansion but resets some information in the universe at the beginning of each expansion phase. It provides a clean slate, alleviating, among other things, Tolman's entropy problem.

7.2 Black Holes

Black holes are small versions of a collapsing universe, regions of space, rather than all of it, falling to ever larger densities. General relativity does not find any limits, the density increases beyond all bounds, leading to a singularity. Before this time is reached, space-time is deformed in drastic ways by large curvature. The singularity is enclosed by a horizon that allows nothing, not even light, to escape. Within the horizon, it is the radial distance from the center that plays the role of time, bound to decrease until it vanishes at the singularity.

With their strong curvature and singularity, black holes provide another testing ground for quantum gravity. Space-time according to general relativity would end when the singularity is reached, denying us a complete picture of space-time. If quantum gravity can resolve the singularity at the big-bang, it might also help us avoid those in black holes, perhaps leading to space-time beyond or within their singular cores.

Compared to cosmological models, black holes show more complexity. In cosmology, we may assume space-time to be almost homogeneous, about the same at all points in space. Around a black hole, the gravitational force does change as we move closer to or away from the horizon, the horizon itself being a distinguished place unlike any others. Black holes and their singularities are not as well-understood as quantum cosmology, a feature that makes them more interesting.

3) Thomas Mann: *The Magic Mountain*.

Black holes present a new issue of their own, related to their complexity. A black hole can come in many different states, even if it is assumed to be of given size. Hawking radiation of a black hole of some mass has a specific temperature, which in equilibrium equals the temperature of the black hole itself. In thermodynamics, we understand temperature as the energy and interactions of microscopic constituents moving with certain velocities, the faster the higher the temperature. We could in principle measure the temperature if we could detect Hawking radiation, but the black hole's constituents lie hidden behind the horizon. Theories can be tested by computing the number of states, namely, all possible configurations with the same total size of the black hole, and comparing with thermodynamical expectations based on the temperature.

Black-hole entropy Black holes are tidy objects. General relativity only allows black-hole solutions of specific forms, characterized by three parameters: the mass, angular momentum, and electric charge. All other properties, for instance, in distortions of the horizon, would not be intrinsic to the black hole, but rather be consequences of how it reacts to surrounding matter, pulled by gravity. General relativity does not tell us what forms a black hole; it does not show us any microstructure.

There are indirect hints for microscopic building blocks. If a quantum field is present near the horizon, it can be excited to higher energy, at the expense of the black hole's mass. The process, Hawking radiation, is analogous to matter in the inflating universe emerging from the vacuum. Black holes radiate, and as Hawking's calculation showed, the radiation looks thermal with temperature related to the mass. Temperature quantifies the motion of microscopic constituents, but we do not see any in general relativity.

Quantum gravity is a candidate for a theory of the microstructure of space-time, including black holes. Strings or loops, in stacked, knotted, interlinked, or other configurations could build black holes, and counting all possible ways to obtain a black hole of a given size should, if these theories are correct, result in a number amounting to the Hawking temperature. Agreement has indeed been achieved, a success claimed by the two main approaches to quantum gravity. The verdict of black-hole states does not speak in favor of either strings or loops.

Counting black-hole states in different approaches to quantum gravity has been an active research area for some decades. With successful results in several approaches, it has become clear that the question is not specific enough to help us distinguish competing ideas. Also, the fuzziness of the horizon, an unsharp demarcation unlike a material surface, spells trouble. One can define the precise place of a horizon in different ways, implying shifts in the actual position. Different definitions have been tried and used for state countings, and agreement with Hawking's temperature has been found in all cases. However, if the horizon is at a different place and the counted states correspond to the enclosed black hole, some properties of the states and their number should change. The fact that different notions of horizon can all lead to consistent results indicates that the properties of quantum space-time treated with such questions are much too generic to tell us anything of direct interest for the specific development of quantum gravity. These questions are important consistency tests, but we must look for something more dramatic, something as violent as a singularity, to make progress.

Black-hole singularity A black hole does not have a central point; its singularity is rather a moment in time. With time and radial distance swapping roles at the horizon, infinite density, or the black-hole singularity in general relativity, is reached at a certain moment in time, when the temporadial distance is zero. If we restrict attention to the region inside of the horizon, black-hole collapse in general relativity is not much different from cosmic collapse: infinite density is reached at a specific moment in time.

Before infinite density is reached, we cross the Planck threshold. General relativity fails; quantum gravity takes over. Scenarios of quantum cosmology then suggest what might happen in black holes, for instance, a nonsingular end state, or a bounce after which infalling matter would spew out of the black hole, perhaps with a core of four-dimensional space deep inside the high-density region. The horizon does not seem to allow matter to re-emerge even if quantum gravity could make it bounce back, but the horizon changes and shrinks due to Hawking radiation. It is a notion derived for classical space-time; quantum space-time at high density has its own rules. The definition of horizons no longer applies as it refers to trapped surfaces based on properties of light propagation. At high density, no light, not even gravitational waves move, and the notion of horizons loses meaning. Quantum gravity might open up black holes, perhaps after waiting long enough for the horizon to evaporate to small size.

Black-hole explosions, with all the collapsed matter visible again, still hot from its high density, would be impressive events by which quantum gravity could be tested. However, before matter can return from the abyss, the horizon must shrink by Hawking radiation. This radiation is weak; for a solar-mass black hole, it is less intense even than the faint cosmic microwave background left from the big bang. For more massive black holes, the intensity decreases because the curvature at the horizon is smaller. The larger mass is more than compensated for by an increased radius. Hawking temperature is inversely proportional to the mass, and it is much smaller than the 2.7 K of the microwave background for all known black holes. Black holes, at present, absorb more energy from cosmic microwaves than they emit by Hawking radiation, making them grow rather than shrink. We would have to wait for several billion years of cosmic expansion and cooling for the first black holes to start evaporating.

Black holes remain theoretical testing grounds, allowing us to probe mechanisms of singularity resolution in situations more complex than isotropic cosmology. Conclusions will remain speculative for quite some time, though they are necessary for a complete and consistent worldview in order to determine how space-time is extended in all regimes.

Multiverse Speculation is addictive. Having started speculating about the form of space-time deep within black holes, of regions yet inaccessible to us, we may go on and ask how different black holes appear together in space-time, branching out through their horizons and high-density regimes. Black holes make space-time quite different from the close-to-homogeneous form assumed in cosmology.

If matter in black holes bounces back and spews out after the horizon has disappeared, the form of space-time does not change much. Some parts may be sealed off for some time, but matter in them still contributes to the total density by the mass and gravity of entire black holes. Even if we can't see what is in a black hole, we know how much mass it contains. A bouncing black hole, from the cosmological perspective, is just another form of compact object, not much different from a neutron star.

If matter does not bounce back or spew back out, it is lost. The horizon would still evaporate, making the black hole and its mass shrink, contributing less and less to the matter density, replaced by Hawking radiation. However, matter that has reached the high-density region, general relativity's singularity, has nowhere to go in our classical space-time. If it does not spew back out, it must remain elsewhere if there is no singularity in quantum gravity; it must enter a place in quantum space-time, not part of our universe, not connected to us by a space-time continuum. Through the high-density phase, matter would appear to split off as a baby universe, a new space-time region holding on to us by nothing more than the threads of quantum space, a delicate umbilical chord soon severed. Even though there is just one quantum space-time, the classical connectivity we expect for our notion of time and causality is lost between our part and these new-born ones. From the point of view of the space-time continuum, the universe appears with disconnected parts; it is a multiverse, multiple generations of universes, all born from the darkness of black holes.

Whether there is a multiverse beyond our universe, and if so what kind, remains wide open and speculative. The scenario just sketched depends on if and how matter might bounce back in a black hole, and whether it can spew out after the horizon evaporates. Unlike simple cosmological models in which one often assumes homogeneity, inhomogeneous collapse is much more complicated. We may think of collapsing inhomogeneous space-time as a collection of homogeneous patches all collapsing at different rates. When there is still some degree of homogeneity and curvature is not large, collapse is rather smooth, with different patches weakly influencing one another's motion. However, when the curvature increases to Planckian values, the densest patches start bouncing, turning collapse into expansion even while less dense ones still collapse. Space-time is torn apart, its pieces changing in opposite ways, some collapsing, others expanding. Collapsing space can fracture into different universes: disconnected regions that start expansion while their surroundings still collapse. A multiverse picture may arise even if no black holes and horizons have formed, but one difficult to analyze. What happens at the turn-around points depends on the equations of quantum gravity, in ways too sensitive to allow clear progress at the current stage of developments.

7.3 Tests

To dam the ever-growing deluge of speculation (and to prevent, as happened often when scientists had to select between different approaches and theories, that decisions become a matter of taste, and debates a matter of polemics) observational insights into the quantum nature of space and time, even if indirect, are needed. Mathematical consistency may place strong restrictions when vast frameworks such as general relativity and quantum physics are to be combined. However, we cannot be sure what the right kind of mathematics is, which of its subfields and structures are best used to match nature. General relativity was lucky to have differential geometry, developed in the nineteenth century to understand properties of surfaces. Quantum mechanics had to await new mathematical methods before its equations could be understood. The different approaches to quantum gravity, foremost string theory, have provided important stimuli for recent mathematical developments, but we cannot be sure that these are the correct tools for nature. Only observations and experiments can help us.[4)]

Our imagination, trained in classical surroundings and sharpened by the use of mathematics, has a poor track record of guessing properties of the microscopic world. Quantum mechanics still appears strange to us; it would not have been found without experiments indicating the failure of classical physics and guiding the way to new laws. Quantum gravity has resisted attempts of direct definitions, combining rules of relativity and quantum theory. It requires new insights, more general principles, and perhaps its own mathematical basis. None of this can be imagined without at least a hint of observations.

Planck scale The needle in the haystack is a problem of scales. There is one tiny needle among a multitude of needle-like stalks of hay. The needle is much smaller than the stack; to find it in a direct manner, we must spend a long time looking at hay stalk after hay stalk. A dedicated search with sensitive probes can proceed faster. If we swipe our hands through the stack until they are pricked by the needle's pin, we avoid looking at every single stalk in the stack. We need sensitive measurements together with a good theory of what we are looking for, that is, we must know that it is a needle, and that our nervous system is a good detector set off when a hand is pricked.

For quantum gravity, we expect scales of the Planck length or density, scales too extreme to allow direct tests of any phenomena the theory could imply. Regions of high density may exist in our universe, but they are hidden behind horizons or by hot and dense plasma as in the early universe. Indirect evidence for the microstructure of space-time might still be found, just as many indications for material atoms

4) Even with observations, we can never be sure that a theory is correct, in the strongest sense of the word. We can only falsify theories that do not agree with observations, and gain more trust in those that remain consistent, without full confirmation. In science, we cannot expect gratification; the best we can do is try to avoid blame. (Sometimes, this statement is true not only for scientists facing nature, but also for scientists facing one another.)

existed well before microscopes of atomic resolution could be developed, and just as we nowadays use sensitive measurements of the energies of hydrogen to determine the radius of the tiny proton within. The unambiguous evaluation of indirect evidence requires a good understanding of the underlying theories, and for this reason the heavy dose of current theoretical investigations is necessary. However, the aim of all theory must be to prepare for the real world, to find the best places to look for evidence, for or against the theory. Our theories are teacher and student at the same time; they devise the toughest tests, and they are the ones who either pass or fail.

To find testable phenomena of quantum gravity, we need magnification effects, provided either by nature or by cunning experimental setups. If we are looking for properties of one tiny space-time atom within the multitude of all space-time, or at least a region of astrophysical or cosmological relevance, the haystack is far bigger than the needle. However, if a theory tells us how space-time is built from atoms, it may become testable if the form of the stack depends on properties of its constituents. We can probe the substructure of the stack, and once we understand it well enough, zoom in on an individual stalk, or the needle. In quantum gravity, even before we find the needle, understanding how the stack is built would be immense progress.

An alternative strategy focuses on the unification aspect of some theories of quantum gravity rather than the one of space-time atoms. Unification implies relationships between the properties of different elementary particles and the fundamental forces, and often predicts new types of particles of large masses. If the relevant scale comes from quantum gravity, masses should be as high as the Planck mass. Even though its value, about two millionth of a gram, sounds light, it is a billion times a billion times as heavy as a proton. Particle accelerators will not reach such energies for a long time, and even cosmic rays won't help. The production of new particles requires high energies, but not many particles need be smashed together; no magnification effects from the number of particles can be expected. Instead, there are hopes for a different kind of magnification, one coming from specific and detailed properties of unified theories. Extra dimensions could be larger than the Planck length, yet small enough to have escaped detection so far. The ratio of their length by ℓ_{Planck} would be a large dimensionless number that might magnify some effects, for instance, by bringing down the masses of particles to values accessible at accelerators. So far, no concrete indications have been found, but hope and interest remains among particle physicists.

Electromagnetic waves Light is our most common messenger. It propagates through matter and space, teaching us much about both. Dispersion in crystals depends on how close the wave length is to the lattice spacing; it gives us clues about the latter. Deflection in the universe depends on curvature; it has revealed properties of space-time. If space-time has a microstructure, light of short wave length should be the first informant we ask.

All electromagnetic waves that have been produced or observed have wave lengths much larger than the Planck length, or energies much below the Planck mass times c^2. Differences to propagation through continuous space-time are tiny when light travels a distance of about its wave length l, as small as the ratio ℓ_{Planck}/l. Even high-energy electromagnetic waves such as X-rays or gamma rays, with wave lengths much shorter than those of visible light, would give tiny ratios. But if light travels a long distance, some effects might add up and accumulate to something measurable. The cosmos provides many distant sources, still visible if they are bright and violent enough.

Light propagates through matter at speeds different from the celerity c, the speed of light in vacuum. If matter has a regular structure like a crystal, the speed depends on the color, or the wave length compared to the crystal's lattice spacing. Colors move at different speeds, and are deflected by different angles when they enter or leave the crystal; a light beam splits up into its colors when moving through a prism. If light travels through vacuum, the microstructure of space-time could play the role of crystal matter. Different colors, or gamma rays of different energies, would propagate at different speeds, in a way that could be computed by theory and tested by observations.

To measure the travel time, we must know when a gamma ray starts at a distant source before it reaches us. Steady light sources don't give us such information, but explosions or short bursts send out all their light in a brief time. The explosive nature also makes them visible from afar; some events, called gamma-ray bursts, send out their energy by gamma rays of short wave lengths. They seem to be ideal candidates to test dispersion effects of electromagnetic waves in atomic space-time.

The satellite Fermi has observed many gamma-ray bursts since its launch in 2008. Much information has been gained about the nature of the bursts, and some of the data have been evaluated to see if rays of shorter wave lengths arrive before or after those of long wave lengths. Delays have indeed been found, some as long as a minute. What these new and detailed observations also indicated, however, was that the bursts seem more complicated than thought before. Some show precursors, events announcing the bursts with less intensity and energy. It remains unclear whether travel delays seen in arriving rays can all be accounted for as related to precursors.

Another effect seen in crystals is birefringence: the color-dependent rotation of polarization of an electromagnetic wave, the direction in which electric and magnetic fields oscillate. If this phenomenon happens for waves in atomic space-time, there should be no polarization of light from distant sources; all waves would have their oscillations turned in different, almost randomized ways. Polarization in star light can be observed, even from distant sources. Implications of quantum space-time on the propagation of light cannot be strong, so far consistent with most theoretical approaches.

Cosmology Cosmological history descends from the highest densities to the smallest ones, from Planckian fulfillment or the classical singularity all the way to near-complete dilution. There should be times, deep in the big-bang phase, when quantum gravity is essential for a correct description. We cannot watch those events because our messengers for observations scatter too much. Perusing patterns shown by electromagnetic waves, in the cosmic microwave background and the galaxy distribution, we observe what happened 380 000 years after the densest phase and later. Messengers with weaker interactions, neutrinos or gravitational waves, might one day give us access to earlier times, but their weak interactions also make them difficult to detect. It will be long before we can use detailed maps shown by those carriers to probe the cosmos.

At times 380 000 years after the densest phase and later, matter and energy in the universe had been diluted to densities far less than Planckian, and even in somewhat controlled earlier phases such as inflation, the density was at most a millionth of the Planck density. Individual corrections in quantum-gravity equations for the expansion rate of the universe or the propagation of structure are tiny. However, the universe evolves for a long time, and if small corrections add up, they may produce sizeable effects. Additionally, instead of just referring to the density, there may be more sensitive effects, depending on the atomic space-time structure of quantum gravity, not bound by the dimensional Planckian estimates. Atomic space-time has at least three scales: the huge Hubble distance, $c/H = ac\Delta T/\Delta a$, the tiny Planck length, and the average size ℓ of spatial atoms at any given time. Three lengths do not determine a unique dimensionless ratio, and if corrections depend on numbers such as $\ell/\ell_{\mathrm{Planck}}$, they may not be small.

Whether such corrections should be present depends on details of quantum gravity, an endeavor fraught with the uncertainty of an incomplete theory. By general principles related to covariance of the theory, or the requirement that physical predictions do not depend on what coordinates we choose, we know that it is not easy to implement a consistent atomic scale. Atomic space would look much like a lattice, but a lattice is not invariant under rotations, let alone Lorentz transformations. For this reason, cosmologists often expect that no significant magnification effects occur, but one possibility is realized in loop quantum gravity.

Symmetry problems of lattices can be circumvented by exploiting one of the features of quantum physics, the superposition principle which allows different states to exist in superposition, realized at the same time for one given system. The universe, or its space at some time, can be in a superposition of lattices oriented in different ways; the combination restores the symmetries that a single lattice would break. And yet, the superposed state remains discrete, some of its properties and predictions depending on the lattice spacing.

Complete lattices are difficult to control with exact mathematical solutions, but good approximations exist and reveal the cosmological dynamics. Modifications of the classical space-time structure can be expressed by a quantum corrected space-time algebra: $T[N_1]T[N_2] - T[N_2]T[N_1] = S[f(N_2\Delta N_1/\Delta x - N_1\Delta N_2/\Delta x)]$. At lower densities, inverse-triad corrections are the main contribution to f, which stays positive (unlike around the Planck density where it is negative, showing that time becomes space) but differs from one. Lorentz boosts, obtained for linear functions N_1 and N_2, do not behave as they do in classical relativity. The speed of light, invariant under the usual transformations, can change; the wave equation invariant under the new algebra is $-\Delta_t^2 u/(\sqrt{f}c\Delta t)^2 + \Delta_x^2 u/(\Delta x)^2 = 0$. If f is not constant but depends on the changing lattice scale of an evolving atomic space, so does the speed $\sqrt{f}c$.

Inverse-triad corrections are significant when the lattice spacing is small, close to where the classical inverse is infinite. For large lattice spacing, on the other hand, discreteness would be visible by deviations from continuum physics. A bounded range of parameters remains consistent with observations; the latest data leave open a window of about four orders of magnitude. We must wait for new and more precise observations before we can rule out the theory, or find clear signatures for it. However, compared with the ratio of the cosmological density to the Planck density, differing by more than ten orders of magnitude, the smaller range gives much more hope.

Sometimes, when they were not busy tending their fields, Mathlas' daughters went to visit him. In spite of the magnitude of his task, he was cheerful, but once his daughters found him depressed. Asked about his mood, Mathlas began: "Ever since I compactified the heavens... " "It's legendary!" they interrupted. Mathlas continued: "Well. Ever since, one question has remained on my mind. I have been standing here year after year, holding the heavenly sphere, thinking it over and over. But I have come no closer to an answer." "What's the question?" the daughters inquired with interest.
"The question is: When I came up with the idea of compactifying the heavens, did I invent the heavenly sphere, or did I discover that the heavens can be compactified for better grasp? At that time, the idea just occurred to me. I had never seen a heavenly sphere before. The gods, with their infinite powers, could handle infinite space. They had no use for a heavenly sphere, but it might have existed as a possibility. Or did I bring the compact sphere into existence when I compactified the heavens?"
His youngest daughter, Queda, fell in: "Oh, I know. I have heard the uncles talk about your problem." "What do they say?" asked Mathlas, curious. "Well ... Physikos says you should just 'shut up and hold that sphere.'" She continued in another direction, trying to distract Mathlas. "You said the gods, with their infinite powers, could handle infinite space. Now, I know that for us mere titans everything has to be finite so we can grasp it. But how were you able to beat the gods if they had infinite powers?"
Mathlas' depression returned: "Well, we never really beat them, I think. We thought we did, but when all of us went on our assigned tasks, we noticed – not right away but after some time – that the gods, or something, must still be there. Or do you think Economastos could steer the universe on his own? And Physikos, he doesn't know his heavens." "He's a drunkard!" burst out Cucida, laughing. "No, no, he's not," Mathlas intervened, "don't be so harsh. Managing energy is a complicated task." "Not as complicated as holding the heavens!" – his daughters were still trying to cheer up Mathlas, but to no avail. "Maybe, maybe. But once, well, once Economastos did get drunk, and, as he told me later, forgot to steer the universe. Even worse, when he realized his oversight, he, on a whim and still intoxicated, tried to make the universe stop." "He is always stirring up trouble." was the daughters' final attempt. Mathlas concluded: "The universe didn't stop. It just kept growing, when he forgot to steer it and when he tried to stop it, it even started accelerating. There must be powers that rule us, impose laws on us. All of us have discovered this in our assigned tasks – all but me. This is why that question keeps torturing my mind."

Acknowledgement

More thanks than can be expressed go to Amy Alderman.

It is a pleasure to thank Holger von Juanne-Diedrich for encouragement, suggestions and some careful reading. In parts, the constructions in Chapters 2–4 are related to the books Gravity: An Introduction to Einstein's General Relativity by James Hartle and Introduction to Quantum Mechanics by David Griffiths, to which one may refer for further details.

Some of the material in this book is based upon work supported by the National Science Foundation under Grant No. 0748336. Any opinions, findings, and conclusions or recommendations expressed in this material are those of the author and do not necessarily reflect the views of the National Science Foundation.

The Universe: A View from Classical and Quantum Gravity, First Edition. Martin Bojowald.

Index

(Words followed by the symbol ◇ indicate topics discussed in boxed passages.)

The Universe: A View from Classical and Quantum Gravity, First Edition. Martin Bojowald.